Ecological Research Monographs

Series Editor: Yoh Iwasa

For further volumes:
http://www.springer.com/series/8852

Sun-Kee Hong · Jianguo Wu · Jae-Eun Kim
Nobukazu Nakagoshi
Editors

Landscape Ecology in Asian Cultures

Springer

Editors
Sun-Kee Hong
Institution for Marine and Island Cultures
Mokpo National University
61 Dorim-ri, Cheonggye-myeon
Muan-gun, Jeonnam 534-729
Republic of Korea
landskhong@gmail.com
skhong@mokpo.ac.kr

Jae-Eun Kim
Institution for Marine and Island Cultures
Mokpo National University
61 Dorim-ri, Cheonggye-myeon
Muan-gun, Jeonnam 534-729
Republic of Korea
ecokimje@yahoo.co.kr

Jianguo Wu
School of Life Sciences,
School of Sustainability,
and Global Institute of Sustainability
Arizona State University, P.O. Box 874501
Tempe, AZ 85287-4501
USA
Jingle.Wu@asu.edu

Nobukazu Nakagoshi
Graduate School for International
Development and Cooperation
Hiroshima University
1-5-1, Kagamiyama, Higashi-Hiroshima
Hiroshima 739-8529, Japan
nobu@hiroshima-u.ac.jp

ISSN 2191-0707 ISSN 2191-0715 (eBook)
ISBN 978-4-431-54084-7 ISBN 978-4-431-87799-8 (eBook)
DOI 10.1007/978-4-431-87799-8
Springer Tokyo Dordrecht Heidelberg London New York

Cover
Front Cover:
Landscape of Beopseongpo in Yeonggwang-gun, Jeonnam Province, Korea (photo by S.-K. Hong; see Chapter 2)

Back cover:
Left: Fishing village on Chuja-do Island, Jeju Province, Korea (photo by S.-K. Hong)
Center: Gonam-myeon, a rural village in Taean-gun, Chungnam Province, Korea (photo by S.-K. Hong)
Right: "Ullimsanbang (Atelier)," in a traditional Korean garden in Jindo-gun, Jeonnam Province, Korea (photo by S.-K. Hong)

Printed on acid-free paper

Springer is part of Springer Science+Business Media (www.springer.com)

Preface

During the last century, ecological research in Asian countries has made many strides in terms of both quantity and quality. In recent decades, landscape ecology as a new interdisciplinary field has been developing rapidly in many Asian countries, mainly following the concepts, principles, and methods outlined by a limited number of textbooks from North America and Europe. However, patterns of landscapes (and seascapes) in Asia are not only spatially heterogeneous, but also have unique characteristics that are driven by different socioeconomic and cultural processes. The spatial patterns of Asian landscapes are strongly related to human activities and their impacts. Anthropogenic patterns and processes have created numerous traditional cultural landscapes throughout the region, and understanding them requires traditional and indigenous knowledge of the landscapes and various patches in them.

Cultural landscapes (seascapes) are a product of the interactions between humans (human history and land use) and natural settings of the land system. Socioeconomic forces are the main drivers of landscape dynamics; their pattern and processes influence the creation and development of cultural landscapes. Thus, cultural landscapes are not static but constantly changing. Also, landscape dynamics are driven by environmental changes including climate change and a suite of natural disturbances. To understand and manage such complex landscapes, interdisciplinary and transdisciplinary approaches, which emphasize the integration between natural and social sciences as well as between multiple landscape functions, are necessary. These approaches are often characterized and advanced by regional foci and contexts. Landscape ecological studies in Asian countries, therefore, provide useful information for ecologists elsewhere to better understand the human-environmental relationship and landscape sustainability.

This book contains chapters that cover a range of topics, including indigenous knowledge of landscapes, ecological impacts of human actions on landscapes, management of landscapes, and land-use policies. An emphasis on human–ecological networks (e.g., traditional knowledge of wind–water geographic theory) seems apparent in Asian landscape ecological studies. Thus, we believe that the term "cultural landscape ecology" captures some of the essential features of the studies focusing on diverse ecological applications in human-dominated landscapes in Asia.

The book consists of three parts. In Part I, "Understanding Asian Cultural Landscapes," authors discuss a number of issues and topics in cultural landscape ecology from multidisciplinary perspectives. Conserving traditional cultural landscapes in a developing Asia is not an easy task. Adequate laws are indispensible to effectively conserve such landscapes. All Asian countries should establish the necessary ordinances and regulations to protect their cultural landscapes. Part II, "Measuring and Managing Patterns and Processes of Cultural Landscapes," focuses on the quantification of the structural pattern of cultural landscapes. It is more complicated to quantify cultural landscapes than natural ones as cultural systems are not only dependent on, but also interact with, the natural system. In Part II, various effects of natural and human disturbances on the characteristics of cultural landscapes are also reported from different countries in Asia. Cultural landscapes result from nature-human interactions, and the loss of certain key features of cultural landscapes may have important, and often undesirable, ecological and sociocultural consequences. Part III, "Concluding Remarks," includes only one chapter, which discusses the unique characteristics and future directions of Asian landscape ecology. Here, the role of an Asian perspective in the science and practice of landscape ecology is explored.

Landscapes around the world have been shaped increasingly by human activities. In order to sustain ecosystems, landscapes, and human society, we must develop a better understanding of how culture and nature interact with one another. To this end, human-dominated landscapes with long traditions, such as those described in this book, may provide valuable lessons. We hope that the reader will find this to be the case.

Sun-Kee Hong
Institution for Marine and Island Cultures
Mokpo National University, Korea

Jianguo (Jingle) Wu
School of Life Sciences, School of Sustainability, and
Global Institute of Sustainability
Arizona State University, USA

Jae-Eun Kim
Institution for Marine and Island Cultures
Mokpo National University, Korea

Nobukazu Nakagoshi
Graduate School for International Development and Cooperation
Hiroshima University, Japan

Acknowledgements

We would like to acknowledge that this work was supported by a National Research Foundation of Korea Grant funded by the Korean Government (NRF-2009-361-A00007) awarded to S.-K. Hong. Some of the chapters were contributions from Hiroshima University projects, including the Global Environmental Leader Education Programme.

Contents

Contributors

Saiful Arif Abdullah (Chapter 4)
Institute for Environment and Development (LESTARI),
Universiti Kebangsaan Malaysia, 43600 Bangi,
Selangor Darul Ehsan, Malaysia

Issouf Bamba (Chapter 11)
Service d'Ecologie du Paysage et Systèmes de Production Végétale,
Université Libre de Bruxelles, CP 169, 50 Avenue F.D. Roosevelt,
1050 Bruxelles, Belgium

Yao Sabas S. Barima (Chapter 11)
Service d'Ecologie du Paysage et Systèmes de Production Végétale,
Université Libre de Bruxelles, CP 169, 50 Avenue F.D. Roosevelt,
1050 Bruxelles, Belgium

Jan Bogaert (Chapter 11)
Service d'Ecologie du Paysage et Systèmes de Production Végétale,
Université Libre de Bruxelles, CP 169, 50 Avenue F.D. Roosevelt,
1050 Bruxelles, Belgium

Farid Dahdouh-Guebas (Chapter 11)
Laboratoire de Complexité et Dynamique des Systèmes Tropicaux,
Université Libre de Bruxelles, CP 169, 50 Avenue F.D. Roosevelt,
1050 Bruxelles, Belgium
and
Onderzoeksgroep Algemene Plantkunde en Natuurbeheer,
Vrije Universiteit Brussel, Pleinlaan 2, 1050 Brussel, Belgium

Forood Azari-Dehkordi (Chapter 5)
Department of Geography, University of California at Berkeley, Berkeley, CA, USA

Wei-Ta Fang (Chapter 13)
Chung Hua University, Hsingchu 30012, Taiwan, Republic of China

Michiro Fujihara (Chapter 17)
Graduate School of Landscape Design and Management,
University of Hyogo, 954-2 Nojimatokiwa, Awaji,
Hyogo 656-1726, Japan
and
Awaji Landscape Planning and Horticulture Academy,
Awaji, Hyogo, Japan

Victor G. Gorshkov (Chapter 10)
Theoretical Physics Division, Petersburg Nuclear Physics Institute,
Russian Academy of Sciences, 188300, Gatchina,
St. Petersburg, Russia
and
Ecological Complexity and Modeling Laboratory,
Department of Botany and Plant Sciences, University of California,
Riverside, CA 92521-0124, USA

Luchman Hakim (Chapter 6)
Department of Biology, Faculty of Mathematics
and Natural Sciences, University of Brawijaya,
Jl. Veteran Malang 65142, East Java, Indonesia

Daisuke Hashimoto (Chapter 16)
Graduate School of Civil Engineering, Kyushu Institute of Technology,
Kitakyushu 804-8550, Japan

Sun-Kee Hong (Chapters 2, 7, and 9)
Institution for Marine and Island Cultures, Mokpo National University,
61 Dorim-ri, Cheonggye-myeon, Muan-gun, Jeonnam 534-729,
Republic of Korea

Shuji Iijima (Chapter 16)
Graduate School of Human-Environment Studies,
Kyushu University, Fukuoka 812-8581, Japan

Dai Isono (Chapter 16)
Department Civil Engineering, Faculty of Engineering,
Kyushu Institute of Technology, Kitakyushu 804-8820, Japan

Keitaro Ito (Chapter 16)
Department Civil Engineering, Faculty of Engineering,
Kyushu Institute of Technology, Kitakyushu 804-8820, Japan

Jian Ji (Chapter 11)
Nanjing University, 22 Hankou Road, Nanjing, Jiangsu 210093, China
and
Chengdu University of Technology, 1 Dongsan Road,
Erxianqiao, Chengdu, Sichuan 610059, China

Hong Jiang (Chapter 11)
Nanjing University, 22 Hankou Road, Nanjing, Jiangsu 210093, China

Yuanmei Jiao (Chapter 3)
The Tourist and Geography Department, Yunnan Normal University,
Kunming 650092, China

Jae-Eun Kim (Chapters 7 and 9)
Institution for Marine and Island Cultures, Mokpo National University,
61 Dorim-ri, Cheonggye-myeon, Muan-gun, Jeonnam 534-729,
Republic of Korea

Nico Koedam (Chapter 11)
Onderzoeksgroep Algemene Plantkunde en Natuurbeheer,
Vrije Universiteit Brussel, Pleinlaan 2, 1050 Brussel, Belgium

Bai-Lian Li (Chapter 10)
Ecological Complexity and Modeling Laboratory,
Department of Botany and Plant Sciences,
University of California, Riverside, CA 92521-0124, USA

Xiuzhen Li (Chapter 3)
State Key Laboratory of Estuarine and Coastal Research,
East China Normal University, Kunming 650092, China

Anastassia M. Makarieva (Chapter 10)
Theoretical Physics Division, Petersburg Nuclear Physics Institute,
Russian Academy of Sciences, 188300, Gatchina, St. Petersburg, Russia
and
Ecological Complexity and Modeling Laboratory,
Department of Botany and Plant Sciences, University of California,
Riverside, CA 92521-0124, USA

Adi Mama (Chapter 11)
Service d'Ecologie du Paysage et Systèmes de Production Végétale,
Université Libre de Bruxelles, CP 169, 50 Avenue F.D. Roosevelt,
1050 Bruxelles, Belgium

Tohru Manabe (Chapter 16)
Kitakyushu Museum of Natural History and Human History,
Kitakyushu 805-0071, Japan

Hiroyuki Miura (Chapter 17)
Awaji Landscape Planning and Horticulture Academy,
Awaji, Hyogo, Japan
and
Okinawa Environment Club, Naha, Okinawa, Japan

Léon Iyongo Waya Mongo (Chapter 11)
Service d'Ecologie du Paysage et Systèmes de Production Végétale,
Université Libre de Bruxelles, CP 169, 50 Avenue F.D. Roosevelt,
1050 Bruxelles, Belgium

Nobukazu Nakagoshi (Chapters 12, 15, and 18)
Graduate School for International Development and Cooperation,
Hiroshima University, Higashi-Hiroshima 739-8529, Japan

Edgar Nyssen (Chapter 11)
Vrije Universiteit Brussel, Vakgroep Elektronica en Informatica,
Pleinlaan 2, 1050 Brussel, Belgium

Mariko Ohnishi (Chapter 17)
Awaji Landscape Planning and Horticulture Academy,
Awaji, Hyogo, Japan
and
Department of City Planning, Itami, Japan

Yoko Ohta (Chapter 12)
Open Network for Nature Conservation, Shimane 694-0064, Japan

Jong-O Park (Chapter 8)
Institution for Marine and Island Cultures, Mokpo National University,
61 Dorim-ri, Cheonggye-myeon, Muan-gun, Jeonnam 534-729,
Republic of Korea

Yoshihiro Sawada (Chapter 17)
Graduate School of Landscape Design and Management,
University of Hyogo, 954-2 Nojimatokiwa, Awaji,
Hyogo 656-1726, Japan
and
Awaji Landscape Planning and Horticulture Academy,
Awaji, Hyogo, Japan

Shigeo Suzuki (Chapter 15)
Faculty of Geo-Environmental Science, Rissho University,
1700, Magechi, Kumagaya, Saitama 360-0194, Japan

Niels Thevs (Chapter 19)
Institute of Botany and Landscape Ecology, Greifswald University,
Grimmer Strasse 88, 17487 Greifswald, Germany

Takashi Umeno (Chapter 16)
Graduate School of Civil Engineering, Kyushu Institute of Technology,
Kitakyushu 804-8550, Japan

Jianguo Wu (Chapter 20)
School of Life Sciences, School of Sustainability,
and Global Institute of Sustainability, Arizona State University,
P.O. Box 874501, Tempe, AZ 85287-4501, USA
and
Sino-US Center for Conservation, Energy, and Sustainability (SUCCESS),
Inner Mongolia University, Hohhot 010021, China

Sampei Yamashita (Chapter 14)
Kyushu Sangyo University, 2-3-1 Matsukadai, Higashi-ku,
Fukuoka 813-8503, Japan

Takakazu Yumoto (Chapter 1)
Research Institute of Humanity and Nature, Kyoto 603-8047, Japan

Stefan Zerbe (Chapter 19)
Faculty of Science and Technology, Free University of Bozen – Bolzano,
Piazza Università 5, P.O. 276, 39100 Bolzano, Italy

Part I
Understanding Asian Cultural Landscapes

Chapter 1
Historical Perspectives on the Relationships between Humanity and Nature in Japan

Takakazu Yumoto

1.1 Introduction

The Japanese Archipelago extends over 3000 km from north to south, and includes subarctic, cool temperate, warm temperate and subtropical climatic zones. These various climatic zones were present even during the dramatic environmental changes of the past 100000 years (e.g., Tsukada 1967). The Japanese Archipelago has been densely populated since the Neolithic Age, and most of the natural environment has been strongly influenced by human activities (Koyama and Sugito 1984). The life patterns of humans have, in turn, been shaped by their use of biological resources in the shape of the fauna and flora they encountered. Moreover, although the Japanese biota is derived from life forms which migrated from the continental mainland when sea-levels were lower, the biodiversity has been augmented by human beings, who have introduced species at various times.

There is still a rich biotic life in the Japanese Archipelago, in spite of intensive intervention by humans in the natural environment, including, for example, an abundance of indigenous species of angiosperm and freshwater fish. This abundance has led to the assumption that human–nature relations in pre-modern Japan were governed by some kind of traditional wisdom that prevented people from exhausting biological resources, or that moderate human activity preserved the abundance of biological resources in Japan.

The question of exactly how stable the coexistence between nature and humans was in the past has not been resolved. How ought one to describe the human–environmental history of Japan? Has there not been, in the Japanese Archipelago, a history of exhausted biological resources and of extinction? If early peoples possessed the wisdom and will to use biological resources in a sustainable way, how commonly were these virtues exercised? Further, if there were cases in which humans exhausted certain biological resources, did any major social changes follow as a result?

T. Yumoto (✉)
Research Institute of Humanity and Nature, Kyoto 603-8047, Japan
e-mail: yumoto@chikyu.ac.jp

S.-K. Hong, et al. (eds.), *Landscape Ecology in Asian Cultures*, Ecological Research Monographs, DOI 10.1007/978-4-431-87799-8_1, © Springer 2011

Although each of these questions has been examined within one historical period or region, or from one disciplinary perspective, no trans-disciplinary examination has been attempted over an area that would represent the whole Japanese Archipelago, or over a time span that encompasses the whole period from the earliest human habitation of Japan to modern times. The objective of this chapter is to review the human role in the vegetation changes witnessed in the Japanese Archipelago since the Neolithic, and to locate the refugia of plants and animals that inhabited the primeval forests.

1.1.1 Human Activities and Vegetation in Pre-History

During the Late Glacial Maximum (LGM, ca. 20000–12000 year BP), a period which was much cooler and drier than the present, the vegetation of Japan would have been unrecognizable to a contemporary observer. The climax vegetation of the southwestern area of Japan is now evergreen broad-leaved forest dominated by evergreen oaks (*Quercus* spp.) and the Lauraceae, but during the LGM, evergreen broad-leaved forest was restricted to the narrow coastal area of southernmost Kyushu, Shikoku and Honshu Islands. Instead, deciduous broad-leaved forest prevailed in southwestern Japan, consisting of beech (*Fagus crenata*), oak (*Quercus crispula*), hornbeam (*Carpinus* spp.), maple (*Acer* spp.) other deciduous species, and five-needled pine (*Pinus koraiensis*) (Tsukada 1967). The northeastern part of Japan, where deciduous broad-leaved forest dominated by *F. crenata* and *Q. crispula* is now found, was covered in coniferous forest, mainly consisting of spruce (*Picea* spp.) and fir (*Abies* spp.) (Kamei et al. 1981).

How has the vegetation in the Japanese Archipelago changed since the LGM? Pollen analyses have indicated that broad-leaved evergreen forests formed across the southwestern part of Japan in the early Jomon period (ca. 5300–3500 BC) (Yasuda and Miyoshi 1998), while Japanese cedar (*Cryptomeria japonica*) dominated the wide, flat lowlands along the Sea of Japan. As the climate subsequently warmed, broad-leaved evergreen tree species and *C. japonica* gradually expanded from their earlier warmer, wetter refugia in the LGM. On the other hand, the northeastern part of Japan began to be covered by deciduous forests consisting of *Fagus*, *Acer*, *Carpinus* and related species (Yasuda and Miyoshi 1998).

There have been many arguments on the subject of the human impact on these early forests. The population density in the early Jomon period was estimated as 1.0 per km^2 in northeast Japan, and 0.5 per km^2 in the southwest (Koyama and Sugito 1984). From the early Jomon, people depended heavily on the seeds of deciduous oak (*Quercus* subgen. *Lepidobalanas*) and chestnut (*Castanea creanata*) as staple foods. Pollen analyses and artifacts excavated from the Sannai-maruyama site (northern Honshu), a Jomon village continually inhabited from ca. 5100 to 3750 year BP, indicate that the *C. crenata* forest appeared after ca. 5650 year BP, and expanded with the appearance of the village. The pure stand, which cannot exist without human intervention, was formed in ca. 4850 year BP and maintained for several hundred years

(Yoshikawa et al. 2006). Signs of horse chestnut (*Aesculus trubunata*) usage also began to appear at the end of the middle Jomon period, when there was also a decrease in the area of the *C. crenata* stand (Yoshikawa et al. 2006). Such evidence shows that Jomon people modified the forest according to their needs.

New crops introduced in the Yayoi period (ca. 1500–900 BC), including rice from southern China, dramatically changed the previous patterns of vegetation and human–forest relations. Forested land was cleared to make way for rice cultivation. In the early Yayoi period, rice was cultivated from Kyushu to Kanto, and prevailed throughout Honshu Island in the middle Yayoi period. Iron tools appear to have been introduced to Japan along with rice cultivation, and were quickly adapted for felling timber. The composition of the forest changed as a result. From the Kinki district to Chubu during the Yayoi-Kofun period (ca. 900 BC–AD 600), the most commonly used wood was that of *Chryptomeria japonica.* People utilized *C. japonica* for various tools and construction. Many forests previously buried by flood or tephra have been excavated in the wet lowlands along the Sea of Japan where natural stands of *C. japonica* cannot be found today. Ancient people seem to have exhausted *C. japonica* trees in such regions.

1.2 Forests in Historical Period

Reconstructing the history of Japanese vegetation from ancient to pre-modern periods relies on combined analyses of pollen, wooden tools and constructions, and historical documents and images. The following discussion is focused on the Kinki district, where such materials are abundant.

The Kinki district was the site of several ancient capitals. By AD 600 the people of Japan were using woodland much more intensively than they did a millennium earlier. Villagers needed cleared land for tilling, and the ruling elite required huge quantities of timber for the construction of monumental buildings such as palaces, temples and government offices (Totman 1989). Massive pillars of hinoki (*Chamaecyparis obtuse*) and Japanese umbrella pine (*Sciadopitys vercillata*) were excavated in Heijo Palace (the Nara period, AD 710–784) and Heian Palace (the Heian period, AD 794–1185/1192), although no natural stands of *C. obtuse* or *S. vercillata* are currently found nearby. At the end of the Heian period, relatively convenient stands of large trees were exhausted, and large timbers were often reused in new constructions.

One of most important woods in early Japan was *Sciadopitys vercillata*. Many coffins excavated in the Kinki district during the Kofun period (ca. AD 0–600) were made of *S. vercillata*. One of the oldest written histories, the "Nihon-shoki," described *S. vercillata* as the most suitable wood for coffins. A third of the pillars in Heijo Palace were made of *S. vercillata*, and a significant ratio of pollen fossils from that time are of *S. vercillata*. However, nowadays stands of *S. vercillata* are found only on very steep mountain ridges in areas remote from Nara city. Earlier people exhausted stands of *S. vercillata* that were located near the plateau.

Ancient urban areas also consumed large amounts of firewood. Evergreen broad-leaved forests in the urban areas of the Kinki district provided the fuel needed to work iron, fire pottery and make salt, and were replaced by pine forests dominated by *Pinus densiflora*. Less impacted forests still remained in mountainous areas in these periods.

In the seventh century, after adopting a Chinese system of governance, the Japanese government claimed ownership of all lands. In practice, many forests remained available for use. In the latter half of the seventh century, however, lords and temples again claimed exclusive land ownership, and governments of this period repeatedly issued decrees forbidding the private use of forests.

A manor system prevailed in the Kinki district in the eighth and ninth centuries, and private use of demesne lands was promoted, in spite of the principle of government land ownership. The manor system initially arose as the government bestowed the responsibility for land management on lords and temples, and compensated them for their expenses in doing so. Later, the lords and temples were allowed to tax certain uses of their demesne land. In the medieval manor system, the lords and temples acquired both judicial authority and military power over the land and people in their demesne lands. Historical documents recount a number of land-based social conflicts in this era, including a famous example of the unauthorized use of private lands in Katsuragawa beside Lake Biwa during the thirteenth and fourteenth centuries. The landowner was a temple called Myo-oh-in. The temple desired to preserve their forest as a sanctuary for meditation and other religious austerities, but villagers nearby claimed the right to harvest trees for firewood and charcoal production.

1.3 Forest Commodities and the Appearance of Grass-Covered Mountains

The typical rural landscapes depicted in Ukiyoe, the Japanese woodblocks of the Edo period, consisted of a combination of pine forests and grass-covered mountains. A pictorial map of Kyoto published in 1884 showed a wide range of surrounding mountains in the north covered by grass and shrubs with few trees, and indicated that pine stands and bamboo thickets occurred only around shrines and temples (Ogura 1992). Older pictorial maps of Kyoto also indicated that no broad-leaved forests existed in the Kyoto Basin after the seventeenth century.

Many photographs from the Meiji, Taisho and early Showa eras (1880s–1940s) confirm that bare mountains prevailed in the southwestern part of Japan. The first topographical maps produced by the government in 1880–1886 also show that grass-covered mountains, which was wasteland, and pine stands occupied wide areas in Japan. When and why were such a wide range of grass-covered mountains established?

The oldest pictorial maps of Kyoto, the Grand Views of Kyoto Capital (Rakuchu-rakugai-zu), were created in the 1520s–1530s. They show the mountains surrounding Kyoto covered by grass and shrubs with few trees, as they were also depicted in the Edo period (Ogura 1992). Pollen analyses undertaken in Mizuro-ga-ike in the northern part of Kyoto established that evergreen oak (*Quercus* subgen. *Cyclobalanopsis*) began to decrease and two-needled pine (*Pinus* subgen. *Diploxylon*) to increase in the seventh century. The same analysis showed that the number of evergreen oaks decreased greatly in the eleventh century, and two-needled pine became dominant in the seventeenth century (N. Sakaki et al. personal communication).

Mountains and hills near villages were important sources of timber, firewood, and grass for fodder and manure. An archeological excavation also proved that in the sixth and seventh centuries several kilns for pottery and roof tiles were located nearby, and old documents survive showing that people in Mizuro-ga-ike village bought the right to harvest grass on Mt. Kibune, which was situated ca. 7 km north of the village. There are a number of pictures from the Edo period showing agricultural workers ploughing twigs and grass into paddy fields as manure. Tokoro (1980) estimated that more than 500 kg of grass is necessary to manure 1 acre of paddy field, and 10–12 acres of grass-covered mountain are needed in order to harvest 500 kg of grass. Tokoro (1980) also estimated that an average farming household needed 135–200 kg of firewood, which is equivalent to 2.5–4.5 acres of woody mountain. Maintaining grass-covered mountains requires regular burning in the early spring. Local governments in the Edo period often issued decrees warning about bush fires and the need to check the spread of fire.

An important aspect of Japanese grassland landscapes is the lack of pasture. From ancient times, governments and some lords kept herds of horses for military use, but there was no widespread grazing of cattle, sheep or goats, and none of the associated landscapes. Farmers kept only a few horses and oxen for ploughing. Very few people had the habit of eating the meat of domestic animals until the modern period, although many people did consume bush meat, such as sika deer (*Cervus nippon*), Japanese hare (*Lepus brachyurus*), wild boar (*Sus scrofa*), Eurasian badger (*Meles meles*), Japanese serow (*Capricornis crispus*), Asian black bear (*Ursus thibetanus*) and Japanese giant flying squirrels (*Petaurista leucogenys*), and they hunted and ate wild birds such as green pheasant (*Phasianus versicolor*), copper pheasant (*Syrmaticus soemmerringii*), Japanese quail (*Coturnix japonica*), several duck species (*Anas* spp.), goose (*Anser* spp.), swan (*Cygnus* spp.) and passerines.

Timber plantations in Japan, consisting mainly of conifers, *Cryptomeria japonica* and *Chamaecyparis obtuse*, were mainly formed after World War II. The cultivation of broad-leaved tree plantations for firewood and charcoal began much earlier, but only a very few coppiced stands of kunugi (*Quercus actissima*) for charcoal production are still maintained, e.g., those in Nose, Osaka Prefecture, and in Motegi, Tochigi Prefecture. Intensive plantations of *Q. actissima* are estimated to have been started in the sixteenth and seventeenth centuries. People harvested the sprouting stems at an interval of 8–10 years as wood for charcoal production. The stands were complete monocultures of *Q. actissima*, and seedlings of *Q. actissima* began to be distributed as a saleable commodity no later than the seventeenth century. People planted seedlings of *Q. actissima* even in the terraced fields, because they regarded it as a crop. A population genetic analysis of *Q. actissima* proved that the genetic diversity of the Japanese populations is very low, and strongly suggested that people introduced *Q. actissima* to Japan from the Korean Peninsula (Y. Tsumura et al. personal communication).

Charcoal making was almost abandoned after the energy revolution of the 1960s, but the woods of *Q. actissima* are still used for oak mushroom (*Lentinula edodes*) culture. Oak mushroom culture began in the 1660s, but the modern style of culture was established in 1943 by Dr. Kisaku Mori.

Until the Meiji period, mountains and hills were managed as commons. In many regions, local rules were established to guide the harvest of grasses and wood. For example, in some cases only a fixed number of people from each family were allowed to harvest manure or firewood during a fixed period of the year. In some regions, only bush knives, not hatchets or saws, were allowed for harvesting firewood. Nevertheless, many forests were over-used in the Edo and Meiji periods, and as a consequence, the newly bare mountains were susceptible to soil erosion. Early industrialization in those periods increased demands for firewood, so that people over-harvested wood from the commons as a saleable commodity. In forests where commons borders were not clear, competition among neighboring communities to harvest wood also accelerated over-use (Chiba 1991).

1.4 Logging and the Wilderness in the Deep Mountains

In spite of such intensive intervention by humans, some primeval forests remained in Japan, and several large mammals, including bears and wolves, survived until recent times. How was this possible?

The Warring States Period (the Sengoku Jidai, 1943–1573) was a time of social upheaval, political intrigue and almost constant military conflict. Huge military forts were repeatedly constructed and destroyed. Many military lords recognized the forests that provided good timber for such constructions as important resources, and controlled them with strict rules and surveillance.

Hideyoshi Toyotomi (1536–1598) dominated the whole of mainland Japan by 1592. He was the first person in Japanese history who was able to demand and receive massive contributions of timber from all parts of the Japanese Archipelago (Totman 1989). His decade-long reign consumed huge volumes of timber. He built and repaired a number of monumental buildings, including Osaka Castle, Fushimi Castle, Kyoto Palace and Hokoji Temple (to house a giant 160-foot Buddha image). Timber poured into the Kinki region from throughout the archipelago, most notably from Kumano in Kii, Hida, Mino, and Suruga to the east, and from deep within the mountains of Shikoku and Kyushu. His government also assumed control of several important timber-producing forests, such as Kiso, Hida, Yoshino and others.

Tokogawa Ieyasu succeeded Hideyoshi, and improved transportation, the single most costly element in acquiring timber. Loggers were evidently cutting far enough into the mountains to make improved transportation worth the effort (Totman 1989).

In order to secure their supply of timber, Tokogawa Shogun and other lords adopted strict forest policies: they assumed direct control of important forests and issued prohibitions on the harvesting of timber in general. They also protected the forests in order to ensure the sources of water supplies. Tokugawa Shogun and several other great lords were fond of falconry. They designated several important habitats for the breeding of the Northern goshawk (*Accipiter gentils*) and Eurasian sparrowhawk (*Accipiter nisus*) as "Osutakayama (hawk-nested mountains)," and prohibited their use by local people.

Although forest resources have been used intensively for centuries, some areas of the Japanese Archipelago's steep topography have remained largely inaccessible. In the Edo period, it is likely that there were areas deep in the mountains which feudal lords could not control effectively. Certain people surreptitiously utilized such areas in deep mountains, including bear hunters and wood turners.

Bear-hunting groups, called "matagi," appear in Edo period local government documents of the Tohoku district. The local government ordered them to provide the gallbladders of Asian black bears, which were considered to be precious medicines, to the central government. Bear hunters were mainly farmers who went hunting only in the winter season. The local government controlled the number of guns and obliged the matagi to donate the gallbladders and furs of harvested bears, but could not check the hunters' activities in the mountains directly. After the Meiji era, bear hunters moved through the archipelago to hunt Asian black bears and Japanese serows. Animal fur was quite a common commodity, because the Japanese military needed a lot of furs for winter fighting against China and Russia at that period.

As for wood turners, or "kijiya," they used large trees of *Zelkova serrata*, *Kalopanax pictus* and *Aesculus trubunata* to make bowls and trays. Once they had exhausted the large trees, they moved to find other places to work. In the Edo period, when travel was severely regulated by local governments, they secured special permission allowing them to enter the higher parts of every mountain all over the Japanese Archipelago. Actually they moved from central Honshu to Akita in the north and to Miyazaki in the south looking for suitable wood. Several documents survive describing conflicts between local people and wood turners for the utilization of wood. However, until the large-scale forest road network was built after World War II, there were huge areas of physically inaccessible places in the deep mountains of the Japanese Archipelago, and apart from bear hunters and wood turners, very few people entered such areas.

1.5 Conclusion

There is a history more than 10000 years of intensive human intervention in the forests of the Japanese Archipelago. The knowledge and skills that humans have developed concerning individual species can be considered to contain both the idea that biological resources should be used in sustainable ways, and the desire to harvest without fear of exhausting the resources. Although ethnological research into Japanese human–environmental history has highlighted such phenomena as public management of lands and resources and environmental preservation through restricted harvest, the cultural and intellectual roots of this philosophy of preservation are still unclear, and have not been convincingly traced to a particular social period or order.

Throughout the period of human habitation, the Japanese Archipelago has been blessed with a relatively warm climate and abundant rainfall, and consequently with abundant biological resources. As a result, the characteristics of the natural environment and human subsistence activities within the Japanese Archipelago varied greatly, as did the relationships between nature and human activities. Under the influence of climatic change and human activities, the distributions of individual species of plants and animals in the Japanese Archipelago and its surrounding landmasses have been constantly changing. Populations have repeatedly divided, expanded, diminished and divided in response to changes in the availability of suitable habitat. Where suitable habitat was unavailable, species became extinct. Until quite recently, there were a surprising number of physically remote and largely inaccessible places in the deep mountains of the Japanese Archipelago. Where they exist today, such places still provide refugia for plant and mammal species which inhabited the primeval forests.

References

Chiba T (1991) A study on bare mountains, 2nd edn. Sosiete, Tokyo
Kamei T, Research Group of Biogeography (1981) Fauna and flora in the Late Glacial Maximum in the Japanese Archipelago. Quat Res (Daiyonki-Kenkyu) 20:191–206 (in Japanese)
Koyama S, Sugito S (1984) A study of Jomon population: computer simulation analysis. Bull Nat Mus Ethn 9:1–40 (in Japanese with English abstract)
Ogura J (1992) History of human and landscapes read from illustrated maps. Yuzankaku, Kyoto (in Japanese)
Tokoro M (1980) A study on pre-modern forestry. Yoshikawakobun-kan, Tokyo (in Japanese)
Totman C (1989) Green archipelago. Ohio University Press, Athens
Tsukada M (1967) Vegetation and climate around 10,000 B.P. in central Japan. Am J Sci 265:562–585
Yasuda Y, Miyoshi N (eds) (1998) Illustrated history of vegetation in the Japanese Archipelago. Asakura-shoten, Tokyo (in Japanese)
Yoshikawa M, Suzuki S, Tsuji S, Goto K, Murata D (2006) Vegetation history and human activities at the Sannai-maruyama site, Aomori Prefecture. Jpn J Hist Bot 2:49–82 (in Japanese with English abstract)

Chapter 2
Eco-Cultural Diversity in Island and Coastal Landscapes: Conservation and Development

Sun-Kee Hong

2.1 Introduction

In the aftermath of the United Nations Conference on Environment and Development, held in 1992, sustainable development has become a shared goal within the international community (Weaver 2005). In this regard, various possible directions for the future have been introduced, one of which has been the goal of bringing about sustainable tourism (Luchman et al. 2007). Sustainable development has been defined as development which meets the needs of present-day tourists and host regions, while protecting and enhancing opportunities for the future. This has been envisaged as leading to the management of all resources in such a way that economic, social, and aesthetic needs can be fulfilled while maintaining cultural integrity, essential ecology processes, biological diversity, and life-support systems (WTO 1992). In addition, various alternative forms of tourism have been suggested to replace the commercially oriented tourism of the 1980s, one of these being the notion of eco-tourism suggested by Hector Ceballos-Lascurain.

The alternative form of tourism known as eco-tourism has also been referred to by other terms, such as green tourism, nature-oriented tourism, soft tourism, and defensive tourism (Abel 2004; Patterson et al. 2004; Luchman et al. 2008, 2009). At its core, eco-tourism represents a form of tourism that revolves around developing an awareness of nature and minimizing the impact on local communities, or what we can define as low-impact tourism (Fig. 2.1). The goals of eco-tourism have been defined as heightening the mutual dependency between natural and cultural resources, and coordinating development and conservation by providing high quality tourism products and experiences in a manner which also involves active

This paper is a modification of the article 'Eco-cultural resources and regional activation of maritime area', which was written in Korean with English abstract and published in the Journal of Korean Society of Rural Planning (2007; 13(3): 61–72).

S.-K. Hong (✉)
Institution for Marine and Island Cultures, Mokpo National University, 61 Dorim-ri, Cheonggye-myeon, Muan-gun, Jeonnam 534-729, Republic of Korea
e-mail: landskhong@gmail.com; skhony@mokpo.ac.kr

S.-K. Hong, et al. (eds.), *Landscape Ecology in Asian Cultures*, Ecological Research Monographs, DOI 10.1007/978-4-431-87799-8_2, © Springer 2011

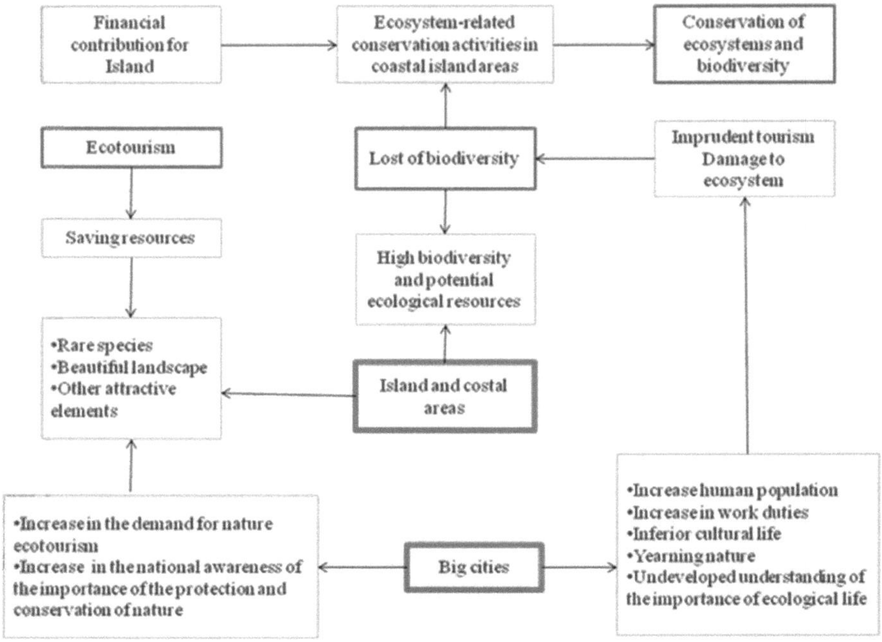

Fig. 2.1 Interaction between eco-tourism and biodiversity. Eco-tourism, which is an environmentally friendly form of tourism, revolves around the sustainable use of local ecosystems and bioresources. It is an alternative form of tourism that contributes not only to the conservation of nature, but also to the economic development of the local community

assessments of local culture (Wunder 2000; Weaver 2005). The marked increase in income and quality of life that has taken place in Korea has been accompanied by a growing desire to enjoy tourism. The ever-increasing number of tourists going abroad has had a negative impact on domestic tourism. Nevertheless, Korea's development of tourist resources continues to lag behind the demand emanating not only from Koreans, but also from foreign nationals staying in Korea for prolonged periods of time, who all desire opportunities to experience Korea's wondrous nature and culture.

While Korea is a maritime country that is surrounded by water on three sides, the great majority of the development which has taken place has been centered on large urban areas. The diplomatic war raging between Korea and its neighbors over what constitute the boundaries of each country's exclusive economic zone (EEZ) has had the effect of greatly increasing the importance of island and coastal areas in Korea (KORDI 2001, 2002). Meanwhile, tidal flats, which are widely perceived as the jewel of maritime resources, can be regarded as the most important ecosystem in terms of both biodiversity and productivity. However, these tidal flats are also regarded as a major cultural resource (Hong 2007; Hong 2010; Hong et al. 2010). Biodiversity and the cultural activities of humans have long enjoyed a close relationship in the case of island and coastal areas. The Earth Summit held in Rio de Janeiro, Brazil, in 1992 was convened to address urgent global problems pertaining to environmental change and the decrease of biological diversity. These problems have also become international socio-economic issues. Approaches to the resolution of biodiversity-related problems that are based solely on natural sciences are now perceived as being outdated efforts associated with the twentieth century. Instead, recent discussions of these matters have revolved around the implementation of interdisciplinary studies that are focused on the socio-cultural aspect of such issues. An international conference was held in Paris in September 1998 on the theme of "Biological and cultural diversity." Later, during a seminar hosted by the National Museum of Natural History (Muséum National d'Histoire Naturelle) in Paris in May 2000 entitled "Biology Education 2000," an effort was made to highlight the socio-cultural values of biodiversity. Japan's Environment Agency has implemented a national project that revolves around an investigation and research into biodiversity. Such a project is designed to restore the gradually disappearing eco-cultural characteristics of traditional villages in island and coastal areas. As part of such ongoing efforts to activate the economies of island and coastal areas, Japan's 100 museums of natural history have undertaken the task of developing the natural and cultural resources found in such areas into tourism resources.

This study seeks to review the possibility of developing eco-tourism resources, with the main focus being on the biological and cultural diversity of Korea's various island and coastal areas. Moreover, an effort is also made to activate regional communities and heighten Korea's maritime competitiveness by exploring some of the methodologies related to the study of sustainable eco-cultural tourism. In addition, this study also analyzes the possibility of developing sustainable eco-cultural tourism in Korea based on an examination of the materials related to natural resource indicators pertaining to the island and coastal areas in the southwest of the Korean peninsula.

2.1.1 The Necessity of Eco-Tourism in Island and Coastal Areas

Maritime countries in South East Asia and Mediterranean islands have sought to foster economic development in their respective island and coastal regions by promoting eco-tourism or green tours, and the development of tourism products that are intricately linked to island and coastal cultures, maritime ecosystems, and biodiversity (Luchman et al. 2007, 2008; Vogiatzakis et al. 2008). Meanwhile, China, whose rapid industrialization has transformed it into an economic powerhouse, has evaluated eco-tourism as a new economic good which makes use of its natural and cultural resources. To this end, the Chinese central government has provided large amounts of money for the promotion of eco-tourism (Zhao et al. 2004). These examples from the island and maritime cultures of other countries have significant implications for the development of Korea's own island and maritime culture policies. This is particularly important because Korea possesses various important species of fish and a cultural life that involves thousands of islands, tidal flats, and coastal sand dunes. Korea also has numerous uninhabited and uninhabitable islands. To this end, the Special Act on the Ecosystem Conservation of Small Islands, such as Dokdo Island, designates 153 islands as "special islands" (Ministry of Environment 2005).

All of these islands boast outstanding natural landscapes or unique biological species. More to the point, many of these islands should be preserved because of their status as habitats for wild animals or wintering sites for migratory birds. However, the majority of these islands have suffered because of poor management, which has resulted in widespread damage to island and coastal area ecosystems, as well as the depletion of fisheries because of reckless overfishing. It is believed that such islands can, when properly managed, become major resources for eco-tourism in the form of ecological cruises.

The development of maritime transportation routes has resulted in the islands along the southwest coast of Korea gradually emerging from their former state of seclusion. In this regard, in the case of areas such as Jeonnam Province's Shinan-gun (*gun* is county in the Korean language), it is expected that the construction of additional bridges will contribute greatly to the invigoration of transportation and commercial activities. Should this come to pass, it is only natural to assume that land-based tourism options will become more popular than ship-based ones. This will lead to an increase in the number of tourism products which revolve around visitors staying in the relevant island or coastal area for extended periods of time. This in turn will create a strong demand for a wider variety of themes and spaces in order to satisfy the natural curiosity of urban visitors. Here, the forest village culture/maritime ecosystem experience which has been offered along Japan's Seto Inland Sea provides a salient example of connecting marine ecosystems to culture via eco-cultural networks and tourism resources (Yanagi 1998; Hong et al. 2007). The activation of a region through tourism requires that the institutional background necessary to develop and manage tourism resources should be in place. To this end, it is necessary to do away with the complicated institutions and regulations regarding the management of cultural properties which have been put in place by various government departments, and establish a body which is responsible for the integrated management of island and coastal tourism. This, in turn, will require holding in-depth discussions on such topics as the design of measures pertaining to the use of eco-cultural tourism to activate regional communities, and the promotion of international maritime cruise tourism as a means to boost international exchanges. Such discussions should be part of the preparation of the proposed Special Act on the Development of the Southwest Coast.

Under such circumstances, it will be necessary to develop the traditional Korean folk and cultural resources associated with fishing villages, with the aim of using such resources as regional eco-cultural resources. Such an exercise should be conducted in conjunction with efforts to ensure the sustainable development of natural resources such as the tidal flats along Korea's southwest coast. These efforts are required as part of wider attempts to promote island and coastal tourism at the international level (Ministry of Maritime Affairs and Fisheries 2001). To this end, it is essential that an ecological investigation should be conducted of the biota, and especially of the natural resources and cultural landscapes, such as natural monuments, so intricately related to the life and culture of island and maritime regions, and that the related environmental problems should be analyzed. Therefore, it is essential to establish comprehensive management methods by which the ecological and cultural capacity of island and coastal areas can be increased.

2.2 Development of Eco-Tourism Resources

2.2.1 *The Link Between Historical–Cultural–Ecological Resources Along the Southwest Coast in Korea*

The process of selecting the islands where eco-tourism should be promoted is a very important one. This is because any hasty development of biological resources in circumstances where the relevant eco-tourism infrastructure has yet to be properly established will inevitably result in tremendous damage being wrought to the ecosystem. Furthermore, as the Korean public's awareness of the importance of the natural environment and its attitude toward tourism continues to lag behind that of the people of advanced countries, concerns have been raised about the implementation of eco-tourism based solely on ecological resources.

Fortunately, Korea's island and coastal areas are endowed with beautiful natural resources, cultural resources that are embedded in these natural resources, and a strong historical background. To this end, it is necessary to develop tourism products for these island and coastal areas that bring together these historical, cultural, and ecological resources. The characteristics of eco-tourism in island and coastal areas in various regions which are considered to feature distinct island and coastal cultures, as well as maritime ecosystems, are summarized in Table 2.1.

There are many examples of trade and cultural exchanges taking place with China through the West Sea. In this regard, vestiges of such trade and cultural exchanges can still be found in the estuaries of the West Sea. In particular, the Shinan area of Jeonnam Province and Taean peninsula in Chungnam Province served as cultural gateways or marine Silk Roads through which a ceramics trade was carried out with the Tang (唐) and Yuan (元) dynasties of China as well as Japan (倭) during the Goryeo Era (高麗時代). In addition, as the coastal ecosystem in this region includes both large-scale tidal flats and salt marshes connected to river estuaries, the area serves as an important sanctuary for migratory birds and as a habitat for salt plants and benthic organisms. The natural sand dunes and beaches found on the islands and coastal areas along the West Coast, as well as the tidal flats that represent the area's core ecological resources, are perceived as being of particularly great value. In this regard, it has been surmised that the natural resources found in the Jeungdo-myeon area of Shinan County in Jeonnam Province (UNESCO Biosphere Reserve site), an area which includes tidal flats and the already developed Ujeon-ri beach and related black pine tree forests, represent prime candidates for the development of eco-tourism (Hong et al. 2006). Furthermore, Jeungdo-myeon also possesses the biggest saltpan in the West Coast area, and is not too far away from the site in the seas off Shinan where the remains of a trade ship have been found. Therefore, because it possesses the historical/cultural/ecological resources needed to satisfy visitors' curiosity, the area has been hailed as an ideal candidate for eco-cultural tourism. The *Beopseongpo* (法聖浦) area of Yeonggwang-gun in Jeonnam Province was the place where Buddhism was first introduced to Korea by the Indian Buddhist monk *Marananta*, who arrived in Baekje (百濟國) via

Table 2.1 Environmental and historical characteristics of island and coastal areas in Korea

Region	Island	Historical characteristics	Cultural characteristics	Ecological characteristics
West Sea region	Ganghwa, Wido, Seonyudo, Heuksando, Jindo, and Gageodo	• Bridgehead in Korea–China trade • Site of conflicts with imperial powers such as the Hideyoshi Invasions of 1592 • Potential for North and South Korean exchanges to activate the fishery industry after reunification	• Gateway for historical and cultural exchanges with China • Development of rural culture	• Ria coast and significant difference between ebb and flow • World-class tidal area, migratory bird sanctuary, plants on tidal flats, and benthic organisms • Habitats for wild animals and plants (sea otters and orchids, etc.) • Changes in the ecosystem caused by large-scale reclamation projects • Serve as estuarine ecosystems for the major rivers
	Tidal cultural zone Tidal flat–island–coastal mosaic			
South Sea and Jejudo Island region	Odongdo, Geomundo, Seukdo, Wando, Bogildo, Geojeodo, and Jejudo Islands	• Trade hub during the ThreeKingdoms Era (Cheonghaejin) • Exchanges with Japan (倭)	• Gateway tomainland culture • Developmentof various species of fish and food culture • Folk culture of the southern provinces	• Marshes, migratory bird sanctuary, and evergreen broad-leaved forest habitat • Home of the biggest *Camellia japonica* forest
	Transboundary region of island and maritime culture Transboundary sphere connecting the east and west cultural and ecosystem zone: human, ecosystem, and cultural exchanges between east and west			
East Sea region	Euleungdo and Dokdo Islands	• Frictions in the Pacific Rim (Russia, Japan and North Korea)	• Linked to highlands in Korea • Development of fisheries and mountainous village culture	• Simple coastline • Changes in the current and increase in the sea surface temperature caused by global warming • Emergence of problems within the marine ecosystem such as the changes in fish species and emergence of the so-called albinic phenomenon
	Pacific Rim Zone Marine resources from the East Sea			

China. The area boasts not only ecologically valuable tidal flats, but is also famous for being the biggest producer of dried croaker in Korea, a distinction which it obtained by combining the croakers caught in the Chilsan fishery and the sun-dried salts produced in the islands along the West Coast (Fig. 2.2). Therefore, the development of coastal tourism in this area should revolve around the promotion of eco-cultural tourism products through which visitors could take in the tidal ecosystem in Beopseongpo, experience the process of manufacturing dried croaker, and explore the relics related to the introduction of Buddhism. As we can see from these examples involving the West Coast area, the economies of island and coastal areas can best be activated, and the natural environment of the relevant communities protected, by developing eco-cultural tourism products which are designed to provide visitors with an opportunity to experience not only the local ecosystem, but also the area's history, culture, and everyday life.

A clear link exists between biodiversity and cultural diversity. The introduction of a foreign organism can change not only the ecological environment of its new habitat, but also the cultural environment of local people. For example, the development of deep-sea and coastal fishing technology inevitably resulted in changes to the marine culture of fishing villages and the lifestyle of fishermen. In the case of a peninsular country such as Korea, the introduction and coordination of foreign organisms in fishing villages and estuaries has become much more commonplace because of the widespread exchanges which take place through island and marine routes. Nevertheless, the culture of fishing villages, which is embedded in the everyday lives of fishermen,, and their social organizations, which have been maintained for thousands of years, have now begun to collapse under the weight of rapid development.

Fig. 2.2 Landscape of *Beopseongpo* of Yeonggwang, Jeonnam Province

Viewed from this standpoint, it is imperative that social and cultural activities are carried out based on the relationship between the natural environment and biological resources (Naveh 2000; Bridgewater and Bridgewater 2004). Recently, a wide range of civic activities relating to the conservation of nature have been undertaken in urban areas. These have included conducting research projects in which citizens have played a pivotal role, citizen-initiated public projects, and bio-oriented villages. However, cultural diversity has been excluded from the bio-oriented activities that have been carried out in some cities (Kiss 2004).

Much like biodiversity, culture also has a systemic structure (Mitchell and Bugger 2000; Jongman 2004). Island and marine systems feature various cultural resources which have been created through interactions between organisms, nature, and humans. Viewed from this standpoint, the numerous tangible and intangible resources which have been created, used, and preserved by the residents of island and coastal areas can be identified as the main elements in terms of cultural diversity. From an eco-cultural standpoint, eco-cultural tourism can be said to revolve around a search for the significance of the biological resources which have long played a role in preserving cultural diversity, and the development of such resources as regional environmental resources. Furthermore, this form of tourism is geared toward the development of eco-cultural conditions that will make it possible to transform the organisms, nature, and cultural resources of islands and coastal areas into Korea's main tourism resources.

2.2.2 Securing Resources

There is a need to develop organisms and cultural elements in a manner which takes into account the links between the relevant area's historical background and biological resources. Korea's coastal areas have played a central role not only in the nation's trade and exchanges with foreign countries, but also in its struggles and conflicts with the latter. These areas have served as a hub through which exchanges have taken place, and cultural elements from China, Japan, and Southeast Asia have been imported. Korea's status as a peninsular country has thus resulted in its islands and coastal areas accepting many aspects of the cultures of other countries. In this regard, it is important that the development of the cultural resources of islands and coastal areas as eco-cultural tourism resources should be carried out in a manner which takes into account their differences from the indigenous life and culture of Korea. Table 2.2 shows the items which, based on a survey and analysis of island and coastal areas, have been identified as potential tangible eco-tourism resources.

In order to determine the current state of core natural organisms (animals, plants, and vegetation), the documentary material compiled by university and national research institutes, as well as that compiled as part of the monitoring of biological resources by government agencies, should be closely examined. In addition, a survey of the biota in tidal flats, tidelands, and estuaries along the southwest

Table 2.2 Index survey of eco-cultural tourism in island and coastal areas and a related case study

Subject	Item	Survey goal
Natural organismsBiota in tidal flats, tidelands, and estuaries	• Habitats for animals, plants, rare species, endangered species, and exotic species • Research into materials related to biological resources in individual island areas	• Monitoring of data related to biological resources • Determination of the ideal spatial distribution in accordance with an analysis by university and research institute-based researchers • Analysis of the use of organisms and the establishment of a conservation strategy
Cultural and landscape elements and natural monuments in island and coastal areas	• Natural monuments, village forests, sacred forests, fish shelter forests, windbreak forests, and old trees	• Analysis of spatial distribution and characteristics • Implementation of the survey of biota • Monitoring of sustainable generative processes and of the state of the management of such processes • Development of long-term management guidelines for natural monuments
Cultural background and characteristics	• Natural monuments in island/coastal and inland areas	• Implementation of a comparative survey • Study of the cultural links between island and inland areas
International exchanges	• Comparison of similar items found in other countries	• Analysis of the international process of conveying life culture

coast should be conducted, and a comparative analysis carried out using existing documentary materials. In this regard, an analysis of the habitats of natural organisms (rare, endangered, and exotic species) needs to be conducted, and their spatial distributions recorded. Such an exercise is required in order to identify the use which can be made of these organisms in the future, and to establish a conservation strategy which revolves around limiting the damage caused by eco-tourism. In addition, comparative analyses of other materials related to biological resources in individual islands and coastal areas must be carried out. The spatial distribution and characteristics of the cultural landscape elements and natural monuments (village forests, sacred forests, fish shelter forests, windbreak forests, and old trees) found in island and coastal areas should also be examined, and a survey of the biota in these areas conducted. Particular attention must be paid to consistent monitoring of the regeneration process of forests, and to the quality of the management of such processes. Such an approach is required in order to develop long-term management guidelines for natural monuments.

Determination of the background and characteristics of every culture requires that the cultural linkages between islands, coastal areas, and inland areas must be examined through a comparative analysis of the natural monuments found in all areas (Korea Forest Service 1999). Some of the forests found in island and coastal areas are of great eco-cultural significance. For instance, the rituals known as *pungeoje* (豊魚祭, the fishermen's ritual prayer for an abundant catch) and *dangje* (堂祭, a communal ritual to the village god) have long been implemented in such forests (Korea Forest Service 1999, Fig. 2.3). The traditional fishing method called *doksal* (a primitive fishing method which involves piling stones in tidal areas and catching the fish which become trapped when the tide ebbs) and forests known as *eoburim* (魚付林, fish shelter forests located near the coastline or on the edge of an island to protect the village from ocean winds and facilitate the gathering of fish, *see* Chap. 18 in this book) can be regarded as part of the traditional fishing culture based on fish habitats which took root in these areas. Such practices and entities are perceived as valuable eco-cultural tourism resources. In this regard, some of the fishing villages along the west coast, and in particular in the Taean peninsula area, have established tourism programs that advertise traditional village forests and *doksal* as part of what they refer to as an opportunity to experience agricultural and fishing villages and eco-tourism (*personal information from* Taean Agricultural Technology Center). The natural beaches found along island and coastal areas are surrounded by pine tree forests and broad sandy expanses. While the species of trees found in coastal forests differ from region to region, the majority consist of black

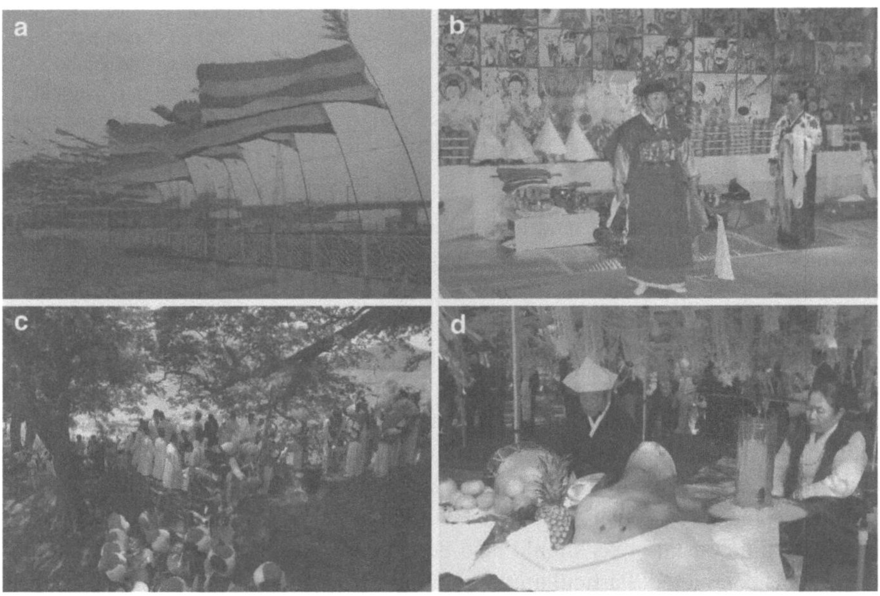

Fig. 2.3 Traditional cultures such as the *pungeoje* (the fishermen's ritual prayer for an abundant catch) and *dangje* (a communal ritual to the village god) in island and coastal areas (**a** and **b**. Youngheung-do in Ongjin-gun, **c**. Beopseongpo of Yeonggwang, **d**. Taean peninsula)

Fig. 2.4 *Eoburim*, *Woosil* and *Doksal* in island and coastal areas (**a**. *Woosil* (stonewall) in Bigeumdo, Shinan County, **b**. *Eoburim* in Mulgun-ri, Namhae County, **c**. *Woosil* (strip forest) in Anjwa Island in Shinan County, **d**. Doksal in Namhae County)

pine (Jeonnam and Jeonbuk Provinces) or red pine (north of the Taean peninsula) (Choung and Hong 2006).

Coastal forests help prevent the erosion of sand dunes by blocking ocean winds, and facilitate the growth of vegetation within such sand dunes (Kim and Hong 2009). In addition, these forests also prevent the ocean winds from damaging crops, and serve as recreational areas for local residents (Hong and Kim 2007). Some of the coastal forests found in Jeonnam Province have been referred to as *woosil*, a term which refers to a stone fence protecting a village from the wind (Fig. 2.4). Similar eco-cultural elements can be found in the islands and coastal regions of China, Japan, and Southeast Asia. To this end, a review of natural monuments, biodiversity, and the international conveyance of life culture should be conducted through gradual on-the-spot inspections, investigations, and analyses of these island and coastal regions, with the results of such a review being used as educational guidelines for potential eco-tourists.

2.3 Investigation of the Social and Cultural Infrastructure

The socio-environmental capacity to implement eco-tourism projects can be assessed by reviewing the general social and cultural resources (such as forest and land-use types, local participation, the location of the village, and transportation systems) of island and coastal areas designated for eco-tourism. The presence of outstanding eco-tourism resources alone should not automatically qualify an area as one which should be developed to this end. This is because ecosystems are very difficult to preserve and manage without the facilities needed to observe and understand them, or the transportation systems required to access the relevant area. Therefore, transportation systems should be established and unpaved roads made in order to facilitate access to the relevant area, albeit in a manner which ensures that the damage to the ecosystems remains minimal. To this end, renting bicycles to tourists so that they can access the sites where the ecosystem can be experienced is perceived as a good measure. A perfect candidate for such a means of transportation is the west coast island of Gogunsan, which, while difficult to access by car, can easily be toured by bicycle. While such a move is naturally motivated initially by the absence of roads capable of accommodating large numbers of automobiles, the use of bicycles can also minimize the human impact on natural resources and ecosystems.

A look at the production structure of island and coastal areas in Korea reveals that the majority have been engaged in agriculture and fishing. While people engage in agricultural activities during the farming season, they often fish in local waters during the off season. While some variations exist, the most common seasonal fish found in specific regions of the West Sea include the blue crab in Ongjin County, and the sea bass, pomfret, croaker, and kingfish in Chungcheong Bay and Shinan County. In this regard, the seasonal nature of agriculture and fishery in these island and coastal areas may have the effect of limiting visitors who desire to participate in such activities. Conversely, eco-tourists can also affect the residents' ability to engage in their regular production activities. A perfect case of such a conundrum is Jeungdo Island in Shinan County. In addition to being home to the biggest sun-dried saltpan in Korea, the island is also home to various kinds of salt plants and inland tidal fish species such as the goby fish. As such, the area attracts many visitors. However, those engaged in the saltpan industry have systematically refused to allow these visitors to enter the saltpan and take pictures. It is also difficult for visitors to experience the wildlife found in tidal waters as the type of clam known as the *Dosinia japonica* is cultivated in the tidal flats where the gobies live. There are many instances of cases where potential eco-tourism sites overlap with those closely related to the economic activities of local residents.

Eco-tourism cannot be carried out without the active participation and cooperation of local residents. Therefore, it is necessary to develop social coordination mechanisms, such as education programs for local residents, which are designed to heighten local participation in island and coastal tourism and promote exchanges with visitors. In this regard, any relevant social and cultural investigation should include a thorough review and analysis of residents' participation prior to the implementation of eco-tourism projects. The materials used as part of this social and cultural investigation should be collected and analyzed before and after the implementation of eco-tourism projects. In addition, this cultural and social material should be reviewed and analyzed on an annual basis in order to keep up with trends in residents' participation in eco-tourism projects.

2.3.1 Case Study

2.3.1.1 Outline of the Case Study Area

The significant tidal variations that characterize the southwest coast of Korea have led to the creation of a complex ria coast and a broad range of tidal zones. Tidal flats, in which various kinds of shellfish and fish species live, can be regarded as one of the most representative tidal zones. The southwest coast of Korea is also dotted with a large number of islands, many of which belong to Shinan County. Consequently, the geographical characteristics of the area and the physical characteristics of the tidal flats have resulted in numerous fisheries and estuaries being formed along the southwest coast of Korea. This, in turn, has led to the formation of small- and large-scale fishing villages up and down the coast. The residents of these fishing villages traditionally carried out the rituals known as *pungeoje* (the fishermen's ritual prayer for an abundant catch) and *dangje* (a communal ritual to the village god) in the local seas and coastal forests. However, the drive for modernization that began in the 1960s ushered in rapid changes in the socio-economic environment of fishing villages and the disappearance of traditional practices and customs. It is therefore necessary to investigate and restore the eco-cultural landscape elements through which this now discontinued traditional culture can be linked to the natural ecosystem. Thereafter, these elements should be developed and used by fishing villages as eco-tourism resources, an effort that should come as part of wider ones to establish an eco-cultural network for island and coastal areas.

2.3.1.2 Survey Method

This study conducted a case survey of the natural resources and village forests found on the main islands along the southwest coast, and in the main coastal areas and estuaries along the west coast. The islands surveyed included those found in Shinan County's Imja-myeon, Jeungdo-myeon, and Jido-myeon, as well as other islands within the county. The index of the natural resources of islands of Shinan County consisted of a survey of geographical characteristics, island types, topography, and climate, and of the main local products and *bohosu* (保護樹, nurse trees). Meanwhile, the survey of traditional forests revolved around the collection of data regarding the current state of the old village forests [*nogeosu*: (老巨樹, old and large trees)] in island and coastal areas. The background and state of the management of village forests, as well as eco-cultural characteristics, were in turn analyzed with the help of village residents.

2.3.1.3 The Results of the Case Survey

The islands of Imjado and Jeungdo attached to Shinan County are located some 20–30 min away by ship from the center of Shinan County, which is itself connected to Muan County. Imjado boasts rich coastal resources, including the longest sandy beach (12 km) in Korea in Daegwang beach. Imjado serves as a tourism resort during the summer season. Meanwhile, the port of Jeonjangpo used to be the largest center in Korea for the processing of salted shrimp. However, one can still see the caves in which these salted shrimp were stored. The island of Jeungdo is the result of a long-term process which involved the reclamation of 99 small islands. Jeungdo, where saltpans were first developed, is still home to the Taepyeong saltpan, which produces the largest amount of sun-dried salt in Korea. The Taepyeong saltpan has long served as an important cultural resource that attracts many tourists to Jeungdo Island. Shinan County has recently begun to develop Jeungdo as the center of Dadohae eco-cultural tourism. Jeungdo includes Ujeon-ri beach and its black pine forest. Jeungdo Island is one of the UNESCO Shinan Dadohae biosphere reserve zones in Korea (Fig. 2.5).

The island town of Jido in Shinan County can easily be accessed from Muan County via a land bridge. Jido's oceanic climate is responsible for its red soil and the outstanding dry field crops such as onions and garlic which are grown there. The main islands in Shinan County are characterized by changes in their landscapes brought about by reclamation projects. Given the difficulties and inconvenience of transportation to and from these islands, and the need to supply food to the mainland, these reclamation projects can be said to have provided the residents with an important socio-economic infrastructure. The rice farming and saltpan activities carried out on these reclaimed lands have also helped to resolve food problems and create income. Nevertheless, the socio-economic environment in the Shinan area has undergone rapid changes that have included the importation of salt, changes to the fishing industries, the decline of agriculture, and a decrease in the population. The rapid social changes that have taken place within fishing villages have resulted in traditional land uses and fishing village culture either collapsing or being significantly transformed. Almost all the traditional village forests situated along the southwest coast, including those which feature the *nogeosu* (large old trees), have either disappeared or been devastated. Although reclamation can be blamed for many of the problems faced by

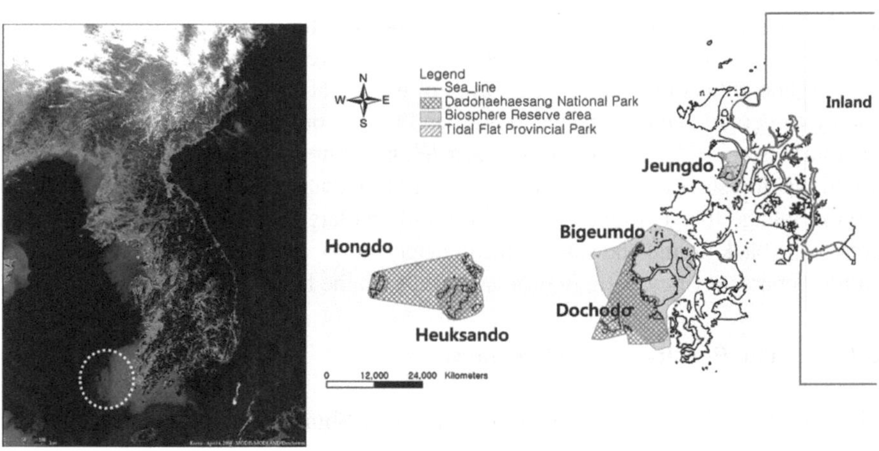

Fig. 2.5 Location and map of the third designated UNESCO Shinan Dadohae biosphere reserve in Korea (Lee et al. 2010)

these traditional village forests today, the onset of their decay can actually be traced back to the absence of responsible management of elements of traditional culture such as the *pungeoje* (the fishermen's ritual prayer for an abundant catch) and *dangje* (a communal ritual to the village god). The economic foundations which made possible the holding of such rituals as the *pungeoje* and *dangje*, which were commonplace until some 30–40 years ago, subsequently evaporated as local populations became increasingly elderly and the fishing industry collapsed. In addition, one finds many instances in which *dangsup* (sacred forests) were intentionally destroyed as new religions began to be introduced to the area from the 1970s onwards.

Coastal forests have also been greatly impacted by the increased footfall. The black pine forests found along smaller beaches such as Daegwang beach in Imjado, Ujeon-ri beach in Jeungdo, and Holtong, Topmeori, and Jogeum-naru beaches in Muan County have been adequately protected and preserved. However, in the case of beaches visited more frequently by tourists, the pressure produced by the feet of tourists has resulted in the roots of black pine trees becoming exposed, and in the extinction of the strip-type forests characteristic of coastal forests. Generally speaking, the disappearance of traditional forest culture has resulted in the traditional village forests found along the southwest coast faring rather badly in terms of the preservation of the forest ecosystem. Recently, efforts have been made by various environmental organizations, such as the Cultural Heritage Administration of Korea, to develop and provide support for elements of the eco-culture of island and coastal areas such as village forests. However, the problems plaguing traditional landscape elements such as the *eoburim* (魚付林, fish shelter forests) and *haeannim* (海岸林, coastal forests), which were so closely related to human life, cannot be resolved by restoration efforts starting solely from the standpoint of natural ecosystems. Instead, the ecological functions of village forests can only be restored when such an exercise is carried out in conjunction with the restoration of traditional culture.

The characteristics of the ecological landscapes of Korea's islands and coastal areas have traditionally been closely related to the agriculture and fishing that lie at the center of the life and culture of local residents. The traditional practices and customs of fishing villages, including *pungeoje*, were also closely related to the everyday use of landscape resources. The development of ports and fishing methods, as well as the use of fisheries (tidal flats), became part of the traditional cultural landscape which was intricately related to the natural resources. Recently, tourism programs that make use of the natural and cultural resources associated with tidal flats as part of an opportunity to experience eco-villages have been carried out in some of the fishing villages along southwest coastal areas such as the Taean peninsula. However, the majority of these programs have been more akin to events designed to increase the income of the village. In some instances, the landscape and ecosystem of fishing villages has actually been destroyed to attract more participants. These problems have for the most part been caused by a lack of specialist–resident participation systems, through which the local culture and ecological resources can be managed in an integrated manner, and the failure to carry out continuous ecosystem monitoring. A more desirable approach would be to establish eco-tourism programs for island and coastal areas that are based on small numbers of people, such as family units, in order to avoid damage to the ecosystem of these island and coastal areas.

This case study of island and coastal areas along the southwest coast focused on ecological resources, and in particular, on the cultural characteristics and the ecological management of traditional villages and coastal forests. A long-term research, monitoring, and protection strategy should be established in conjunction with the biota and vegetation that represent the main natural resources used for eco-tourism, as well as for the eco-cultural landscape elements so intricately related to these resources. Such a strategy should, in turn, be employed when developing the tourism resources of island and coastal areas in Korea, including those along the southwest coast.

2.4 Conclusions

The island and coastal areas of Korea are currently plagued by a lack of the infrastructure needed to conduct simple eco-tourism based on observing and experiencing the biota and the local ecology. To this end, better use should be made of the tangible and intangible resources possessed by individual islands and coastal areas, such as local history and culture, and great care should be taken over the development of related resources which can be linked to eco-tourism. The development of such eco-tourism resources requires that basic research into natural resources and the investigation thereof should be carried out by specialists capable of conducting studies on topics related to natural history, such as indigenous knowledge, folk customs, the social environment, and the history of people's everyday lives (Hong 2010). It is essential that the items to be researched are selected by such specialists.

After an investigation of ecological resources such as the biodiversity and unique topography of island and coastal areas conducted by ecology specialists, it is also necessary to prepare assessments of what constitutes an appropriate site for eco-tourism based on discussions with specialists in the field of natural history and the cultural/social sectors. At this point, after preparing such an assessment of land-use types, vegetation, and the natural landscape, sites appropriate for eco-tourism can be selected based on an analysis of their spatial–ecological characteristics using Geographic Information System technology (Nakagoshi and Hong 2001).

Once the island and coastal areas that mesh with the goals of eco-tourism have been selected, it becomes necessary to analyze the cultural and social infrastructure, as well as the tourism resources of the relevant island and coastal area, and to plan touring methods that are based on the basic research which has been conducted (Weaver 2005). Desirable guidelines for eco-tourism should revolve around educating visitors to use and experience the resources which can be found in the relevant area in an appropriate manner. In other words, *blue*(seascape or water body)–*green*(landscape or green space)–*human*(anthropogeneous landscape) networks (Hong 2007) linking together urban–rural–coastal–island areas can be created where visitors from the mainland are able to experience and learn about ecological resources with the active cooperation of the residents (cooperation with the residents to create green tours through which visitors can see, feel, and learn about the relevant resources).

The successful conduct of such community-based eco-tourism requires a thorough understanding of the characteristics of island and coastal areas, the maximization of limited resources, and visitors' cooperation with the residents who are actively participating in the relevant activities (Mitchell and Reid 2001; Kiss 2004).

Islands and coastal areas have only a very limited number of bioproductive resources. In addition, unlike terrestrial ecosystems, island and coastal ecosystems feature ecological characteristics such as isolation and openness. As a result of these characteristics, one finds many instances in which previously stable ecosystems in island and coastal areas have been damaged as a result of visitors introducing foreign species (Gössling 1999). The ecosystems of island and coastal areas are extremely sensitive to external disturbances (Vogiatzakis et al. 2008). As the dissemination of introduced organisms (exotic plants and pets, etc.) or disease (pine wilt disease, *see* Chapter 16 in this book) can lead to environmental problems which can greatly damage sustainable eco-tourism, steps must be taken to avoid the spread of such foreign influences in the first place. There are presently 14 different laws regarding the use, development, and protection of Korean island and coastal areas (Ministry of Environment 2005). There is a direct link between such laws and the ability to promote eco-tourism in island and coastal areas. More to the point, the smooth implementation of eco-tourism programs using the biodiversity and culture of island and coastal areas will require consistent application of such laws, and the integrated management of administrative organizations related to islands (Patterson et al. 2004).

Acknowledgment This research was supported by the National Research Foundation of Korea Grant funded by the Korean Government (NRF-2009-361-A00007).

References

Abel T (2004) Ecosystems, sociocultural systems, and ecological economics for understanding development: the case of ecotourism on the island of Bonaire, N.A. PhD Thesis, University of Florida

Bridgewater P, Bridgewater C (2004) Is there a future for cultural landscapes? In: Jongman RHG (ed) The new dimensions of the European landscape. Springer, Dordrecht, pp 193–200

Choung HL, Hong SK (2006) Distribution patterns, floristic differentiation and succession process of *Pinus densiflora* forest in South Korea: a perspective from nation-wide scale. Phytocoenologia 36(2):213–229

KORDI (Korea Ocean Research and Development Institute) (2001) Studies on inventories and a sustainable use of tidal flats in Korea. Ministry of Marine Affairs and Fisheries. BSPM 118-00-1370-3. 1214 p

KORDI (Korea Ocean Research and Development Institute) (2002) Studies on inventories and a sustainable use of tidal flats in Korea. Ministry of Marine Affairs and Fisheries. BSPM 16100-1465-3. 885 p

Gössling S (1999) Ecotourism: a means to safeguard biodiversity and ecosystem functions. Ecol Econ 29:303–320

Hong SK (2007) Linking man and nature landscape systems: landscaping blue-green network. In: Hong SK et al (eds) Landscape ecological applications in man-influenced areas: linking man and nature systems. Springer, Dordrecht, pp 505–523

Hong SK (2010) Island ecology on biological-cultural diversities and human adaptation in seascapes. J Ecol Field Biol 33(2):115–120

Hong SK, Kim JE (2007) Village forests of estuary in West Sea: forested landscape management for conserving resort area. Isl Cult 29:441–473 (in Korean with English abstract)

Hong SK, Park JW, Yang HS (2006) Ecological characteristics of black pine forest as ecotourism resource-Jeungdo, Shinan-gun, Jeonnam. Isl Cult 28:223–244 (in Korean with English abstract)

Hong SK, Nakagoshi N, Fu B, Morimoto Y (2007) Landscape ecological applications in man-influenced areas: linking man and nature systems. Springer, Dordrecht, 535 p

Hong SK, Koh CH, Harris RR, Kim JE, Lee JS, Ihm BS (2010) Land use in Korean tidal wetlands: impacts and management strategies. Environ Manage 45:1014–1026. doi:10.1007/s00267-006-0164-3

Jongman RHG (2004) Landscape linkages and biodiversity in European landscape. In: Jongman RHG (ed) The new dimensions of the European landscape. Springer, Dordrecht, pp 179–189

Kim JE, Hong SK (2009) Landscape ecological analysis of coastal sand dune ecosystem in Korea. J Kor Soc Environ Restor Tech 12(3):21–32

Kiss A (2004) Is community-based ecotourism a good use of biodiversity conservation funds? Trends Ecol Evol 19:232–237

Korea Forest Service (1999) Green tourism and mountain village activation. Korea Forest Service, Seoul

Lee HJ, Cho KM, Hong SK, Kim JE, Kim KW, Lee KA, Moon KO (2010) Management plan for UNESCO Shinan Dadohae Biosphere Reserve (SDBR), Republic of Korea: integrative perspective on ecosystem and human resources. J Ecol Field Biol 33(2):95–103

Luchman H, Hong SK, Kim JE, Nakagoshi N (2007) Nature-based tourism in small islands adjacent to Jakarta City, Indonesia: a case study from Seribu Island. J Kor Wet Soc 9(1):31–46

Luchman H, Hong SK, Kim JE, Nakagoshi N (2008) Tourism and cultural landscape at the Tengger, East Java, Indonesia: the implications for ecotourism planning. Kor J Env Eco 22(3):207–220

Luchman H, Kim JE, Hong SK (2009) Cultural landscape and ecotourism in Bali Island, Indonesia. J Ecol Field Biol 32(1):1–8

Ministry of Environment (2005) Report for conservation of natural environment of coastal and islands area. Korea Environmental Institute, Seoul, 168 p

Ministry of Maritime Affairs & Fisheries (MOMAF) (2001) MOMAF-PEMSEA regional workshop on Sihwa, management strategy and regional initiatives for coastal environmental management. March 2001, Seoul. 318 pp

Mitchell N, Bugger S (2000) Protected landscape and cultural landscapes: taking advantage of diverse approaches. George Wright Forum 17(1):35–46

Mitchell RE, Reid DG (2001) Community integration: island tourism in Peru. Ann Tour Res 28:113–139

Nakagoshi N, Hong SK (2001) Vegetation and landscape ecology of East Asian Satoyama. Glob Environ Res 5:171–181

Naveh Z (2000) The total human ecosystem: integrating ecology and economics. Bioscience 50(4):357–361

Patterson T, Gulden T, Cousines K, Kraev E (2004) Integrating environmental, social and economic systems: a dynamic model of tourism in Dominica. Ecol Modell 175:121–136

Vogiatzakis IN, Pungetti G, Mannion AM (2008) Mediterranean island landscapes: nature and cultural approaches. Springer, Dordrecht, 372 p

Weaver DB (2005) Comprehensive and minimalist dimensions of ecotourism. Ann Tour Res 32:439–455

WTO (1992) Agenda 21 for the travel and tourism industry.

Wunder S (2000) Ecotourism and economic incentives- an empirical approach. Ecol Econ 32:465–479

Yanagi T (1998) Nature and environment of Seto Inland Sea, Japan. Kobe Shinbun Publisher, Japan, 244 p

Zhao B, Kreuter U, Li B, Ma Z, Chen J, Nakagoshi N (2004) An ecosystem service value assessment of land-use change on Chongming Island, China. Land Use Policy 21:139–148

Chapter 3
Ethnic Culture and Nature: Interactions in the Hani Terrace Landscape

Yuanmei Jiao and Xiuzhen Li

3.1 Culture, Nature, and Landscape

A landscape is a complex higher-order relationship system with many subsystems (Forman and Gordon 1986; Forman 1995) which together (and also by virtue of the geography) form a recognizable part of the earth's surface. It is formed and maintained by the mutual actions of abiotic and biotic forces as well as human activities (Naveh and Lieberman 1984; Pickett and Cadenasso 1995; Moss 1997). Thus, landscapes can be viewed in three ways: (1) a perception of landscape or scenery; (2) a heterogeneous mosaic whose attributes are pattern, function, and ecological processes; and (3) a landscape ecosystem (Antrop 1997, 2000; Brandt 1997; Wiens 1999).

A landscape is a combination of nature and culture, with humans playing the key role. Humans cannot make, but can only use, protect, or destroy, nature and affect the landscape through their culture. Figure 3.1 illustrates the fact that any kind of transition can exist from pure natural landscapes to urban landscapes (Nassauer 1995, 1997; The Dornach Landscape Document 2000).

An agricultural landscape is primarily the result of the interaction between the natural features of the region and the activities and mentality of the farmers. This interaction, or the traditional management regimes, have resulted in rich (agri) cultural landscapes with high esthetic values (Hendriks et al. 2000). Over the course of this century, the combined impact of population, technological change, and economic development has greatly increased the magnitude of human impacts on landscapes (WCED 1987; Vitousek et al. 1997; Papport et al. 1998; Herzog et al. 2001). Some cultural landscapes are jeopardized and are subject to protection. For example, terraced hillsides, coastal reclaimed areas, and irrigated arid lands (Bertollo 1998, 2001), in which the component ecosystems have developed over many centuries in response to anthropogenic management and are now widely

Y. Jiao (✉)
The Tourist and Geography Department, Yunnan Normal University, Kunming 650092, China
e-mail: ymjiao@sina.com
X. Li
State Key Laboratory of Estuarine and Coastal Research, East China Normal University, Kunming 650092, China

S.-K. Hong, et al. (eds.), *Landscape Ecology in Asian Cultures*, Ecological Research Monographs, DOI 10.1007/978-4-431-87799-8_3, © Springer 2011

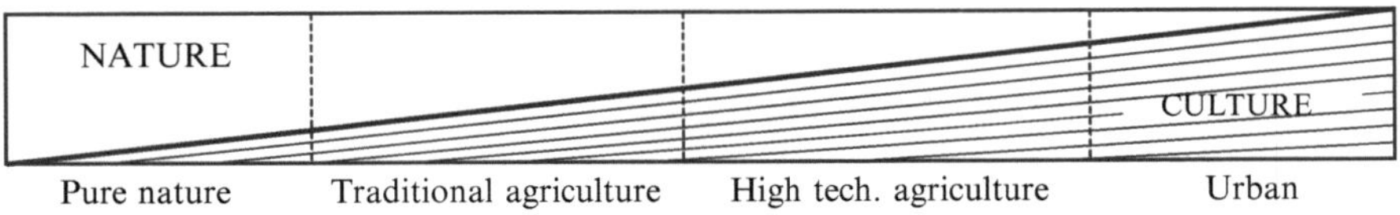

Fig. 3.1 Landscape as a combination of nature and culture (Zonneveld 1995)

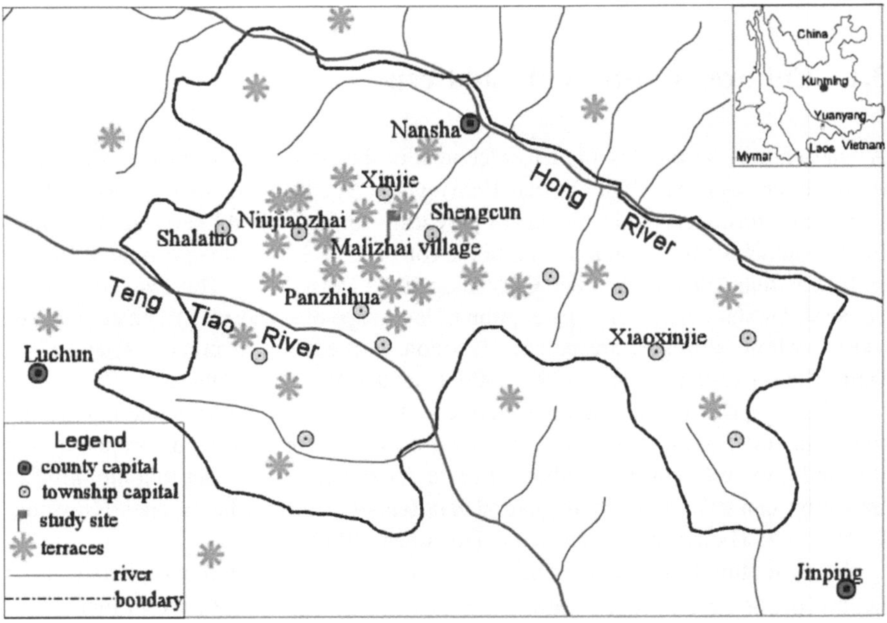

Fig. 3.2 Map of Yuanyang County

recognized as of major conservation importance (Bergland 1991; Berks et al. 1988; Bunce et al. 1998; Selman 1994).

Yunnan Province, an important world tourist destination, lies in southwestern China. Yunnan is a mountainous and frontier area with a population of 25 ethnic groups, and is famous for its rich biodiversity, cultural diversity, and mineral resources, as well as for soil erosion, landslides, and mudflows (Li et al. 1998; Zhong et al. 1999). The population of the Hani ethnic group is more than 1.2 million, and they are the fourth largest race in Yunnan Province. They are mainly distributed in the subtropical mountainous region between the Red River and Mekong rivers, especially in Honghe Hani autonomous prefecture. Through long periods of occupation and productivity, they created the famous Hani Terrace landscape and maintained its stability and sustainable development for more than

1300 years. The harmonized, terraced agro-cultural landscape of the Hani ethnic groups is an ideal landscape that is productive, clean, harmonious, and sustainable (Palketti 1999), and has important natural, cultural, and esthetic value. However, at present, owing to population growth and the alteration of social systems and consciousness, this system has lost its sustainability and the balance that once existed between culture and nature. This paper analyzes the harmony and conflicts between Hani culture and nature, and investigates its interacting mechanisms through field surveys in Yuanyang County, the core region of the Hani ethnic group (Fig. 3.2).

3.2 Semi-Directive Interview Methods

A list of topics for discussion has been designed to suit the study's subject and goals. The deputy head of Shengcun School, Mr. Li, who has 15 years teaching experience at the school and is very familiar with the local Hani ethnic groups, assisted in selecting villages to visit and people to interview. These interviewees included the Hani people in Malizhai and Luomadian, and the Quanfuzhuang villages of Xinjie township in Yuanyang County. The location and time of the interviews were at the interviewee's convenience. The interviewees included the "Beima" or "Mopi," who is the guardian and facilitator of traditional Hani culture and ethics, and also the village chief, with whom to discuss the history, culture, lifestyle, and other relevant topics. For group interviews, we selected men, women, and elders to discuss their plant species, cropland locations, harvesting methods and taboos, management of forests, water, land, and livestock power, and especially their management and maintenance of terraces and their current problems of paddy agriculture. Pictures and examples are used in the field survey. In addition, statistical information from the government is utilized.

3.3 Nature and the Hani Terrace Landscape

3.3.1 Location

Yuanyang County lies between 102°27′–103°13′E and 22°49′–23°19′N. The region occupied by the Hani ethnic groups has historically been very isolated, allowing for the preservation of their culture over many centuries. Owing to its steep topography and distance from major cities (334 km from Kunming, the capital of Yunnan Province, and 146 km from Gejiu, the capital of Honghe district), external communications are minimal and transport to the region is difficult on the unmade roads (The Editors Committee of Honghe District 2007).

3.3.2 Topography and the Hani Terraces

Yuanyang County is located on the southern part of Ailao Mountain. Due to the presence of the Hong River in the north and the Tengtiao River in the south, the landform of Yuanyang County is relatively higher in the central section and lower in the southern and northern areas. The higher central mountains restrict the cold airflow from northern China as well as the warm, humid airflow from the southeast and southwest monsoonal currents from the tropical Pacific and Indian oceans. This creates high topographic rainfall, resulting in the phenomenon of "high mountains, high waters." In addition, due to the deep river valleys and the fragmented earth surface, soil erosion is unavoidable. To adapt to the steep slopes and make full use of the abundant rainfall, the Hani people developed the terrace system that changed the slope gradient and preserved the mass of water and soil, thereby combining production with protection (Fig. 3.3).

3.3.3 The Subtropical Monsoon Climate and the Hani Paddy Terraces

Yuanyang County has a subtropical monsoon climate. Owing to its high altitude and the influence of the northern and southern airflows, this area has a unique climate with distinct dry and humid seasons. The average temperature is about 16.4°C (Table 3.1). Impacted by the southeastern wind coming from Northern Bay and the southwestern wind coming from the Bay of Bengal in the Indian Ocean from May to October, the region is suitable for the growth of rice, which requires both high temperatures and high humidity during the growing period. The increase in rainfall with elevation in mountainous areas results in an increase in the altitude at which rice can be grown. Locally this can reach 1 950 m, but this is highly dependent on the terraces.

Fig. 3.3 Pictures of the Hani terrace (Photos by Jiao Yuanmei)

Table 3.1 The climate of Yuanyang County

Item	Spring months (3–5)	Summer months (6–8)	Autumn months (9–11)	Winter months (12–2)	Dry season months (11–4)	Rainy season months (5–10)	Yearly total
Sunlight (h)	584	364	371	452	971	800	1 770
Sunlight (%)	33	21	21	25	55	45	40
Radiation (therm.cm^2)	38	31	26	25	57	63	120
Precipitation (mm)	287	685	312	114	308	1 090	1 398
Percent of annual rainfall (%)	21	49	22	8	22	78	100

Data source: Editors of Honghe District Records, Honghe District Records, Sanlian Bookstore of Living, Reading and Knowledge, August, 1994, Beijing: P4–9.

3.3.4 Soil and the Formation of Hani Terraces

Granite, curdle rocks, and a red covering of Trias are widely distributed in Yuanyang County. With the abundant precipitation and thick vegetation, the region is rich in shallow groundwater and spring water to irrigate the Hani terraces. The soil types on the terraces are effloresced from flat rock, sandstone, shale, and granite, and its quality is so dense and clinging that it can easily conserve water and fertilizer. The Hani ethnic groups utilize the paddy soil to build ridges around the fields, thus enhancing the function of water conservation and maintaining the presence of terraces for centuries.

3.3.5 Forests and the Maintenance of Hani Terraces

Through the assimilation and maintenance of precipitation by the subtropical evergreen broad-leaved forests, the infiltration time of water is prolonged, and the groundwater surfaces as springs with a high flow rate, thereby providing irrigation water for the terraces. In addition, the forests retain the water and soil and decrease the rate at which they are lost, thus maintaining the use of terraces for thousands of years. Therefore, the forests have been a key function in preserving the terraces for many centuries (Jiao 2000).

3.4 The Hani Culture and Terraced Landscapes

3.4.1 Culture and the Cultural Landscapes of Hani Terraces

Culture is the sum of the material and mental civilization created by an ethnic group in a particular period and environment (The Editors Committee of Chinese Cyclopedia 1990). The method of production can be regarded as a reflection of the culture in traditional social and economic conditions. The paddy agriculture of terraces is the production model developed by the Hani people, and their culture was generated during the development of terrace agriculture. From their basic necessities of life, wedding and funeral customs, festivals and celebrations, religions and cults, to their attitude to life, morality, and esthetic consciousness, come from the practices of terrace paddy agriculture. Therefore, the Hani culture is commonly referred to as the Hani terrace culture (Bai 1997). The core of Hani terrace culture is the terraced landscape, which is composed of forest ecosystems, village cultural systems, and terrace ecosystems (Jiao 1999).

3.4.2 The Components and Spatial Structure of Hani Terrace Landscapes

The components of Hani terrace landscapes are the forest ecosystems, the village systems of Hani culture, and the terrace ecosystems, and these are distributed on the mountain slopes along contour lines (Fig. 3.4) (Wang 1998).

Among the three subsystems, the forest ecosystem is a relatively steady system formed by the interaction of the forest community and its environment. It includes the forest community (plants and animals) and the environment (soil and atmosphere). The village system of Hani culture is a human ecosystem composed of Hani people, the farming technology of the terraces, and their physical and mental culture. The terrace ecosystem is a human-controlled paddyfield ecosystem composed of the paddy, the terrace animals, the microorganisms, and their environment.

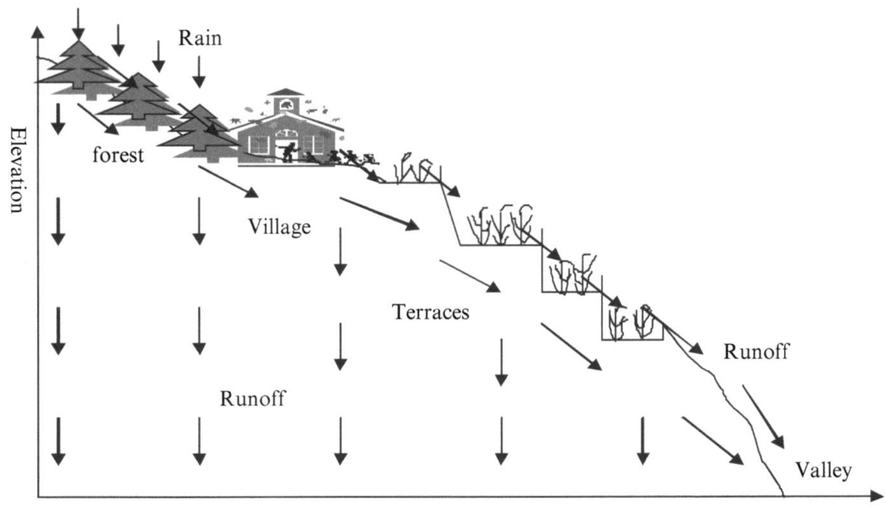

Fig. 3.4 The spatial structure of a Hani terraced landscape

3.4.3 The Function of the Hani Terraced Landscape

The basic energy of Yuanyang Hani terrace landscapes comes from the sun. Naturally, the solar energy and the rain enter the forest and the terrace ecosystems and are transformed into organic mass and energy via green plants, which then provide sustenance to wild animals, human beings, and livestock. The forest ecosystem provides clean water, firewood, vegetables, meat, fruits, and other living produce to the village system of Hani culture. The terrace ecosystem provides rice, fish, vegetables, fodder, and building materials to the village system of the Hani culture, which is the center of the flow of mass and energy. Through its special function of control and modulation, the village system manages the whole landscape and the input of mass and energy from the other two ecosystems, and constantly maintains its development. In addition, it maintains the stability of terrace and forest ecosystems, and enables them to provide useful mass and energy via its outputs of human and animal power (Fig. 3.5).

3.4.4 The Traditional "Village Division Measure" of Hani Ethnic Groups

Hani terrace agriculture is the most productive type of traditional mountain farming. Its high productivity encouraged the increase in the Hani population, and resulted in the development of Hani terracing around the villages. When the walking distance exceeds 1 day, or the increased population cannot get enough food from an existing terrace, a group of people who have strong kindred relationships will move away from the "mother village" to another place to build a "son village," thereby maintaining the Hani terrace landscape for centuries (Wang 1987). This "village

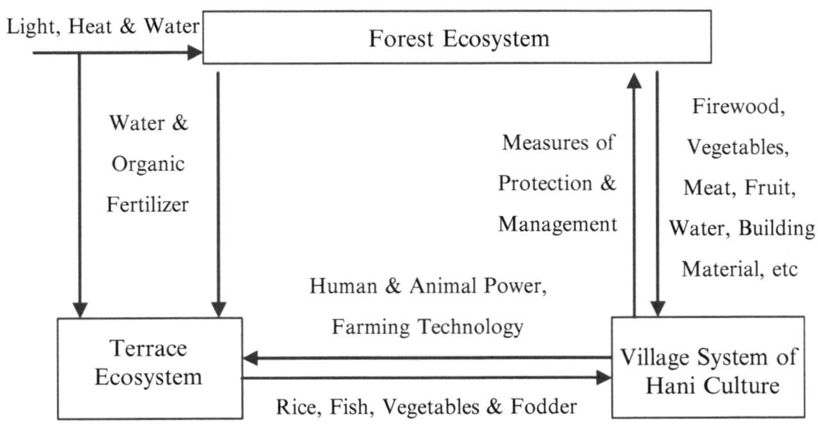

Fig. 3.5 The function of Hani terrace landscapes

division measure" is the Hani people's adaptation to the natural environment. It spreads the population burden on the environment and ensures the stabilization of the Hani terrace landscapes.

3.5 The Conflict Between Hani Culture and Nature

3.5.1 The Obstruction of Nature and Culture

Owing to the obstruction by the Hong river in the northwestern, northern, eastern, and southeastern areas of Yuanyang County, and the proximity of two poor frontier mountainous counties (Jinping County and Luchun County) on the southern and western sides (Fig. 3.2), communication between Yuanyang County and the outer world has historically been very difficult during the rainy season, and the area can only be reached by sampan ferry over the Hong river in the dry season. In addition, the county is famous for outbreaks of malaria as a result of its subtropical climate, and this deterred outlanders from moving to the area. Furthermore, the deep valleys in the high mountains and the sparse population led to communication difficulties within the county and the development of close-knit communities within the Hani ethnic groups. This was enhanced by the differences in language, trappings, festivals and celebrations, wedding and funeral customs, affiliations and nature worship, and the self-sufficient family economy (Sun 1991).

3.5.2 Low Productivity

Compared with modern agriculture, the productivity of terrace agriculture is very low. For example, in 2006, the average food output of China was 5 225 kg/ha, in Yunnan Province it was 4 078 kg/ha, while in Yuanyang County it was 4 025 kg/ha. The Quanfuzhuang village in Yuanyang County was even lower at 3 423 kg/ha. Furthermore, the average land area per unit of labor is low. In 2006, it was only 0.03 ha in Quanfuzhuang village, while the Chinese average was 0.09 ha, which resulted in a low average output per hour of labor. The use of chemical fertilizers is very low, with the main fertilizers used being organic. The total amount of chemical fertilizer used in 2006 was only 32.9 kg/ha in Quanfuzhuang village. In addition, low-level technology is generally employed.

3.5.3 Tension of Human–Land Relationships

The Hani terrace agriculture can provide few employment chances for industries and services owing to its highly self-sufficient family economy. Any increase in agriculture depends mainly on the expansion of land and labor, and results in the reclamation of marginal land and an increase in the stress put on the land (Table 3.2) (Yang 1990).

Table 3.2 The population and field area in Luomadian village in Yuanyang County from 1985 to 1999 Units: people, hectares

Year	Population	Labor	Crop land	Paddy	Corn	Grain	Soya	Vegetable	Banana taro	Tea
1985	363	158	20.6	10.7	4.9	2.5	1.8	0.4	0.4	0.1
1988	388	189	24.0	10.7	6.6	6.0	3.4	–	–	0.1
1991	368	183	24.5	10.7	6.6	4.6	4.3	–	–	0.1
1993	378	187	27.7	10.7	2.2	1.6	9.3	1.3	0.4	0.1
1995	379	192	26.9	10.7	5.2	1.5	5.9	1.6	0.9	0.1
1997	388	202	33.7	10.7	5.5	1.2	6.6	1.5	0.8	0.1
1999	402	238	37.9	10.7	6.4	1.6	9.3	1.6	1.3	0.1

Data source: The statistics of Luomadian village in Yuanyang County, 1984–1999.

Table 3.2 shows that in Luomadian village in Yuanyang County, from 1985 to 1999, the population growth rate was 7.2% per year, but the labor growth rate was 33.3%. Owing to the lack of investment, technology, and information, the majority of the surplus labor remains in the countryside. The only way to survive is to reclaim new fields, which has resulted in a 1.84-fold increase in field area over a 15-year period. However, the area of paddy terrace has remained at 10.7 ha during this period because of the limited moisture and heat at this elevation. The dry land area is 2.42 times that of 15 years ago, and the main crops grown on the new land are soybean, corn, vegetables, and banana taro. Furthermore, the number of households has increased from 70 in 1985 to 90 in 2006.

3.6 Interactions of Hani Culture and Nature

The Hani culture has developed in a closed environment, and there has been little communication with the world outside. This has made the Hani terrace landscape a typical stable ecosystem. The village is the basis of Hani culture. One village always forms a terrace landscape. When the output of the Hani terrace in one village cannot support all the people, a "village division measure" will be adopted in order to achieve a balance between humans and the land.

Today, however, the harmonious relationship between the Hani culture and the natural landscape has changed dramatically owing to improvements in communication and medical treatment, and the influence of national policy. This has caused variations in the traditional Hani culture as well as numerous problems, for instance, the decrease in forest areas and the increase in water shortages, the decrease in the amount of cropland per capita, the poverty of people, and the degradation of the environment. A new relationship between the Hani culture and the environment has been created (Fig. 3.6).

The population of Yuanyang County has increased too much due to improvements in health and the flexible national birth policy in ethnic and frontier areas. Under the current policy and the control of land delimitation, Hani people cannot balance the relationship between increased population and the terrace system with the "village division" technique. People now remain in their own village, leading

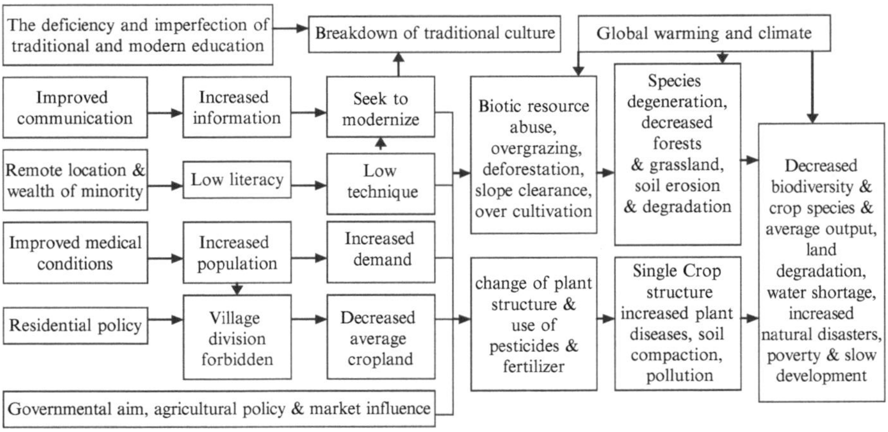

Fig. 3.6 The relationship between Hani culture and the environment

to a decrease in the average amount of cropland year after year. The low technology resulting from remote locations and multi-minorities has intensified the food crisis because of the decrease in cropland per capita. On the other hand, owing to the impact of governmental aid and the market economy, and in order to obtain an economic income, the poor Hani people have adopted many techniques to resolve their problems, such as the use of cross-breed animals and rice varieties, new technology, and the enhancement of organic fertilizer, which has positive benefits for the environment. At the same time, they have altered the structure of land use (land cover) unconsciously during land reclamation. Biotic resource collection and forest exploitation have led to a decrease in forested areas, the aggravation of soil erosion, a shortage of water, and the decline of biodiversity. Large mono-cropping of high-yield varieties has resulted in a loss of agro-biodiversity and the decline of average output.

3.7 Conclusion

The landscape is the product of culture and nature. Through use and management of the local environment, Hani ethnic groups have created the terrace landscapes that have significant natural, esthetic, and historical values. Nowadays, due to changes in communication, technology, consciousness, and the impacts of policy, the traditional Hani terrace landscapes have altered, and this has caused an imbalance in the traditional human and environment relationship. In order to avoid the complete collapse of these traditional and sustainable agricultural practices and Hani culture, measures need to be taken to improve the current conditions.

Acknowledgments This study was financed by the National Natural Science Foundation of China, No. 40401022 and No. 40861023, and the Natural Science Foundation of Yunnan Province, No. 2004 D0016Q.

References

Antrop M (1997) Landscape, planning and landscape ecology. Lecture given to the international Ph.D. course 'Landscape Ecology and the Dynamics of Agricultural Landscapes', November 2 to 7, Roskilde University, Department of Geography and Development Studies

Antrop M (2000) Background concepts for integrated landscape analysis. Agric Ecosyst Environ 77:17–28

Bai YB (1997) The source and structure of Hani culture. In: The corpus of the first international conference of Hani culture. Yunnan Nationality Press (in Chinese)

Bergland BE (1991) The cultural landscape during 6000 years in southern Sweden – the Ystaas Project. Ecological Bulletin 41. Copenhagen, Muksgaar, p 495

Berks HH, Berks HJB, Kaland PE, Moe D (1988) The cultural landscape: past, present and future. Cambridge University Press, Cambridge, p 521

Bertollo P (1998) Assessing ecosystem health in governed landscapes: a framework for developing core indicators. Ecosystem Health 4(1):33–51

Bertollo P (2001) Assessing landscape health: a case study from Northeastern Italy. Environ Manage 27(3):349–365

Brandt J (1997) The histories and goals of landscape ecology. Papers presented at the international Ph.D. course "Landscape Ecology and the Dynamics of Agricultural Landscapes", 2–7 November, Dragerup, Denmark

Bunce RGH, Bell M, Farino T (1998) The environmentally sensitive area legislation in the United Kingdom and its potential application to the Picos de Europe mountains in northwest Spain. Environ Conserv 25(3):219–227

Forman RTT (1995) Land mosaics: the ecology of landscape and regions. Cambridge University Press, Cambridge, p 36

Forman RTT, Gordon M (1986) Landscape ecology. Wiley, New York

Hendriks K, Stobbelaar DJ, Mansvelt JDV (2000) The appearance of agriculture: an assessment of the quality of landscape of both organic and conventional horticultural farms in West Friesland. Agric Ecosyst Environ 77:157–175

Herzog F, Lausch A, Muller E, Thulke H, Steinhardt U, Lehnann S (2001) Landscape metrics for assessment of landscape destruction and rehabilitation. Environ Manage 27(1):91–107

Jiao YM (1999) Study on the cultural ecosystem of Hani terrace. Hum Geogr 14(S1):56–59 (in Chinese)

Jiao YM (2000) The formation of high gradient terraces in southwest China – a case study of Hani terraces in southern bank of Hong River. Econ Geogr 20(4):94–96 (in Chinese)

Li XZ, Xiao DN, Wang GJ (1998) The fragile karst landscape in southwestern China. Ambio 27(3):246–247

Moss MR (1997) Lecture given to the international Ph.D. course, "Landscape Ecology and the Dynamics of Agricultural Landscapes", 2–7 November, Roskilde University, Department of Geography and Development Studies

Nassauer JI (1995) Culture and changing landscape structure. Landsc Ecol 10(4):229–237

Nassauer JI (1997) Placing nature – culture and landscape. Island Press, Washington, DC

Naveh Z, Lieberman AS (1984) Landscape ecology: theory and application. Springer, New York, p 356

Palketti MG (1999) Using bioindicators based on biodiversity to assess landscape sustainability. Agric Ecosyst Environ 74:1–18

Papport DJ, Gaudet C, Karr JR, Baron JS, Bohlen C, Jackson W, Jones B, Nainan RJ, Norton B, Pollock MM (1998) Evaluating landscape health: integrating societal goals and biophysical process. J Environ Manage 53:1–15

Pickett STA, Cadenasso ML (1995) Landscape ecology: spatial heterogeneity in ecological systems. Science 269:331–334

Selman P (1994) The ecology and management of cultural landscape. In: Proceedings of a conference organized by the International Association for Landscape Ecology (UK), September, 1993. Landscape Issues 11(1)

Sun GS (1991) Antiquity. Mystery. Width – the source of Hani culture. Yunnan People's Press, Kunming (in Chinese)
The Dornach Landscape Document (abridged version) (2000) Get connected to your place! A discussion document prepared for and during the international conference "the culture of the European landscape as a task" at the Goetheanum, Dornach, Switzerland 6th–9th September
The Editors Committee of Chinese Cyclopedia (1990) Chinese Cyclopedia. Geography Chinese Cyclopedia Press, Beijing (in Chinese)
The Editors Committee of Honghe District (2007) The general situation of Honghe Autonomous Prefecture of Hani and Yi ethnic groups. Yunnan Peoples Press, Kunming (in Chinese)
Vitousek PM, Mooney HA, Lubchenco J, Milillo JM (1997) Human domination of earth's ecosystems. Science 277:494–499
Wang QH (1987) Search on the "regime of connected place-name" of Yuanyang Hani ethnic groups. Yunnan Soc Sci 3:85–88 (in Chinese)
Wang QH (1998) The ecology and the spatial structure of Hani peoples' living. Yunnan Soc Sci 2:71–74 (in Chinese)
WCED (World Commission on Environment and Development) (1987) Our common future. Oxford University Press, Oxford, UK
Wiens JA (1999) Toward a unified landscape ecology. In: Wiens JA (ed) Issues in landscape ecology. IALE Press, USA
Yang HS (1990) Agricultural ecology. Agriculture Press, Beijing (in Chinese)
Zhong XH, Zhang WJ, Luo J (1999) The characteristics of the mountain ecosystem and environment in the Gongga Mountain region. Ambio 28(8):648–654
Zonneveld IS (1995) Land ecology. SPB Academic Publishing, Amsterdam, The Netherlands

Chapter 4
The Characteristics of the Cultural Landscape in Malaysia: Concept and Perspective

Saiful Arif Abdullah

4.1 Introduction

A cultural landscape, occasionally known as a rural landscape, is an ecosystem consisting of the most long-term human-modified landscapes (Farina 2000). It is a combination of cultural and ecological values, where their quality and function have been mediated by the close inter-relationship between humans and nature (Naveh 1995; Buckley et al. 2008). The human–nature interaction affects the characteristics of the cultural landscape, but the interaction is also dependent on the culture of the society as well as the condition of the initial landscape at any given time (Farina 2000). In addition, cultural landscapes have their own structural and functional complexities, fragility, and resilience (Farina 2000). Therefore, the characteristics of cultural landscapes are different from one region to another.

In European countries, cultural landscapes dominate a considerable proportion of the land area, particularly in countrysides with special characteristics (Plieninger et al. 2006). In this continent, cultural landscapes result from a combination of extensive and integrated use of natural resources, which occurred in different overlapping stages (Vos and Meekes 1999). They are mainly characterized by agricultural landscapes (Rabbinge and van Diepen 2000; Mander et al. 2004), and the special features or elements include hedgerows, terraces, pollarded trees, vineyards, open fields, woodlands, scattered small farms, and castles (Austad 1988; Farina 2000; Oreszczyn 2000; Oreszczyn and Lane 2000). It is a heterogeneous mosaic with a high biological diversity (Gomez-Limon and Fernandez 1999), and some has an aesthetic value which favours tourism and recreation, for example, the cultural landscapes of Montado in Portugal (Pinto-Correia 2000) and the Flanders region in Belgium (Antrop 1997).

Asian cultural landscapes also have special characteristics and features. However, they are not extensively identified and assessed, except in a few countries

S.A. Abdullah (✉)
Institute for Environment and Development (LESTARI), Universiti Kebangsaan Malaysia, 43600 Bangi, Selangor Darul Ehsan, Malaysia
e-mail: saiful_arif2002@yahoo.com

S.-K. Hong, et al. (eds.), *Landscape Ecology in Asian Cultures*, Ecological Research Monographs, DOI 10.1007/978-4-431-87799-8_4, © Springer 2011

such as Japan and Korea. In these countries, agricultural landscapes appeared as a premier characteristic of cultural landscapes (e.g. Nakagoshi and Hong 2001; Hong 1998). This resembles the situation in European countries, but the features are absolutely different owing to differences in cultural and ecological influences. In Japan, the cultural landscape is known as *Satoyama*, which is characterized by areas of agricultural fields and households surrounded by mountains (Takeuchi 2001). The special features include the terraces of paddy fields, the secondary vegetation, particularly pines, and the mixed deciduous species (Hong et al. 1995; Kamada and Nakagoshi 1996). In Korea, the cultural landscape features a combination of high mountains and the gentle slopes of low mountains, houses, and plains of agricultural fields and streams (Hong 2001; Nakagoshi and Hong 2001). Interestingly, one thing that is common to cultural landscapes in Japan and Korea is that the spatial pattern of their features is strongly related to the theory of Feng-shui (Nakagoshi and Hong 2001). In this theory, people determined the setting of the features based on cultural attributes such as religion, social structure, economics, politics, and bio-geo-ecology (Forman 1995; Zonneveld 1995).

Nevertheless, beginning with the industrial revolution in European countries during the eighteenth century, and followed by rapid economic growth after World War II coupled with population growth, dramatic changes have been seen in the cultural landscapes of the world (Antrop 1997). These come hand-in-hand with rapid urbanization processes and changes in agricultural practices and development (Rabbinge and van Diepen 2000; Kim et al. 2007). All these factors have put cultural landscapes under threat. For example, the impact of human disturbance has caused a disruption in the ecological processes (e.g. Hong et al. 1995; Kitazawa and Ohsawa 2002), a degradation of aesthetic and cultural values, and in some cases a loss of identity (Antrop 1997). Because the central premise of a cultural landscape is the feedback loop of interactions between humans and nature (Nassauer 1995), cultural landscapes are an important entity which must be considered for sustainable development. Therefore, identifying cultural landscapes and assessing their characteristics and features, as well as understanding their relationships with the socio-economic and cultural background, will show that human impact is pivotal in achieving a balance between the environment and its development.

In the Southeast Asian region, however, the concept and perspective of cultural landscapes have not received much attention, although there are many areas which fit the premises of a cultural landscape. In this region, the most renowned cultural landscape is the rice terraces of the Philippine Cordilleras, which have been listed as a World Heritage Cultural Landscape (UNESCO World Heritage Centre 2003). In fact, there are many areas in Southeast Asian countries with the potential to receive this recognition, but the lack of a reliable definition, identification, and assessment of a cultural landscape in the context of each country may hinder their efforts to achieve such a status. This recognition is necessary and pivotal for the conservation and/or preservation of traditional values of natural resource use, culture, and socio-economics. Therefore, in this chapter, the concept and perspective of cultural landscapes in Malaysia, which is one of the most highly developed countries in Southeast Asia, is discussed. This is to provide a basis for developing a definition of a cultural landscape in a Malaysian context. To provide a detailed perspective and concept of a cultural landscape in Malaysia, an example, the Merbok estuary in the Malaysian peninsula, is described.

4.2 Cultural Landscapes in Malaysia: Concept and Perspective

Malaysia is a tropical country with two regions, peninsular Malaysia and the States of Sabah and Sarawak in Borneo, better known as Malaysian Borneo (Fig. 4.1). Peninsular Malaysia and Malaysian Borneo represent about 40% (132 928 km^2) and 60% (197 605 km^2), respectively, of the total land area of the country. This country is not only endowed with a high biological diversity of flora, fauna, and complex ecosystems, but it also has a wide variety of traditional values formed by various ethnic groups with different cultural, religious, and socio-economic backgrounds. A factor behind these traditional values is the migration of people, particularly from the Indo-China continent and the Malay Sundaland, which is one of the main aspects of Malaysian history. At present, the Malay, Chinese, and Indian people form the three major ethnic groups in this country. The Malays represent the largest proportion of the total population (60%), followed by Chinese (20–30%) and Indian (10%). The earliest migration to this country by the Malays occurred in 1000 BC, whereas the Chinese and Indian migration was in the thirteenth century, when their traders established regular contacts in peninsular Malaysia (Jomo et al. 2004).

Fig. 4.1 Geographic location of Malaysia (peninsular Malaysia (A) and Malaysian Borneo (B, Sabah; C, Sarawak))

From the nineteenth century to the mid-twentieth century, the major economic activity was different between ethnic groups. Agriculture was the main economic activity among the Malay community. The Chinese were primarily involved in trade and business, while the Indians were engaged as workers in rubber and oil palm estates and other major crop plantations. This scenario influenced the formation of the characteristics and features of human settlement landscapes in Malaysia. The settlements of the Malay communities are mostly found in rural sites. Their agricultural activities included paddy and slash-and-burn agriculture, but this mostly occurred in Sabah and Sarawak, in small-scale vegetables farms and orchards. However, this was only for their daily subsistence and not for trade. Although the Chinese dominated the urban settlements, they can also be found in rural areas. This rural community was mostly engaged in vegetable farming and trade. Meanwhile, Indian community settlements were usually confined within the plantation estates where their employer provided the houses.

The rural, or traditional, villages of the Malay and Chinese are located adjacent to the forested areas of mountains or foothills. Because of this proximity, the communities, particularly the Malays, also depend on forest resources such as rattan, fruits, bamboo, and fish for their daily subsistence. In fact, this is still the practice of villagers, and has been handed down throughout the generations. The use of these resources on a daily-subsistence basis shows that the sustainable harvesting of forest resources is very important. Therefore, the concept of the sustainable use of forest resources has long been applied by the villagers to ensure that those resources are available for their livelihood. Although this is probably not true for slash and burn agriculture, which is known to cause severe soil erosion, degradation, and nutrient losses (e.g. Lim and Douglass 2000), the concept might also be applied to other land uses, for example the setting, design, and size of their houses, farms, and orchards. Hence, there is harmonization between natural and human activities, which reflects the close inter-relationship between humans and nature.

Rural villages can also be found along the coastal areas or river banks downstream. The communities depend on marine or aquatic resources for their daily subsistence. Nevertheless, agriculture remains as part of their economic activities because the relatively flat topography and soil of the coastal area is suitable for agricultural land use, particularly paddy and coconuts. This led to what we can see at the present time, i.e. that the large paddy fields and coconut plantations are the major elements that compose the landscape of the coastal margins, for example, in the Kedah, Selangor, and Perak States of peninsular Malaysia. The association between people and forested areas also exists in coastal villages. Mangrove forest is the predominant natural landscape, as well as other swamp vegetation such as nipa. These provided resources for the people and played an important role in ecological function and balance. Furthermore, this natural ecosystem is also important as a breeding site for various aquatic species, so the sustainable use of forest resources is vital for the peoples' livelihood. The major use of mangrove is timber for charcoal production. Before the mid-twentieth century, the use of charcoal as a fuel source was common among rural inhabitants in this country. However, at present, the use is very minimal and is restricted to particular purposes.

This perspective suggests that factors such as geography, climate, cultural and socio-economic background, and religion might also affect the characteristics and features of cultural landscapes in Malaysia. Geographic factors have divided cultural landscapes into two categories: (1) those located in mountainous or foothill sites close to forested areas, and (2) those located along the coast and/or river banks downstream. Therefore, the first category can be described as a mountainous cultural landscape, whereas the second category is a coastal cultural landscape. In both cultural landscapes, agriculture has been established as the major human activity or land use, and the harmonization between humans and nature is shown by the integrated use of natural resources. Agricultural areas are generally surrounded by, or adjacent to, natural forest. In the mountainous cultural landscape, vegetable farms and orchards are the premier features, with small streams flowing nearby. The traditional houses are generally scattered, and located adjacent to, or within, the farm or orchard. The villagers also depend on forest resources as part of their subsistence. However, in the coastal cultural landscape, paddy fields and coconut plantations are the main features. The houses are scattered or else located along the river banks, or are congregated in one locality on the coast to form fishing villages. People in this cultural landscape also use forest resources, particularly mangrove, in their daily lives, but this use might not be as extensive as that of the villagers in mountainous cultural landscapes. This is because mangrove resources and other swamp forests are relatively less abundant than the forest resources of the mountainous areas. Furthermore, the richness and abundance of marine or aquatic resources provide other alternative natural resources for the coastal villagers.

However, over the centuries, the cultural landscapes in Malaysia are not unusual in having been encroached upon by various anthropogenic activities, which are driven by socio-economic development. Because of this, in some cases Malaysian cultural landscapes have been totally changed into other landscape types, for example, urban landscapes. The possible indicators that show that the urban areas were once a cultural landscape are the clusters of traditional Malay or Chinese houses. Nowadays, these traditional houses are in the midst of new or modern houses and buildings. Some have been renovated but maintain their original structure and design, whereas others are left abandoned. Despite facing the challenges of rapid developmental progress, quite a number of areas in Malaysia still show the characteristics and features of a cultural landscape. Furthermore, the traditional values regarding natural resource use are still practiced among the rural peoples, especially the older generation. Yet developmental activities are continuing to influence the socio-economics of the inhabitants. As a result, the long-term traditional landscape features of paddy fields, vegetable farms, and orchards are expanding, and this is essential to meet the demands of socio-economic progress and population growth. Some of the expansion is part of the implementation of economic development policies, under the auspices of government agencies, in order to improve the socio-economic conditions of the rural people. At the same time, non-traditional landscape features have also emerged in the cultural landscape, such as oil palm and rubber plantations, and aquaculture. This has been brought about either by the government or by private agencies, and the local people are usually employed as labourers.

4.3 Merbok Estuary: A Coastal Cultural Landscape

The Merbok estuary (5°40′N, 101°25′E), is an example of a coastal cultural landscape and is located in the State of Kedah on the Malaysia peninsula (Fig. 4.2). This estuary lies between the town of Sungai Petani to the east and a coastal site to the west, and between the foot of the Gunung Jerai Forest Reserve to the north and the Muda river to the south. The Merbok river is the major river flowing through the area, and many small streams form its tributaries. The width of the Merbok river is about 20 m in the upper reaches and 2 km at the mouth, and the depth varies between 3 and 15 m (Ong et al. 1991). Being located close to Thailand, this area specifically, and the State of Kedah in general, have a tropical monsoon climate with two distinct seasons: dry and warm from January to April and wet from May to December. The average monthly temperature is between 21 and 32°C, and the annual rainfall is between 2 000 and 2 500 mm. The topography is generally flat and the vast tract of this estuary is made up of unconsolidated Quaternary sediments (Khoo 1996). The main soil type is Keranji, which is favourable for the growth of mangrove and other swamp vegetation.

Historical records and archaeological evidence indicate that the present configuration of the shoreline of the estuary has been the same for almost 1 400 years (Khoo 1996). Geological studies have shown that the Merbok estuary was formerly a bay before the regression of the sea during the Late Holocene (Soo 1976). Research has shown the estuary to be the earliest site of human settlement in the present Kedah State. This has been confirmed by archaeological and written

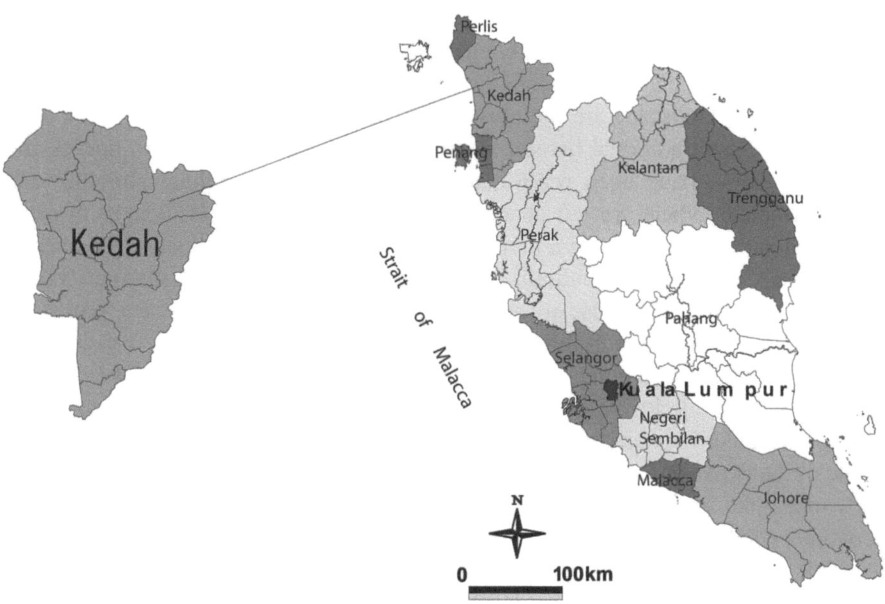

Fig. 4.2 The location of the Merbok estuary, in the State of Kedah, in peninsular Malaysia

records from Chinese, Arab, and Indian sources (Khoo 1996). Furthermore, it has been reported that until the fourteenth century, the estuary was an important port and trading center for traders from China, India, and the middle-east (Braddell 1950, 1980 stated in Khoo 1996). However, the regression of the sea level has caused the Merbok estuary to lose its prosperity and standing as an important center for trade and services (Khoo 1996).

It is apparent that factors such as geological events, the historical migration of people, culture, socio-economic background, and possibly religion have played important roles in determining the characteristics and features of the cultural landscape. Nowadays, the setting of this landscape portrays harmonization between humans and nature, where most inhabitants are still practicing the traditional values in their use of land and their life-style, although they are exposed to various basic necessities of modern life such as telecommunication and electricity. Here, traditional values in the use of land mean that most villagers are still farmers and fishermen, and have a moderate life-style in accordance with their earnings. The landscape is a heterogeneous mosaic, where the land use/land cover is a combination of natural and man-made elements (Fig. 4.3). Agricultural land (agricultural and vegetation areas) dominate the area (62.5%), and other land use/land cover represents about 2–21% of the total land area (Table 4.1). Generally, the features that compose the cultural landscape can be grouped into three categories: traditional, historical, and non-traditional features.

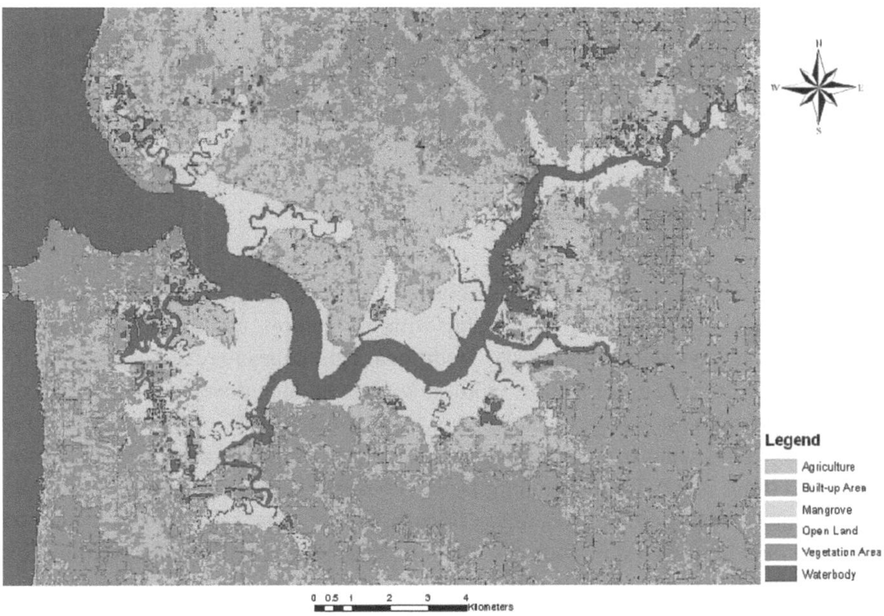

Fig. 4.3 Land use/land cover of the Merbok estuary in 2001

Table 4.1 Proportion of land use/land cover of the Merbok estuary in 2001

Land use/land cover	Area (ha)	%
Agriculture	6532.4	26.3
Built-up area	5041.7	20.3
Mangrove	3660.1	14.7
Open land	633.2	2.5
Vegetation area	9007.6	36.2
Total	24 875	100

Note. Agriculture includes paddy fields and aquaculture areas, whereas vegetation includes plantations of oil palm, rubber, and coconut

Fig. 4.4 Mangrove forest is a natural traditional feature of the Merbok estuary (photo by S.A. Abdullah)

4.3.1 Traditional Features

These refer to customary or long-established features. This means that the local people have developed a close relationship with the features, and this spirit is handed down from one generation to the next. Traditional landscapes may include both natural and non-natural features. In this cultural landscape, mangrove forest (Fig. 4.4) is the predominant natural feature, and covers approximately 15% of the total land area (Table 4.1). This mangrove contains one of the largest ranges of mangrove species in the world (Ong 2003). Furthermore, it is also known as a nursery area for many species of aquatic life, and thus has high ecological value, especially for the livelihood of the local fishermen. At present, the major use of mangrove resources is timber for charcoal production, where the process is implemented by the local people but the timber harvesting is controlled by the Department of Forestry (Fig. 4.5). Paddy fields are a non-natural traditional feature that cover a vast area of the estuary (Fig. 4.6), and along both sides of the Merbok river (see Fig. 4.3). However, in the case of paddy fields, the land is not necessarily owned by the local people. In this context, the landlords are outsiders, and they hire their paddy plots to the local people. Here, the issue of land ownership is of great concern because it may influence the traditional use of the land and the future sustainability of the cultural landscape. Malay villages and traditional houses make up the third component of the traditional features. This component also includes the fishing villages. These features are associated with paddy fields. They are scattered throughout the estuary, but are mostly located within, or at the periphery of, the paddy fields. However, the fishing villages can be found on the coastal fringes at the mouth of the Merbok river.

Fig. 4.5 Timber harvesting of mangrove for charcoal production in the Merbok estuary (photo by S.A. Abdullah)

Fig. 4.6 Paddy fields are a non-natural traditional feature that characterize the cultural landscape of the Merbok estuary (photo by S.A. Abdullah)

4.3.2 Historical Features

Historical features refer to all historic sites that have been discovered from archaeological evidence. In this estuary, the historical features refer to all the remnants and ruins of the basements of temples and stone stupa structures. There is archaeological evidence that Hinduism and Buddhism, which were brought by traders from India and China, were practiced by the population of the estuary during early settlements in the fourth and fifth centuries (Nik Hassan 1980 stated in Khoo 1996). However, there are not as many of these features as in the renowned historical site of Lembah Bujang, which is located adjacent to the north side of this estuary.

4.3.3 Non-Traditional Features

The encroachment of anthropogenic activities has created man-made features in the cultural landscape. Some man-made features appear to have been established for a long time, with little or no sign of expansion or change, and people have become accustomed to their existence even though they are considered to be non-traditional features. In the Merbok estuary, the main non-traditional features are the agricultural plantations, such as rubber, oil palm, and coconut. These agricultural plantations cover the largest proportion of the land area of the estuary (Table 4.1), and are considered to be important for the function of the human–nature ecosystem.

4.4 Conclusion

This chapter has presented a baseline concept and a perspective of the cultural landscape in Malaysia. Basically, agricultural fields are the main characteristic, but features may differ according to their geographic location either near the mountains or on coastal sites. There are several factors involved in the process of establishing the characteristics and features of a cultural landscape in any country. These factors include geological events, early human settlement by people from other continents, and the cultural and traditional backgrounds of various ethnic groups, including religious and socio-economic development. However, during the process their importance might be different, and each might be responsible for creating or determining different categories of cultural landscape features, i.e. traditional, historical, and non-traditional features.

In the context of the Merbok estuary, geological events played an important role in shaping the configuration of the estuary and eventually determined the existence of natural traditional features. Early human settlement and cultural background are inter-related factors where they are responsible for the establishment of long-term close interactions or harmonization between humans and nature. This is related to the use of the land by the locals that created non-natural traditional features such as paddy fields, which they depend on for their livelihood. The harmonization is also shown by the distribution pattern and design of the traditional houses. Socio-economic development is a prevailing factor that may imposes dramatic changes on the cultural landscape. This factor contributed to the creation of man-made features as compared with the non-natural traditional landscape. The establishment of man-made features was mainly influenced by external drivers, particularly when related to the implementation of land developmental policies. An example of such features is the plantations of oil palm, rubber, and coconut. However, these have now been there for a long time and have not varied in size, so the people are used to their existence and acknowledge that they have an important ecological function in the human–nature ecosystem. Thus, long-term human–nature interaction has occurred, which is the main premise when defining the features of cultural landscapes.

At present, there are many locations in Malaysia which have the characteristics and features of a cultural landscape. Although there are three major ethnic groups living in this country, the Malay villages possibly constitute the largest community, with the Chinese group second, and a very small Indian community. This situation contributes to the socio-economic structure and cultural background of each ethnic group. The proportion or distribution of the types of cultural landscape may also differ between States in Malaysia, since they are also influenced by the socio-economic status of each state. The highly developed states may have a poorer cultural landscape compared with that in less developed states.

With this baseline concept and perspective, further relevant research is needed to develop a solid definition of cultural landscapes in Malaysia. Two research components are needed to achieve this goal. Firstly, research into the identification and characterization of a cultural landscape. This includes studies on the geological events and historical background of the cultural landscape, which can reveal the unknown properties that have created the cultural landscape. The second research component is an assessment of features or elements that characterize the cultural landscape. This is in order to provide details of the types or categories of feature in the cultural landscape. Nevertheless, at the same time, research related to an assessment of the human impact on the cultural landscape is also necessary. This assessment is pivotal, because the rapid economic development of human activity is continuously encroaching into the cultural landscapes. To conclude, this chapter should be seen as a platform toward studies of the cultural landscape in Malaysia and other Southeast Asian countries. This is in order to include cultural landscapes as an entity for sustainable development.

Acknowledgments This study is part of a research project entitled "Landscape fragmentation and heterogeneity of mangrove landscape in peninsular Malaysia" funded by Research University Grant UKM-GUP-ASPL-08-06-212, courtesy of the Ministry of Higher Education Malaysia.

References

Antrop M (1997) The concept of traditional landscapes as a base for landscape evaluation and planning. The example of Flanders Region. Landsc Urban Plan 38:105–117

Austad I (1988) Tree pollarding in Western Norway. In: Birks HH, Birks HJB, Kaland PE, Moe D (eds) The cultural landscape: past, present and future. Cambridge University Press, Cambridge

Buckley R, Ollenburg C, Zhong L (2008) Cultural landscape in Mongolian tourism. Ann Tour Res 35:47–61

Farina A (2000) The cultural landscape as a model for the integration of ecology and economics. Bioscience 50:313–320

Forman RTT (1995) Land mosaics: the ecology of landscapes and regions. Cambridge University Press, Cambridge

Gomez-Limon J, de Lucio Fernandez JV (1999) Changes in use and landscape preferences on the agricultural-livestock landscapes of the central Iberian Peninsula (Madrid, Spain). Landsc Urban Plan 44:165–175

Hong SK (1998) Changes in landscape patterns and vegetation process in the far-Eastern cultural landscape: human activity on pine-dominated secondary vegetation in Korea and Japan. Phytocoenologia 28:45–66

Hong SK (2001) Factors affecting landscape changes in central Korea: cultural disturbance of the forested landscape system. In: Van der Zee D, Zonneveld IS (eds) Landscape ecology applied in land evaluation, development and conservation: some worldwide selected examples. ITC, Netherlands, pp 131–147

Hong SK, Nakagoshi N, Kamada M (1995) Human impacts on pine-dominated vegetation in rural landscapes in Korea and western Japan. Vegetation 116:161–172

Jomo KS, Chang YT, Khoo KJ (2004) Deforesting Malaysia. The political economy and social ecology of agricultural expansion and commercial logging. Zed Books, New York, 253 pp

Kamada M, Nakagoshi N (1996) Landscape structure and the disturbance regime at three rural regions in Hiroshima Prefecture, Japan. Landsc Ecol 11:15–25

Khoo TT (1996) Geomorphological evolution of the Merbok estuary area and its impact on the early state of Kedah, northwest peninsular Malaysia. J Southeast Asian Earth Sci 13:347–371

Kim JE, Hong SK, Nakagoshi N (2007) International trends of rural landscape researches for land management and policies. In: Hong SK, Nakagoshi N, Morimoto Y (eds) Landscape ecological applications in man-influenced areas: linking man and nature systems. Springer, the Netherlands, pp 489–504

Kitazawa T, Ohsawa M (2002) Patterns of species diversity in rural herbaceous communities under different management regimes, Chiba, central Japan. Biol Conserv 104:239–249

Lim JNW, Douglas I (2000) Land management policy and practice in a steep land agricultural area: a Malaysian example. Land Degrad Dev 11:51–61

Mander Ü, Pallang H, Ihse M (2004) Development of European landscapes. Landsc Urban Plan 67:1–8

Nakagoshi N, Hong SK (2001) Vegetation and landscape ecology of East Asian 'SATOYAMA'. Global Environ Res 5:171–181

Nassauer JI (1995) Culture and changing landscape structure. Landsc Ecol 10:229–237

Naveh Z (1995) Interactions of landscapes and cultures. Landsc Urban Plan 32:43–54

Ong JE (2003) Plants of the Merbok Mangrove, Kedah, Malaysia and the urgent need for their conservation. Folia Malaysiana 4:1–18

Ong JE, Gong WK, Wong CH, Din ZHj (1991) Characterization of a Malaysian mangrove estuary. Estuaries 14:38–48

Oreszczyn S (2000) A systems approach to the research of people's relationships with English hedgerows. Landsc Urban Plan 50:107–117

Oreszczyn S, Lane A (2000) The meaning of hedgerows in the English landscape: different stakeholder perspectives and the implications for future hedge management. J Environ Manage 60:101–118

Pinto-Correia T (2000) Future development in Portuguese rural areas: how to manage agricultural support for landscape conservation? Landsc Urban Plan 50:95–106
Plieninger T, Höchtl E, Spek T (2006) Traditional land-use and nature conservation in European rural landscapes. Environ Sci Policy 9:317–321
Rabbinge R, van Diepen CA (2000) Changes in agriculture and land use in Europe. Eur J Agro 13:85–100
Soo SW (1976) Semi-detailed soil survey of the Sungei Merbok area, Kedah. Division of Agriculture, Ministry of Agriculture, Soil Survey Report 8
Takeuchi K (2001) Nature conservation strategies for the 'SATOYAMA' and 'SATOCHI', habitats for secondary nature in Japan. Global Environ Res 5:193–198
UNESCO World Heritage Centre (2003) Cultural landscape: the challenges of conservation. World Heritage papers 7. UNESCO World Heritage Centre, Ferrara, Italy
Vos W, Meekes H (1999) Trends in European cultural landscape development: perspectives for a sustainable future. Landsc Urban Plan 46:3–14
Zonneveld IS (1995) Land ecology. SPB Academic Publisher, The Hague

Chapter 5
Introducing Geo-Cultural Landscapes in Iran

Forood Azari-Dehkordi

5.1 Introduction

"Cultural landscapes" are those where the use of the land reflects an amalgam of environmental possibilities (such as gradients, climate, and soil fertility) and human endeavor. The process of creating a cultural landscape has produced classic landscapes which are acknowledged to be as important a heritage as fine historic buildings and vernacular settlements. Noting that concern for such landscapes is now universal, and has moved away from its former Old World focus, Phillips (1998) has confirmed a growing international awareness of the link between cultural and natural diversity, and the vulnerability of both to outside processes. This awareness is paralleled by a widespread reaction against the ways in which the global economy and technological advances have created increasingly standardized and homogenous environments. Cultural landscapes are thus no longer seen as a sectoral, elicit, "Western" topic (Selman 2006), but rather as arenas for multifunctional planning across a wide range of environments.

The definition of a cultural landscape recognizes the role of human construction and imagination in creating and interpreting units of the environment that nevertheless possess a functional as well as a visual coherence. This definition assumes that a fundamental feature of the landscape is its distinctive "character," which has resulted from a complex pattern of actions and interactions, manifest in both the historical legacy and contemporary dynamics. It implies that distinctive places are frequently the outcome of a fortuitous combination of natural and human factors.

Thus, cultural landscapes are "synoptic" spaces where human and nonhuman elements are fused in a physical and social entity laden with individual and collective associations. In this regard, Phillips (2002) has referred to the cultural landscape as comprising: (1) nature plus people, (2) the past plus the present, and (3) physical attributes (scenery, nature, history, heritage) plus associative (social and cultural) values. However, in Iran several of these cultural landscapes have been

F. Azari-Dehkordi (✉)
Department of Geography, University of California at Berkeley, Berkeley, CA, USA
e-mail: fazari@berkeley.edu

S.-K. Hong, et al. (eds.), *Landscape Ecology in Asian Cultures*, Ecological Research Monographs, DOI 10.1007/978-4-431-87799-8_5, © Springer 2011

places of worship. Places of worship are those natural and historical monuments that are perceived by the local population as sacred, and as having spiritual and religious values (Zandanova 2007).

Despite, some major exceptions, such as the course of the Silk Road and the origin of one of the major centers of civilization in the ancient world, few cultural landscape studies (e.g., Nüsser 2001) have been carried out in the Middle East, and such studies are especially rare in Iran. Therefore, the purpose of this chapter is to introduce the extent of Takht-e Soleyman as a geo-cultural landscape. It is made up of geological elements that were considered by the ancient people dwelling in the area to be the holy character of the land, and a place to worship, and after the Arab invasion of Iran the elements of the area were protected by the people because of their sacred landscape characteristics.

5.2 Geo-Cultural Landscape Character

Toward the end of the twentieth century, there was an emerging consensus that landscapes could most consistently and helpfully be described in relatively non-judgmental terms, based on recordable features that contributed to their distinctiveness. Thus, an analysis and explanation of a landscape's character became the keynote, rather than an evaluation of its scenic beauty (Selman 2006).

The character of a cultural landscape derives from (1) a *combination* of factors such as geology, landform, vegetation, land use, fields, and human settlement patterns, (2) its past, present, and/or future context, and (3) the inter-relationship between biophysical and cultural factors. Thus, cultural landscape characters can be seen as an expression of the way in which the natural and cultural elements of terrestrial ecosystems combine to create unique places with specific ecological and economic, as well as social, functions and values.

Therefore, *characterization*, i.e., a way of identifying areas of distinctive character, classifying and mapping them, and describing and/or explaining their character, yields landscapes that are single and unique areas that may capture a "sense of place" for people. In this regard, *character* refers to a distinct, recognizable, and consistent pattern of elements in the landscape that makes one landscape different from another, rather than better or worse. Characteristics are elements, or a combination of elements, which make a particular contribution to a distinctive character. Finally, the term *elements* stands for individual components which make up the landscape, such as trees, stones, or walls.

5.3 Cultural Landscapes in Iran

Ancient settlements in mountainous landscapes followed both linear and punctuated patterns. Linear patterns, such as the distribution of caravan settlements or *caravanserai* (inns), were constructed along commercial routes. One important route was the Silk Road (around 8 000 km), which was developed during the Safavieh Dynasty (1501–1736) as a main governmental road network.

However, the natural landscapes of Iran have been extensively used and developed by humans for centuries. 9000 years ago, einkorn wheat (*Triticum boeoticum*) and 10,500 years ago sheep were domesticated in Iranian and Plateau (Wright, 1976; Makhdoum 2003), and the generation and modification of cultural landscapes followed. After that, humans modified landscapes and generated new ones according to their activities.

Animals transport themselves through the landscape and distribute the effects of their feeding, depositing waste materials, pollination, and seeds across environmental spaces (Reiner and Driese 2004). In Iran, nomadic cattle herders have used the Iran Plateau for centuries, and because of the movement of their animals they have to a large extent determined the plant distribution in this region. For example, the mountain meadows, steppes, and xerophytic plant communities represent sufficient food potential for cattle, and have been used as summer pastures. Extensive phytogeographical studies of alpine flora in Iran (Akhani 2007; Noroozi et al. 2008) provide evidence for the development of the vegetation coverage in this cultural landscape.

However, the most important factor for human settlement and landscape building in this arid region, as in any other part of the world, is the availability of potable and agricultural water. A large spring that has made a lake is a major factor, and a tendency to worship in sacred places in ancient times has also influenced the development, configuration, and preservation of these places. Takht-e Soleyman, as an important cultural landscape in Iran, is an example of the existence of a geo-cultural landscape.

5.4 History of Takht-e Soleyman

Takht-e Soleyman was created around 3 000 years ago. The site of Takht-e Soleyman was destroyed in AD 627 by the Byzantine army of Heraclius I in a counter-attack for the Sassanian invasion of the Roman territories. The Byzantines destroyed the fire temple and took away its treasures, which were offerings by Sassanian kings. The site fell into disuse and was subsequently abandoned. Archeological excavations have revealed some traces from the eighth and ninth centuries AD, but it was not until the fourteenth century that Takht-e Soleyman regained its importance as the palace of Ilkhan Aba-Qaan. The Mongols reoccupied and restored the main buildings of the Sassanians, such as the fire temple and the big western *iwan* (ICHTO 2008) [iwan: (Persian: *eyvān*) is defined as a vaulted hall or space, walled on three sides, with one end entirely open].

The Ilkhanid period was undoubtedly one of the most prosperous occupations of the site after the fall of the Sassanian empire. Architectural remains discovered in the course of excavations have revealed masterpieces of Middle Persian art and architecture. After the demise of the Ilkhans, from the end of the fourteenth century, the site of Takht-e Soleyman, although still partly occupied, never recovered its previous importance. Inundating flood water from the lake gradually washed out the mortar from the joints in the lower parts of the fire temple. The cavities which now took the place of mortar between the bricks caused irregular sinking of the courses. All this resulted in vertical cracks and fissures in the brick masonry. The dome of the fire temple must have fallen down in this way, as well as that of the west *iwan*.

The site of Takht-e Soleyman was discovered in 1819 by the British traveler Sir R.K. Porter. In 1831, Colonel W. Monteith, in his perilous journey to western Iran, visited Takht-e Soleyman. Then in 1838, Sir H. Rawlinson correctly identified the large pile of ruins in the center of the Takht as the fire temple. The site was also visited by A. Houtum-Schindler in 1881, A.V.W. Jackson in 1903, and A.F. Stahl in 1907. Each one left an informative description of the ruins. E.F. Schmidt surveyed Takht-e Soleyman from the air in the summer of 1937, taking the first valuable photographs of the site, but it was not until October 1937 that it was thoroughly investigated by A.U. Pope and D.N. Wilber for an architectural survey by the American Institute for Iranian Art and Archaeology. The site was then explored by the Swedish archeologists H.H. von der Osten and B. Almgren in 1958 on behalf of the German Archaeological Institute. H.H. von der Osten subsequently led the first campaign of excavation with R. Naumann. A serious program of study in 14 campaigns of work at Takht-e Soleyman and its surroundings has been carried out by the German Archaeological Institute under R. Naumann, and then under D. Huff (ICHTO 2008).

5.5 The Takht-e Soleyman Area

Takht-e Soleyman is the remains of an ancient castle located in Afshar, Western Azerbaijan, North of Takab (47°30′E, 36°40′N), Iran (Fig. 5.1). The Throne of Soleyman site has been designated a UNESCO world heritage site (No. 1077/2003). The area is well known for five cultural landscape characters, namely, Takht-e Soleyman (literally means the "Throne of Solomon"), Zendane Soleyman (literally means "Solomon Prison"), Stone Dragon (also Solomon Stick), Belqeis Fortress, and Fire Temples.

5.5.1 Takht-e Soleyman

Takht-e Soleyman has been given many names throughout history. It was called "Gaznak" or "Azargoshnasb" (the Fire Temple) by ancient Iranians, "Gazka" by the Romans, and "Shiz" by the Arabs. The name of the site was Athur Gushnasp (Azarghoshnasb) or the "place of fire of the warriors" during the Sassanian period (AD 224–641). The site is mentioned by an Islamic historian as Shiz, and Saturiq during the Ilkhanid period in the fourteenth century AD. The name Takht-e Soleyman, which is used today, was given to the site in medieval times. According

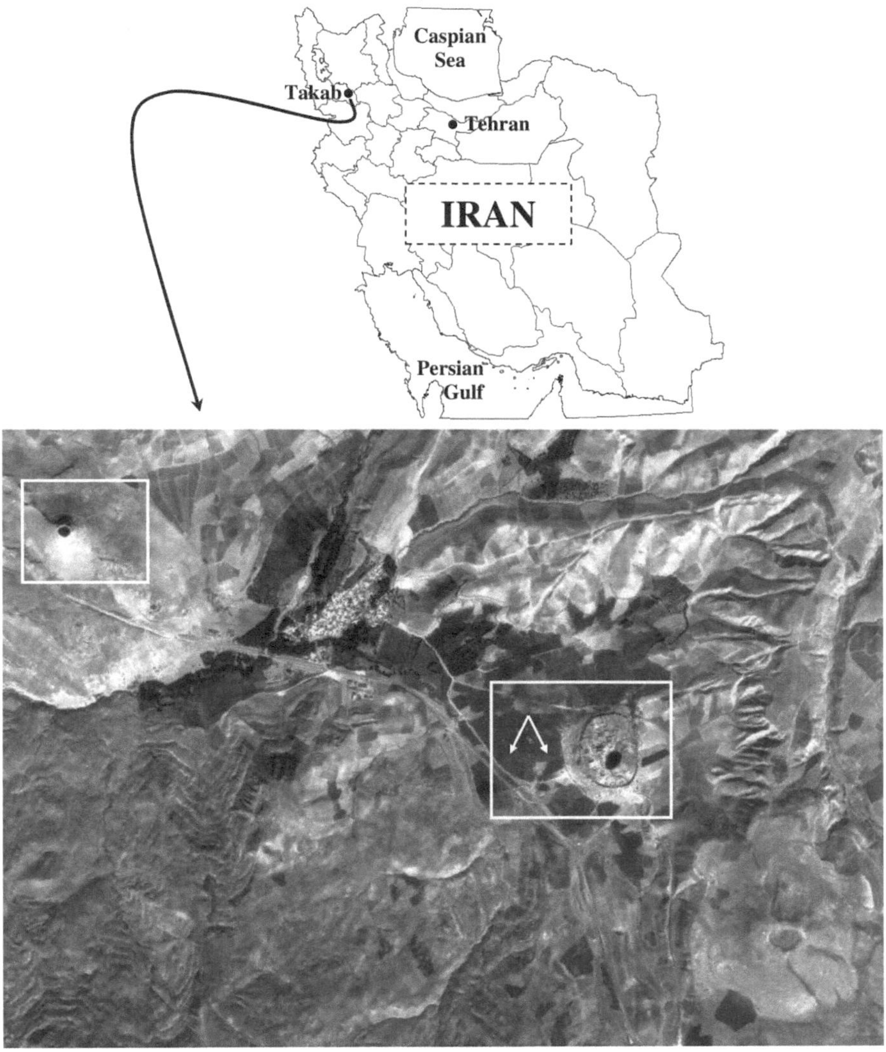

Fig. 5.1 Location of Takht-e Soleyman near Takab, Iran. The air photograph of 1997 shows the historical site and its vicinity. While northwest is on left, the left box engulfs mount Zendan-e Soleyman (Solomon Prison). The arrows give the location of Stone Dragon (Fig. 5.9) and the Right Box is enlarged in (Fig. 5.2)

to ancient texts and legends, Solomon possessed supernatural powers such as talking to animals or flying on carpet.

In fact, Takht-e Soleyman was constructed as a castle around 3 000 years ago. In ancient times and in the Achamenid (538–331 BCE) and Parthian (250 BCE to 224 CE) Empires (Bouni 2008; Lendering 2008), this castle was situated on the strategically important main thoroughfare from Ekbatan (Hamadan) to Armenia and Syria. During the time of its use, this castle served as a fire temple and was the most

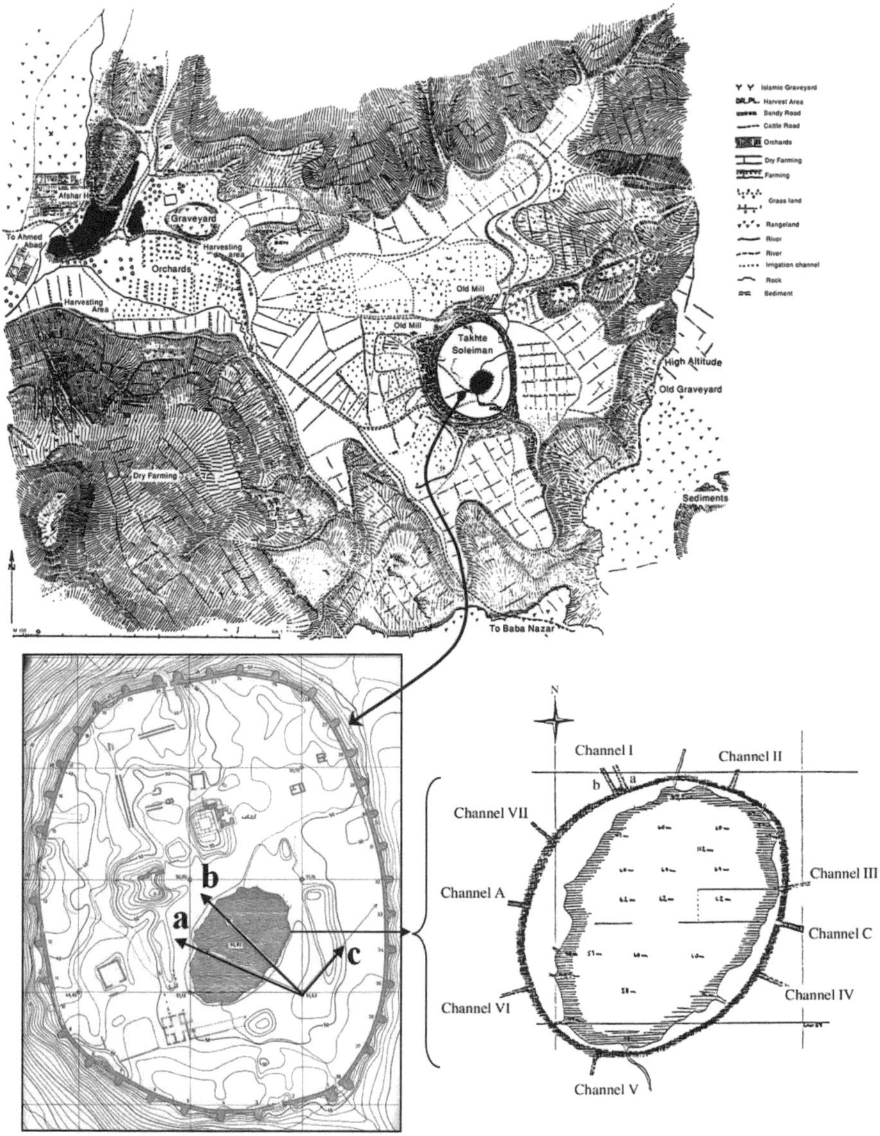

Fig. 5.2 Layout and topographic map of Takht-e Soleyman or Azargoshnasb fire temple, a Zoroastrian holy complex in Iran. In the bottom left the "a" arrow points to the Zendan-e Soleyman, "b" points to the tumulus, and "c" points to Belqeis Fortress (see Fig. 5.1) (Source: von der Osten and Naumann 1961)

important pilgrimage site for Zoroastrian kings, high priests, and priests. Takht-e Soleyman was reduced to ruins by the Roman Emperor Heraclius I in 627.

The Takht-e Soleyman site is a stratified semi-conical structure made up of calcareous Pliocene sediments (Asadi and Hale 2001) that is 400 m (to the North) by 320 m (to the West). It is surrounded by an ovoid stone wall (hedgerow) with a perimeter of 1 120 m. It has an elevation of 50 m from the bottom of the western side of the valley, and is about 10 m in height from the eastern side (Figs. 5.1 and 5.2).

A small lake is located inside the castle, with a depth of 64–110 m and an area of 80×120 m. In the spring floods, the whole surface appears conspicuously as brilliant deep-white stone. For decades, many outlets from this spring were used to irrigate agricultural lands (Fig. 5.3). However, today only two permanent water discharge outlets exist (Fig. 5.2, bottom right). In spite of these two water outlets, the flat stone surface looks like a surface of calcareous sediments with no vegetation coverage.

5.5.2 *Zendan-e Soleyman*

The hollow mountain known as the Zendan-e Soleyman is situated 3 km to the west of Takht-e Soleyman. The height of this conical mountain is 97–107 m above neighboring land, and a hole approximately 80 m deep, with an opening of approximately 65 m diameter, can be seen from above (Fig. 5.4).

Studies (Kleiss and Boehmer 1965; Damm 1968) have shown that this crater, which dates back to the Pliocene geological period, and had then contained lakes with the springs streaming into the lakes, is actually the last evidence of a dormant volcano. The springs flowing from the middle of the basin into the Agh Darreh valley consist of warm water containing many minerals, which formed the mountainous mass of Takht-e Soleyman and the Zendan-e Soleyman. The sedimentary layers of the springs are part of the natural Pliocene landscape. Travertine layers of the Zendan-e Soleyman prison have piled up on Miocene rocks, tightly pressing the mortar to the southern edge of the basin (Kleiss 1971).

Flat conical lime tuff has piled up in the upper part of the valley. However, it determines the structure of the last periods of the valley. An attempt to determine the age of the mountain and prison precisely according to the piled mineral mass and other hard materials was not satisfactory because it was not known what changes the springs had undergone in terms of temperature, water flow, dissolved sediments, and gas. The appearance of the prison and the springs of Agh Darreh today shows the process of these changes. The drying up of the water flowing from the mountain was not sudden. Instead, the water level gradually decreased over a long period of time, and the reason for this was the wet sediments of the lower layers where they had been more compacted.

Around the cone-shaped opening of the prison mountain, there are signs of a holy temple dating back to the first millennium BC (Fig. 5.5). According to the historical objects found, the remains of the architectural monuments around the opening of the prison mountain are connected to the *manas* ruling in this region of Iran from 830 to 660 BC. It seems that the holy place flourished for as long as the prison mountain lake contained water, and after the water had dried up, the place was deserted as a temple, although some of its architectural units were used for guarding the castle for while thereafter (Oehler 1962).

Fig. 5.3 Views of an outlet and vestiges around Takht-e Soleyman-site lake and the remains of Khosrow's Iwan (right)

Fig. 5.4 Southern views of mount Zendan-e Soleyman. The left view is visible to the northern part of Takht-e Soleyman and the recreation path of visitors is depicted in the right picture

Fig. 5.5 The remains of holly temple near mount Zendan-e Soleyman

This place has terraces and chambers for the residence of pilgrim Zoroastrian priests, as well as a holy courtyard with three entrances, a minor side door in the western side, a sloped stairway on the southeastern side, and the main gate along the uphill path which ran diagonally from east to west (Naumann 1967).

Another point to be noted about Zendan-e Soleyman is that the existing gas and spring outlets at the bottom of the well are still active. The walls of the gas spring have subsided into a hole inside the mountain, but this powerful gas-bearing spring, which is placed high on the mountain, has enough force to throw out any unwanted stones and keep its mouth open and flowing with water. In fact, we cannot yet distinguish the spring inside the opening from a height of 80 m because the ruins and sedimentary layers which collapsed into the hole hide it from sight. In general, the prison mountain, which is based on the piled sediments, if sectioned from 60 m below the summit from the back of the mountain has a structure which is very similar to the present condition of Takht-e Soleyman and a sedimentary bench of almost the same size. The locals thought that rebellious devils have been imprisoned by the order of the Solomon.

5.5.3 Belqeis Fortress

The ruins of Belqeis Fortress (Takht-e Belqeis in Farsi) (Fig. 5.6) were inspected with scientific precision in 1959 by von Der Osten, and were studied and documented in 1966 and 1969 by D. Huff (1974). The Belqis double mountain, which has two closely located summits, is 3200 m high and is northeast of Takht-e Soleyman. On the highest part of this mountain, there are the remains of fortifications which date back to the Sassanian era. The general specifications were provided by Huff (1974), and have a close architectural/historical relationship with the monuments of Takht-e Soleyman because of their geographical and chronological position.

Fig. 5.6 Part of the architectural remains of Belqeis Fortress (left) and its preliminary plan (right)

The Belqeis fortress is enclosed in an area approximately 60×50 m. According to a preliminary plan, the enclosing wall includes nine citadels and there are also possibly another four which are not yet known because of insufficient scientific excavation.

In front of the structure, there is a multistory terrace including long shelter-like chambers with dome-shaped ceilings. The architecture and type of bricks have Sassanian dimensions, and the yellow-colored sandstone pieces as well the marks left from the work of Sassanian stone carvers makes it more than likely that the structure dates back to this era (Naumann 1967, 1977).

In their general design and their veranda, hall, and the round porch, these fortifications not only might have been an Iranian palace, but could also be considered to have been installations of a fire temple. Architecturally and geometrically, the direction of the fortifications, which face toward the ancient Takht-e Soleyman complex, puts both structures in such a close relationship that it becomes strongly possible to think of the fortifications as a Sassanian fire temple. In addition, both architecturally and historically, there is also a belief among the common people that there is a mythological relationship between the Takht-e Soleyman works and the Belqeis Fortress.

5.5.4 Fire Temples

After having passed the *iwan* (Figs. 5.3 and 5.7) at the center of the complex, room "A" is a domed, square room on the main axis. Naumann (1977) believed that in this cruciform room the holy fire was exposed during services, and pilgrims could circulate in the four vaulted corridors which surrounded the middle room and see the fire. In the adjoining cross-shaped room "B" the place for the fire in the center, in the form of a square basin, is still preserved (Fig. 5.8). It seems that in the four holes in the corners there once stood stone altars, and one of them was discovered in this room. The stone altars possibly held a platform, as the figure is coined on one of Sassanian coins. Four of these 90 cm height altars were discovered in different places in this small cross-shaped room (B) (Fig. 5.8, left). This was certainly the holiest place in the temple. Because the fire was kept burning when it was not lighting in room A during a religious service.

However, some scholars believe that the holy fire was never exposed in room A because then it could be seen by anybody and was accessible because of its axial position, which is prohibited according to the Avesta. The two rooms mentioned above (A and B) in the center of the temple were built of fired brick only. In all other parts of the great building a well-cut limestone is used, and only vaults and cupolas in this part were made of fired brick. On the East side of the fireplace there are two similar vaulted rooms (C and D) (Fig. 5.8). Besides the possibility of fire treasury, a copy of the Avesta was kept in these rooms where Sassanian Kings also used them to prepare and dispatch their gifts out. In the niches of these rooms, holes for fixing wooden shelves are still visible. A vaulted corridor connects the rooms.

5.5.4.1 West Temple

According to the excavator, the western complex (on the west side of corridor Ko1) there is a fire temple. It consists of a series of rooms in the south connected to a group of buildings of which two basilica halls are quite impressive. The halls PB and PA, each with three naves separated by two rows of rectangular or round pillars, were situated one after the other. On the same axis there is an inner pavilion with a square room and three niches (probably a dining room). The pillars are all made of gypsum covered with fired bricks. The halls were decorated with stuccoes and reliefs. The two basilica halls through room PC led to a cross-shaped room (PD) that was also a place for the fire or a shrine. On the west of this group lies a bigger vaulted cross-shaped room with a similar function as that of room A in the eastern complex.

Fig. 5.7 View of Iwan Khosrow (top) and its location in Takht-e Soleyman site (right)

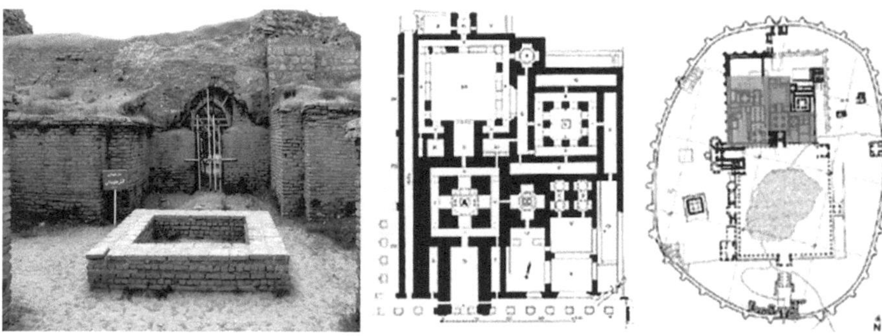

Fig. 5.8 Room B (*left*), main fire temple (middle), and locations of other fire temples in site (*right*)

5.5.5 Stone Dragon

In the southwestern side of the external area of Takht-e Soleyman, about 150 m away from the enclosing wall, there are remains an exceptionally winding path appealing to the eye of a sedimentary wall related to water stream with a length of 300 m and approximately 2 m height (Fig. 5.9). This stone conduit named by the locals, the Stone Dragon. It has been "petrified" by the sediment of lake water that was transferred to the residential region in the Sassanian era. The local people of Shiz said to Arab invaders that dragon has turned to stone upon the prophet Solomon's order.

5.6 Takht-e Soleyman as a Geo-Cultural Landscape

Cultural landscapes are comprised of worship complexes coupled with the landscape environment. The worship complexes (of natural and anthropogenic origins), in turn, include objects or sites of worship, and inherent structures and attributes. The characteristics of the Takht-e Soleyman site were originally a consequence of geological changes in this landscape that attracted people to swarm pilgrim activities around the lake and to build a worship complex within a landscape that has specific geological elements of a landscape. However, its conversion to a Zoroastrian fire temple can be classified as an anthropogenic designation to worship complex.

The reasons for it remaining as a sacred cultural landscape throughout history may vary. When the Arabs conquered and occupied Iran, most of the Iranian ziggurats and Zoroastrian temples were demolished and replaced with mosques. This was particularly the case with fire temples, as the worship of fire is considered blasphemous in Islam. However, local residents were often able to circumvent the newly introduced religious requirements and maintain sacred places such as the major ziggurats by giving them Islamic names. This is how sanctity can preserve and protect a historical monument. The values of this geo-cultural landscape are so

Fig. 5.9 Stone Dragon. The left picture gives the ending point

outstanding, in spite of being hidden, that despite the contradictions between current dominant political issues and the ancient religions of Iran, the site was brought to the attention of UNESCO and was finally nominated as a world heritage site in 2003.

5.7 Concluding Remarks

Before UNESCO knew about the location of Takht-e Soleyman, the holiness of the sacred complex intrigued the Shiz residents to protect this geo-cultural landscape by converting the original names of the elements in the sacred landscape (which were established in the Zoroastrian era) to Islamic names. Thus, the name "Azargoshnasb Fire Temple" was converted to "Takht-e Soleyman," the mountain with a major fire temple was changed to "Belqeis Fortress", and "Zendan-e Soleyman" was re-named to keep invaders from destroying the Holy courtyard and Zoroastrian pilgrim fire temples in that mountain. It is interesting to know that according to Islamic beliefs, Belqeis was the beloved queen of the prophet Solomon.

Other stories have also been told by Shiz people to safeguard this religious complex. For example, by telling the story that the devil is "imprisoned by Solomon under the mount Zendan-e Soleiaman," and describing the conversion of a dragon to a petrified-conduit of water (the stone dragon). Therefore, a nation managed to preserve their cultural landscape heritage from complete destruction by Arab invaders. In fact, the landscape of the present Takht-e Soleyman complex remains because of introducing disguised-names of cultural landscape elements that were understandable and sacred landscape elements to invaders.

References

Akhani H (2007) Diversity, biogeography, and photosynthetic pathways of Agrusia and Heliotropium (Boraginaceae) in South-West Asia with analysis photogeographical units. Bot J Linn Soc 155:401–425

Asadi HH, Hale M (2001) A predictive GIS model for mapping potential gold and base metal mineralization in Takab area, Iran. Comput Geosci 27:901–912

Bouni A (2008) History of Iran. Achaemenid, Persian-Syria, pp 538–331

Damm B (1968) Geologie des Zendan-e Soleiamn und seiner Umgebung. Sudstlisches Balqash Gebirge Nordwest Iran, Wiesbaden

Huff D (1974) Sasanidisch-Fruhislamische Ruinpltze im Belqis Massiv in Azerbeidjan, Arch ologisches Mitteilungen aus Iran, Neue Folge, Band 7, Berlin

ICHTO (Iran Cultural Heritage, Handicraft and Tourism Organization) (2008) Takht-e Soleiman. Retrieved 2008-12-31 from www.ichto.ir

Kleiss W (1971) Zendan-e Soleiamn. Die Bauwerke, Beitr ge zur Archologie und Geologie des Zendan-e Soleiamn, Wiesbaden

Kleiss W, RM Boehmer (1965) Takht-i Soleiamn und Zendan-e Soleiamn. Die Grabungen auf dem Zendan-e Soleiamn. Arch ologischer Anzeiger, Heft 4, Berlin

Lendering J (2008) History of Iran. Parthian Empire. Retrieved 2008-12-12 from Iran Chamber Society from www.iranchamber.com/history/parthians/parthians

Makhdoum MF (2003) Fundamental of land use planning. University of Tehran, Tehran
Naumann R (1967) Takht-i Soleiamn and Zindan-e Soleiamn. Various excavations in 1959. In: Pope AU (ed) A survey of Persian art, vol. XIV. Tokyo
Naumann RE (1967) Takht-i Soleiamn. Ausgrabung des Deutsches Archologischen Instituts in Iran. Katalog der Ausstellung, Munchen
Naumann R (1977) Die Ruinen von Takht-i Soleiamn und Zendan-e Soleiamn und Umgebung, Deutsches Archologisches Institut, Abteilung Teheran Fuhere zu Archologischen Pltzen in Iran, Band II, Berlin
Noroozy J, Akhani H, Breckle SW (2008) Biodiversity and phytogeography of the alpine flora of Iran. Biodivers Conserv 17:493–521
Nüsser M (2001) Understanding cultural landscape transformation: a re-photographic survey in Chitral, eastern Hindu Kush, Pakistan. Landsc Urban Plan 57:241–255
Oehler H (1962) vorberichte uber die Grabungenim Jahre 1960 und 1961. Archologischer Anzeiger, Heft 4, Berlin
Phillips A (1998) The nature of cultural landscapes: a nature conservation perspective. Landsc Res 23:21–38
Phillips A (2002) Management guidelines for IUCN category V areas: protected landscapes/seascapes. IUCN, Gland, Switzerland
Reiner WA, Driese KL (2004) Transport process in nature: propagation of ecological influences through environmental space. Cambridge, Cambridge University Press, 302 pp
Selman P (2006) Planning at the landscape scale. Routledge, New York, 213 pp
von der Osten HH, Naumann R (1961) Takht-e Soleiman. Iran Organization for Cultural Heritage, Tehran (translated to Persian by F.N. Samiee, 1995) 194 pp
Wright Jr HE (1976) The environmental setting for plant domestication in the Near East. Science 194:385–389
Zandanova BA (2007) Worship places as elements of cultural landscapes (as exemplified by the "Tunkinsky" national park). Geogr Nat Resour 29:191–194

Chapter 6
Cultural Landscapes of the Tengger Highland, East Java

Luchman Hakim

6.1 Introduction

Cultural landscapes represent a situation that is associated with the local culture and traditions. They are widely distributed throughout the world, and are very diverse in terms of structure and function. This is a result and a manifestation of the world's cultural diversity. Cultural landscapes, with their tangible and intangible values, are important to humanity (Plachter and Rössler 1995), and therefore the conservation of cultural landscapes is crucial in order to achieve human prosperity. Nevertheless, rapid and uncontrolled development is often reported as leading to environmental degradation and threatening the cultural landscapes. For instance, many sacred forests have a variety of functions, and many traditional management practices have disappeared. Throughout the world, many agricultural landscapes have been abandoned. According to researchers, there has also been a decline in knowledge and respect for traditional values in many areas (Heywood and Watson 1995; Dudley et al. 2005).

While such landscapes are abundant in Indonesia, the term cultural landscape is rarely discussed in scientific literature. Consequently, literature and knowledge about Indonesian cultural landscapes are rarely found. The criteria needed for assessing cultural landscapes are absent. The diversity and conservation issues regarding Indonesian cultural landscapes are significant. Biologically, Indonesia is home to 10% of the world's angiosperms (flowering plants), 12% of its mammals, 16% of its amphibians and reptiles, 17% of its birds, 25% of its fish, and 15% of its insects, making Indonesia a leading biodiverse country. This archipelago not only has huge biodiversity, but also possesses high socio-cultural diversity (Hakim 2004; Sani and Hanun 2004). The interaction between natural resources and people has been recognized, and has led to the creation of many cultural landscape forms. These have been given little attention in the past. The implication of these issues is

L. Hakim (✉)
Department of Biology, Faculty of Mathematics and Natural Sciences, University of Brawijaya, Jl. Veteran Malang 65142, East Java, Indonesia
e-mail: lufehakim@yahoo.com

S.-K. Hong, et al. (eds.), *Landscape Ecology in Asian Cultures*, Ecological Research Monographs, DOI 10.1007/978-4-431-87799-8_6, © Springer 2011

that extensive study about cultural landscapes in Indonesia is needed to allow such resources to become sustainable and therefore enhance human prosperity.

This chapter aims to describe the recent status of cultural landscapes in East Java, Indonesia. East Java is an example of abundant cultural landscapes, but attention to their existence is rarely discussed. East Java has huge biodiversity and is considered to be important as a world biodiversity conservation site. Local people's perspective of nature includes much indigenous knowledge, and many practices which maintain the land in a sustainable manner, protect biodiversity, and ensure prosperity. However, recent rapid development has influenced urbanization, changed land use uncontrollably, and allowed the regions to become densely populated. Nowadays, East Java is one of the most important provinces in Indonesia in terms of political and economic perspectives (Whitten et al. 1996; Hakim 2004; Sani and Hanun 2004). This population increase potentially threatens the biodiversity and cultural landscapes, and therefore seeking a strategy to overcome these problems is urgently needed.

This discussion will focus on the Tengger Highlands in East Java (Fig. 6.1) as a case study. Administratively, most of the highland area has been protected, and it became an integral part of the Bromo Tengger Semeru National Park. The park was founded in 1982 and covers an area of about 50276 ha, stretching from the Tengger Highlands in the north to Mt. Semeru in the south. The park extends to four regencies, namely Malang, Pasuruan, Probolinggo, and Lumajang (Fig. 6.2). The park is

Fig. 6.1 East Java and the distribution of national parks, namely Bromo Tengger Semeru (A), Meru Betiri (B), Baluran (C), and Alas Purwo (D)

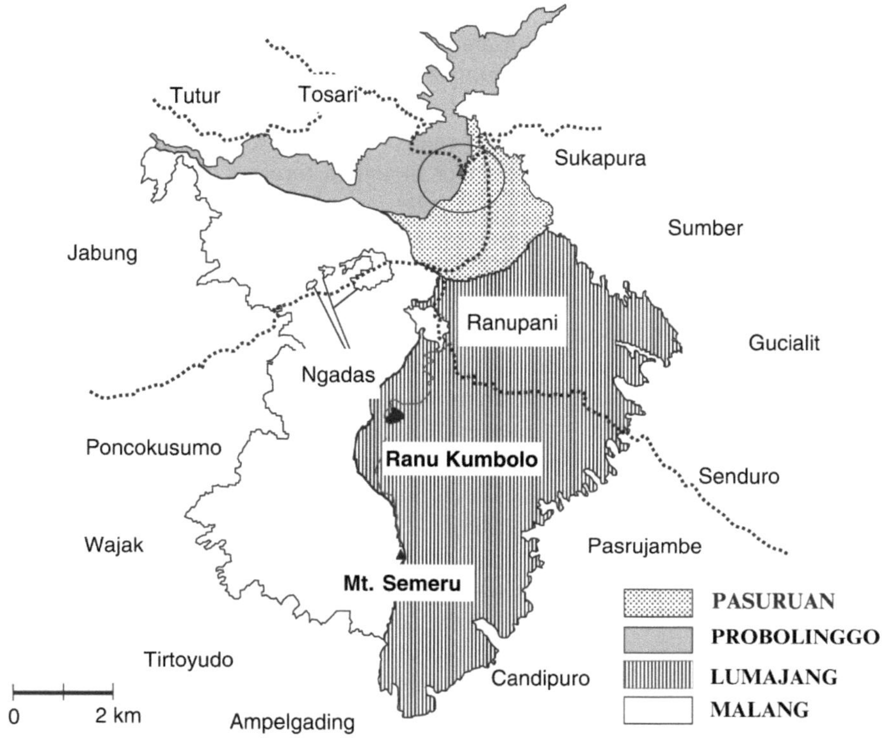

Fig. 6.2 The study area (Bromo Tengger Semeru National Park). *Dotted lines* indicate access to Mt. Bromo-Tengger caldera (the *area* in the *circle*)

known as a nature-based tourism destination, and is visited by many international and domestic tourists (Cochrane 2006; TNBTS 2003).

6.2 Biodiversity and Local People

Located in the eastern part of Java Island, Tengger Highland is part of a wide volcanic area which extends throughout Java, Bali, and the Lesser Sundas Islands. The recent amazing landscape of Tengger Highland is associated with the geological history of the Indonesian archipelago. In the past, the eastern part of Java Island was one of the most active volcanic areas in the world, and one recent big eruption created a vast sand–sea caldera and several peaks in the highlands, including Mt. Jatur (2705 m), Mt. Ider-ider (2617 m), Mt. Pundak Lembu (2826 m), Mt. Lingker (2278 m), Mt. Keciri (2627 m), Mt. Pananjakan (2272 m), Mt. Batok (2470 m), Mt. Kursi (2581 m), Mt. Watangan (2601 m), Mt. Widodaren (2650 m), and Mt. Bromo (2392 m) (UNDP and FAO 1980; TNBTS 2003). It is extremely difficult to identify the number of species and their classification accurately, but there are strong indications that the biodiversity is very rich. The highlands have numerous types of ecosystems and are home to many rare and endemic species. Nevertheless, many of them have recently been threatened by rapid habitat degradation, illegal hunting and harvesting, and other human activities. Ecologically, the forest vegetation is classified into two categories: the lower and upper mountain forest types. The vegetation of the upper mountain forest is characterized by an abundance of *Casuarina junghuhniana*, *Albizia* sp., and *Acacia decurrens*. Very little has been reported about the ecology and biodiversity of the upper mountain forest. *Vaccinium varingifolium* is the most common plant on the slopes and crater walls of Mt. Bromo. The edelweiss *Anaphalis* sp. grows on Mt. Penajakan, where populations are able to survive in poor soil conditions (Whitten et al. 1996). Edelweiss has a special value during the *Kasodo* ceremony as an "offering" plant. Unfortunately, the populations in many mountain areas in East Java, including Tengger Highland, are heavily threatened by exploitation and illegal harvesting. Other species found in the upper mountain zone are *Tithonia diversifolia*, *Dahlia* sp., *Rubus idaeus*, *R. rosaefolius*, *Eupatorium rotundifolium*, *Ricinus communis*, and *Brugmansia* sp. (Hakim et al. 2008).

The highland is famous for the existence of a wide sand–sea caldera called the Tengger Caldera. Surprisingly, biodiversity information about this ecosystem is rare. Physically, the northern part of the caldera has relatively poor vegetation owing to the substrate conditions. The part dominated by a sandy surface has high substrate porosity, like a desert. Small populations of *Imperata cylindrica* grow patchily in very dry environments. Conversely, the southern part consists of fertile soil which allows more vegetation to grow (UNDP and FAO 1980). Herbarium records from the Department of Biology, University of Brawijaya at Malang, show that several species grow there, including *I. cylindrica*, *Foeniculum vulgare*, *Pennisetum* sp., and several ferns species. Some species, such as *C. junghuhniana*, *A. decurrens*, and fern trees, grow scattered along the caldera's wall. The top of the caldera wall in the south is a habitat for numerous non-native species such as *Lantana camara*, *E. rotundifolium*, and *Belamcanda* sp. From the perspective of biodiversity conservation strategies, the clearing of non-native plant species and reducing their dispersal is urgently needed (Cronk and Fuller 1995; Heywood and Watson 1995). In East Java, some national parks have experienced great pressure from non-native plant species that can alter the habitat (Whitten et al. 1996; Hakim 2004; Hakim et al. 2005).

In the lower mountain forests, *Ficus* sp., *Vitex* sp., *Artocarpus* sp., *Erythrina* sp., and *Syzigium* sp. are dominant. These species often form a dense canopy and thereby a special microclimate which allows orchids, ferns, and mosses to grow. A high diversity of rattans and climbers form a complex horizontal structure, while a diversity of herbs and shrubs form complex vertical structures. The understory encompasses diverse species such as *Alocasia* sp., *Colocasia* sp., *Amorphophallus* sp., *Cyathea* sp., *Pandanus tectorius*, and *Musa* sp. According to Whitten et al. (1996), several endemic species were recorded in the past, including *Plectranthus steenisii*, *Melastoma zollingeri*, *Solanum alpinum*, *S. anacamptocarpum*, *S. viscidissimum*, *Viola javanica*, and *Malaxis tenggerensis*, but no further information is available. The lower mountain forests are sources of food, medical material, and charcoal, which offer numerous benefits to local people. The high diversity of plants in the lower mountain zones is a potential genetic resource.

There are no comprehensive data available regarding animal diversity. Wildlife is rare in the caldera (Whitten et al. 1996), but mountain forests are reported to be inhabited by primates, herbivores, birds, and insects. According to a Field Report of UNDP/FAO in 1980, the highland has at least 80 bird species from 34 families. In Mt. Penanjakan, recorded species include *Spilornis cheela*, *Gallus gallus*, *Lanius schach*, *Streptopelia chinensis*, and *Ichthyophaga* sp. The most frequently recorded bird in the caldera ecosystem is *Mirafra javanica*, which nowadays is under pressure owing to habitat disturbance.

The highland is inhabited by local people who are known as *Wong Tengger* (Tenggerese). They are one of the sub-ethnic groups of Javanese Hindus, who were isolated after the rise of Islam in Java in the late eighteenth century (Hefner 1999). Culturally, they have rich folk beliefs, and these support the wise use of natural resources. The local people's perception of nature apparently contributes to the recent huge increase in pristine highland forest. For the Tenggerese, the whole highland is sacred, especially places that play an important role in their beliefs. The local culture and belief system states that Mt. Bromo and Tengger Caldera are sacred sites and therefore should be protected from potential disturbance. The site is believed to be the home of deities (*Dewata*), and has a special association with the origins of the earth (Whitten et al. 1996; Hakim et al. 2008). This is a fundamentally important component in this cultural landscape.

Tenggerese mythology prescribes certain rituals for deities and their ancestors in the calderas and Mt. Bromo. This perception is expressed during *Kasodo*, the main ceremony among the Tenggerese, in order to celebrate their origins, and takes place every 270 days. This ceremony opens with the inauguration of honoured members of Tenggerese society, and is followed by a traditional performance. At midnight, the inauguration of new priests is performed at the *Poten* (a temple) on the north side of the sand–sea of the caldera. The *Kasodo* consists of a huge ceremony on the crater of Mt. Bromo, into which offerings are thrown, and in a temple at the foot of Mt. Bromo and Mt. Batok. At the time of the *Kasodo*, it is estimated that an additional 20000–25000 people enter the park (Hefner 1999; Whitten et al. 1996). During the performance and celebration, numerous plant species are used as honourable plants. These include *Areca catechu*, *Cocos nucifera*, *Zea mays*, *Rosa* sp., *Musa* sp., *Cananga odorata*, and *Michelia champaca*. Many vegetables and fruits are also used as offering plants. These are mostly collected from the gardens of Tenggerese homes (Hakim et al. 2008).

Some parts of the caldera are governed by taboos. In the north, for instance, *Poten*, groves, and a single area dominated by a stone called *Watu Kutho* are among the sacred sites which have a special relationship to the culture. According to the local people, *Watu Kutho* is the unseen kingdom that controls the caldera. In this kingdom, speaking evil, disturbing the vegetation and stone features, throwing waste, and doing immoral acts are prohibited. The worship of nature is a significant practice among Tenggerese in the caldera. Groves have potential biodiversity value and urgently need an inventory of species. The humid conditions in these groves support the growth of ferns, herbs, and grass. In some countries such as India and Ghana, groves have been recorded as having high biodiversity value, and importantly the sacred values that are attached to groves have become key components in conserving the biodiversity of these environments (Amoako-Atta 1995; Chandrashekara and Sankar 1998).

In general, it can be said that the Tengger Highland has huge biodiversity, and represents a unique cultural landscape with tangible and intangible values. However, in order to move towards the sustainable use of these resources, completing a biodiversity database is crucial. In Bromo Tengger Semeru National Park, these issues have been planned, but apparently have not been implemented. It seems that the problem comes from a lack of human resources, database management, and stakeholder support. In terms of protection issues, the large size of the park relative to the limited number of staff and park rangers is claimed to be a factor limiting the protection of the park's resources (Subagiandi Pers.com). The focus of the management must be to maintain the physical integrity of the park and mitigate human impact.

6.3 The Ecology of Rural and Upland Farming

Prior to recent agricultural development, the rural and upland areas in Tengger Highland had hardly been developed. Initial agricultural practices probably appeared after the fourteenth century. There are few data regarding the past status of agriculture here, but it has been reported that several plants, such as *R. communis* (Jarak) and *Allium* sp., have been cultivated and sold by the Tenggerese since 1785. In Java, the Jarak tree has been cultivated for a long time to provide biofuel material. In the eighteenth century, Europeans introduced potatoes and peppers to the highland farmers. These were followed by other vegetables, which were mainly planted to supply European and Chinese demands. While Tengger produced numerous agricultural crops, the local people reported consuming corn as a main food (Hefner 1999).

As in many developing countries, the green revolution has become the prime theme of agricultural development. In Indonesia, it was introduced in the 1970s with the objective of increasing rice production. The government introduced high-yielding plant varieties, provided pesticides and chemical fertilizers, and promoted modern equipment and machinery. Since rice production has become a central theme of the nation's development policy, paddy fields have received a lot of attention, but small orchards in remote areas were ignored. In these areas agricultural development was stagnant, particularly in areas where paddy fields were absent. In the case of the Tengger Highland, which has small patchy fertile areas, the limited infrastructure was an additional barrier to applying any green revolution program. Consequently, agricultural practices and the social systems of farmers have been little affected by any green revolution concepts. Until the beginning of the 1990s, this situation allowed Tenggerese indigenous knowledge to be conserved, and communities were little affected by modernization (Hefner 1999).

In Tengger, most agricultural land is located and distributed on particularly steep slopes (Fig. 6.3). According to Whitten et al. (1996), traditionally these farmers did not cultivate the steepest slopes, but nowadays such land has mostly been cultivated. This change indicates rapid human penetration onto land which is unsuitable for agricultural use. The reasons for this are not yet completely understood. Some scholars consider that limited fertile land, an increasing population, and land conversion due to community settlements are the prime reasons (Hefner 1999; Hakim and Nakagoshi 2007; Subagiandi Pers.com). Nowadays, the steepest slopes have been planted with crops which include potatoes (*Solanum* sp.), garden leeks (*Allium porrum*), and cabbage (*Brassica* sp.), as are widely found in Ngadas and Ranupani, the only two villages located inside the park. Agricultural practices in these villages are also problematic. Planners argue that steep cultivated land should be discontinued, but such a policy seems difficult to implement. Generally speaking, the cultivation of the steep slopes has led to soil erosion and environmental hazards. However, the question of where the farmers could be moved to must be answered. In addition to cultural reasons from the Tenggerese, the central government does not have the funds or proper planning laws to move the farmers out.

The need to conserve the land in a sustainable manner could be achieved by the terrace system, which has been practiced in some villages in the Tengger Highland. Around the world, terrace farming has evolved as an adaptation to extreme topography. Terraces have been created as an adaptation to utilize hilly and sloping environments (Whitten et al. 1996). Recent research findings suggest that such techniques have benefits for soil conservation, particularly in minimizing the rapid surface run off of irrigation water, reducing landslide risk, and mitigating soil erosion (Lansing 2000). In Tengger, some farmers in some places have developed a terrace system as an adaptation to living on such steep land, and so far it seems to have been successful in conserving the land. Farmers have adopted an agroforestry system where

Fig. 6.3 Agriculture on the steepest slopes

Toona sureni, *Melia azedarach*, *Bambusa* sp., *Gigantochloa* sp., *Swietenia mahagoni*, *Albizia falcata*, and *C. junghuhniana* have been planted for garden demarcation or inside the garden itself. Such plant species provide numerous benefits that people in Tengger want and need, such as civil construction material, firewood, fruit, resins, and medicines.

As well as being a dominant feature of steep agricultural landscapes, the landscape of the Tengger Highland is famous for its wide apple, *Malus domestica*, orchards, and particularly those found in Nongkojajar, Poncokusomo, and Gubuk Klakah villages. The combination of soil, temperature, and humidity has led to the cultivation of apples in the Tengger Highland. Apples were introduced from Europe in the eighteenth century, and in Indonesia they were successfully cultivated in Batu and Tengger. These areas are administratively parts of the Malang regency, and a strong image of Malang as an apple city has been created. In every discussion about Malang, people ask and talk about apples. Apples have successfully constructed a strong image of place, and put Malang on the national map (Hakim and Nakagoshi 2007). In discussing the management of steep land, the use of a terrace system to plant apple orchards is interesting. This is not only an excellent model of how people can adapt to steep land, but shows the possibility of applying a terrace system in the Tengger Highland in order to achieve maximum crop production and conserve the environment. These facts imply that apple orchards should be protected as an integral part of cultural landscape conservation.

The Tenggerese live in small settlements where their properties are rich in biodiversity. Biodiversity is particularly rich in their home gardens, and reflects a human tradition to maintain a domestic environment. Hakim and Nakagoshi (2007) recorded at least 153 plant species growing in home gardens. The tradition of managing a garden by planting several plant species is a significant strategy in biodiversity conservation in highland environments. According to researchers, the diversity and beauty of home gardens provide a brand image along tourist corridors, and therefore become a crucial component in destination planning and development (Baud-Bovi and Lawson 2002; Gunn and Var 2002; Akbar et al. 2003). In Tenggerese gardens, the species grown have at least one of several functions, including ornamental plants, medical plants, economic plants, and wild species. Ornamental plants are the most important species, and include *Zantedeschia aethiopica*, *Canna* sp., *Rosa* sp., *Belamcanda chinensis*, *Crinum asiaticum*, *Gladiolus grandiflorus*, *Hydrangea macrophylla*, and *Dahlia pinnata*. According to Raunkiaer's life form classification, chamaephytes are dominant, followed by cryophytes, hemicryptophytes, and phanerophytes in that order. This composition reflects a significant difference from lowland home gardens in Java, where phanerophytes dominate and gardens show a multilayer structure (Fernandez and Nair 1986; van Steenis 1987). However, our previous research showed that non-native plants species were abundant in Tenggerese home gardens. These have become a significant threat to the ecosystem because the presence of non-native plant species leads to a decline in native habitat and environmental disturbance (Cronk and Fuller 1995).

6.4 Landscapes in Danger

While a legal basis for protection and conservation was declared two decades ago, stress on the highland ecosystem increases continuously. In Tengger, humans significantly alter the habitat and pollute the soil and water. This phenomenon can be seen in Lake Ranupani, where the ecosystem has been heavily disturbed. Recently, several activities which put the ecosystem of the lake in danger have been recorded, such as bathing and washing using detergent, and extracting water using machinery. In the past, alien fish species were introduced to the lake and they potentially threaten native species. The lake is visited by tourists and Hindus on pilgrimage, and their activities potentially pollute the lake's ecosystem. The pollution of wetlands represents a loss in potential habitat for fish and other aquatic species, thus disrupting conservation programs. The diversity and composition of the original wetland species are relatively unknown. This means that protected areas do not provide complete and integrated data for conservation needs. As a component of the cultural landscape, lake conservation is crucial. Park management should be able to minimize negative activities and mitigate potential threats.

The highland is also facing problems related to forest fires and their severe impact. According to official data, the caldera and its surrounding area were a fire hot spot during 2000–2006. Fire was frequently recorded in the dry season, from August to October. Fire frequently occurs in grassland ecosystems and in some areas dominated by shrubs. In August 2006, for instance, fire burnt grassland and shrubs in 140 ha in Wetangan-Laut Pasir, 75 ha in Bantengan, 45 ha in Mt. Kursi, 30 ha in Adasan, and 25 ha in Rudjak. More than 3 000 Casuarina trees were burnt in the middle of September 2006. The number of affected areas and burned trees would be higher if the records included fire accidents in Wonokitri, Ngadas, Sukapuro, and other areas in the highland (TNBTS 2007). These figures indicate that forest fire is the most significant threat to highland biodiversity.

Hunting is another crucial problem and in the Tengger Highland it is not a new phenomenon. Traditionally, hunting was the most important livelihood activity because it provided food, and particularly protein, for the family. In recent decades, however, hunting has taken place in every part of the highland ecosystem. Increases in hunting are commonly the consequence of economic pressures, increased numbers of rural dwellers, the establishment of new roads to access forests, and other causes (Subagiandi Pers.com). According to official documents, the most important game animals are Munjak deer, Timor deer, and wild pig. Some species, such as birds, are caught and sold by local people (TNBTS 2007). The decline of birds and other wildlife not only has a potential effect on ecosystem integrity, but also changes the Tenggerese cultural landscape. This is because wildlife is a crucial component of the landscapes. The loss of wildlife means the loss of some seed dispersal agents that could alter the vegetation structure and tree survival. In tropical countries, this concept is becoming important because a diversity of wildlife plays a role in enhancing the reproduction and dispersal success of a number of plants (Heywood and Watson 1995; Turner et al. 2001).

Tourism poses a potential disturbance to the highland. Incidents which include vandalism, illegal flora harvesting, solid waste accumulation, pollution, and habitat disturbance are frequently reported (Hakim 2004). Increasing numbers of tourists to the caldera and Mt. Bromo have led to a significant growth in transportation services (Subagiandi Pers.com). In these cases, the greatest threats to cultural landscape integrity are major infrastructure projects that are carried out without being subjected to an environmental impact assessment. To date, the impact of cars, jeeps, motorbikes, and horses are rarely evaluated, but there is some indication that sand–sea areas are being affected. For instance, roads to the caldera and Mt. Bromo crater have been heavily polluted by horse dung and oil. During big events such as the *Kasodo* ceremony, thousands of tourist vehicles were recorded, and this is ecologically beyond the road's carrying capacity. In addition to the physical disturbance, tourism has an impact on the local people (Walker et al. 2000). Many local people in developing countries such as Indonesia have been shown to be susceptible to such changes. The impact of socio-cultural changes has frequently been reported, and includes changes in social systems and the disappearance of local culture and tradition. Modernization potentially reduces the number of farmers, and there is then an increase in barren agricultural land. In the future, it seems that a code of conduct should be developed.

With the development of mass tourism and modern agriculture, the number of young farmers has decreased and traditional agricultural land has gradually been abandoned. This situation is ironic in regard to promoting cultural landscape conservation. As in many other parts of the world, the agricultural landscape in Tengger Highland needs the farmers. The importance of farmers for the maintenance of a cultural landscape cannot be ignored. The loss of farmers is a potential factor in land-uses changes and therefore affects the cultural landscape. This implies that to preserve the cultural landscape, the number of farmers should be increased, but because of their poor status, low wages, decreasing garden productivity, and numerous other problems, it could be difficult to implement conservation programs here. Therefore, an economic initiative to increase human prosperity is needed. Several options, such as promoting sustainable nature and society-based tourism, for instance, should become part of a well-designed strategy.

6.5 Tourism in Cultural Landscapes

In Indonesia, tourism has been recognized as one significant factor in increasing economic prosperity, while at the same time it has the potential to support conservation programs. This means that promoting tourism in Tengger Highland will be key to achieving a conservation agenda (Hakim et al. 2008). This is consistent with the vision of central government that tourism should contribute to national development. Recently, there have been ambitious targets to increase international tourist arrivals to 8.5 million in 2012. Many nature-based tourism destinations have become targets for this increased tourist trade, including many cultural landscapes such as the Tengger Highland.

While the beauty of the Tengger Caldera and Mt. Bromo was recognized in the eighteenth century by Europeans, tourism developed and flourished in the 1970s when this sector was promoted as a key to development. In Tengger, tourism development in the 1980s was supported by the development of new infrastructure and accommodation, as well as transport systems. It has been reported that national and regional economic benefits from tourism have been significant, but the economic benefits at the local level are few (Cochrane 2006; Hakim et al. 2008). Conversely, at the local level, tourism leads to environmental and social disturbances.

In accordance with planning for sustainable tourism in the Tengger Highland, our previous research reached several important conclusions. Firstly, most policy making in connection with tourism planning and management was controlled by central government, with little community participation. In other words, top-down planning was dominant, rather than bottom-up (Hakim et al. 2008). In modern conservation, such a top-down approach has been recognized as one of the weaknesses of planning. The modern conservation approach argues that local voices and participation are crucial for drawing up proper plans in order to achieve the sustainable use of resources (Heywood and Watson 1995; Dudley et al. 2005). Therefore, future planning should promote the involvement of local people.

Secondly, tourism planning and development in the 1970s only focused on the caldera and Mt. Bromo. These plans included poor fundamental assessments of the site's carrying capacity, and poor environmental impact assessment. Rural areas and their huge potential attractions have received little attention. Tourism development was object-oriented, but neglected the human dimension. These situations fundamentally illustrate how planners fail to maximize the potential value of the Tengger Highland as an integral cultural landscape. Tourism development in Tengger must be able to provide positive impacts to the relevant parties, but such impacts on most of the local society have been very few to date. Areas nearest the tourism objectives grow quickly and become urbanized, and the areas nearby become less developed (Hakim et al. 2008). According to researchers, the imbalance in development also has negative consequences such as increasing social conflict. In many developing countries, such disparities have been recognized as a factor contributing to vandalism, social conflict, and social harassment (Walker et al. 2000). In the case of the Tengger Highland, therefore, these would be counter-productive to the conservation objective.

Thirdly, the issue of authenticity has been ignored, and is less understood in general. Recently, the issue of authenticity is emerging as a crucial component to enhance tourism competitiveness. Authenticity is often applied in heritage tourism and ethnic/tribal-oriented attractions (Fyall et al. 2005; Swarbrooke 2006). In this regard, reports about a destination's authenticity conclude that a destination should be able to provide an authentic experience (or tourist experience) regarding the object of interest. It is relevant that planning achieved sustainable tourism in the Tengger Highland, including strengthening the authenticity of the cultural landscape. Fundamentally, planning scenarios should provide a mechanism to allow traditional and indigenous knowledge to exist in perpetuity.

Fourthly, programs related to social development are poor. For instance, training was delayed or too short, and the development of skills was not always sufficient to support tourism development. The park project focused more on the park's attractions than on community development. The contribution of local people and their culture has been relatively neglected. For a long time, the landscape of the Tengger has been viewed as a natural phenomenon, and the human culture dimension has been neglected. For better planning in the context of sustainable tourism in Tengger, more studies and analyses of local people, local culture, and their participation in possible strategies are recommended.

The above points imply that document planning for highland conservation should be improved. Park authorities and communities should view the Tengger Highland as an integral cultural landscape. In Tengger, the potential for sustainable tourism comes from a combination of several components, such as natural richness, local culture, and social support (Hakim et al. 2008). Consequently, the planning perspectives for tourism development should be able to promote local people's involvement in a park's conservation strategy. These should be formulated using an integrated approach. New planning should be able to protect biodiversity effectively, while at the same time promote and encourage the local culture as a key toward proper management.

6.6 Concluding Remarks

This chapter has evaluated the recent status of the Tengger Highland, and the conclusions can be summarized as follows. One, this study confirms that the Tengger Highland is a cultural landscape where landscape, culture, and humans coexist. There are two fundamental characteristics of Tengger in accordance with its classification as a cultural landscape, namely intangible and tangible values. Intangible values relate to the fact that the highlands are strongly associated with Tenggerese faith and beliefs. This means that their socio-cultural organization has built a local identity and thereby enhanced the prosperity of the local people. The fertile land provides opportunities for agriculture, producing a diversity of crops and stimulating the local economy. Two, the biodiversity is diverse, but few data are available regarding biological and ecological systems. This is a weakness of the current conservation program. Nevertheless, the local belief system and traditional knowledge are apparently significant factors in enhancing and supporting conservation. This traditional knowledge benefits land protection and mitigates human impact on the environment, and therefore supports biodiversity conservation. Three, steep agricultural land dominates the Tengger Highland and has become a landmark of Tenggerese villages. Suitable agricultural practices in these steep lands, such as agroforestry and the development of terraces, enhance land conservation. Four, there are some problems, ranging from pollution, forest fires, and illegal hunting to vandalism, which threaten the cultural landscape. Forest fires have become a crucial issue because of their ability to reduce biodiversity. Five, tourism in the Tengger Highland has the potential to support conservation programs. However, so far, it has featured very little in planning and development. Tourism planning has been dominated by a top-down approach and little participation by the local people. In addition, much attention has been paid to the caldera and Mt. Bromo, but little to the other areas. Planning has failed to define destination authenticity, and programs related to the development of social capacity have been poor.

The overall result of this study suggests that the Tengger Highland, whether as a religious or tourism destination, is a complex system which needs attention and integrated planning and development. Only the integration of the components of the ecosystem will allow the traditional way of life of the local people and their numerous needs to be sustained, and therefore appropriate planning should accommodate the numerous uses of local resources. This means that among the traditional religious practices, tourism and nature conservation should coexist with planning and development.

Acknowledgements The author gratefully acknowledges the contributions of Bagyo Yanuwiadi, Jati Batoro, Endang Arisoesilaningsih, Sasmito Djati, and Fariana Prabandari to this research project. Special thanks go to Mr. Subagiandi, the former Head of Bromo Tengger Semeru National Park, and his staff for providing invaluable information, discussions, and support during our field survey.

References

Akbar KF, Hale WGH, Headley AD (2003) Assessment of scenic beauty of the roadside vegetation in northern England. Landsc Urban Plan 63:139–144

Amoako-Atta B (1995) Sacred groves in Ghana. In: von Droste B, Plachter H, Rössler M (eds) Cultural landscapes of universal value. Gustav Fischer Verlag in cooperation with UNESCO, Jena, pp 80–95

Baud-Bovi M, Lawson F (2002) Tourism and recreation: handbook of planning and design. Architectural Press, Oxford

Chandrashekara UM, Sankar S (1998) Ecology and management of sacred groves in Kerala, India. For Ecol Manage 12:165–177

Cochrane J (2006) Indonesian national parks: understanding leisure users. Ann Tour Res 33(4):979–997

Cronk QCD, Fuller JL (1995) Plant invaders: the threat to natural ecosystems. Chapman and Hall, London

Dudley N, Higgins-Zogib L, Mansourin S (2005) Beyond belief: linking faiths and protected areas to support biodiversity conservation. World Wide Fund for Nature, New York

Fernandez ECM, Nair PKR (1986) An evaluation of the structure and function of tropical home gardens. Agric Syst 21(4):279–310

Fyall A, Garrod B, Leask A (2005) Managing visitor attraction. Elsevier, Oxford

Gunn CA, Var T (2002) Tourism planning: basic, concepts and cases. Roudledge, New York

Hakim L (2004) Dasar-dasar Ekowisata. Bayumedia Press, Malang (in Indonesian)

Hakim L, Nakagoshi N (2007) Plant species composition in home gardens in the Tengger highland (East Java, Indonesia) and its importance for regional ecotourism planning. Hikobia 15(1):23–36

Hakim L, Leksono AS, Purwaningtyas D, Nakagoshi N (2005) Invasive plant species and the competitiveness of wildlife tourism destination: a case of Sadengan feeding area at Alas Purwo National Park. J Int Dev Coop 12(1):35–45

Hakim L, Hong SK, Kim JE, Nakagoshi N (2008) Tourism and cultural landscape at Tengger, East Java: the implications for ecotourism planning. Korean J Environ Ecol 22(3):207–220

Hefner RW (1999) Geger Tengger: Perubahan sosial dan perkelahian politik. LKIS, Yogyakarta (in Indonesian)
Heywood AH, Watson RT (1995) Global biodiversity assessment. UNEP – Cambridge University Press, Cambridge
Lansing F (2000) Foucault and the water temple. Crit Anthrop 20(3):309–318
Plachter H, Rössler M (1995) Cultural landscape: reconnecting culture and nature. In: von Droste B, Plachter H, Rössler M (eds) Cultural landscapes of universal value. Gustav Fischer Verlag in cooperation with UNESCO, Jena, pp 15–18
Sani RR, Hanun SA (2004) State of the environment in Indonesia 2004. Ministry of Environment Republic of Indonesia, Jakarta
Swarbrooke J (2006) The development and management of visitor attractions. Elsevier Butterworth Heinemann, Oxford
TNBTS (2003) A study for nature-based tourism development of Bromo Tengger Semeru National Park. Ministry of Forestry PHKA – Bromo Tengger Semeru National Park, Malang (in Indonesian)
TNBTS (2007) Annual report. Ministry of Forestry PHKA – Bromo Tengger Semeru National Park, Malang (in Indonesian)
Turner MG, Gardner RH, O'Neill RV (2001) Landscape ecology in theory and practice: pattern and process. Springer, New York
UNDP and FAO (1980) Bromo-Tengger/Gunung Semeru proposed national park management plan 1981–1985. Field report of UNDP and FAO National Park development project, Bogor
van Steenis CGGJ (1987) Flora untuk Sekolah di Indonesia. Pradyaparamitha, Jakarta (in Indonesian)
Walker JL, Mitchel B, Wismer S (2000) Impact during project anticipation in Molas, Indonesia: implication for social impact assessment. Environ Impact Assess Rev 20:513–535
Whitten T, Soeriaatmadja RE, Afiff SA (1996) The ecology of Java and Bali. The ecology of indonesia series, vol II. Periplus Ltd., Singapore

Chapter 7
Traditional Forests in Villages Linking Humans and Natural Landscapes

Sun-Kee Hong and Jae-Eun Kim

7.1 Introduction

The sustainable management of natural resources and the effective use of various resources found in a particular area play an important role in the creation of an environmentally friendly local life culture capable of coexisting with nature (Antrop 1997; Nassauer 2005; Hong 2007). In this regard, forests have long been viewed as a traditional form of eco-culture, where the communal aspects of the village are combined around the agricultural lifestyle of the residents, with the latter's social and public values (Volker 1997; Choung and Hong 2006; Kim et al. 2006). Forests also play important public functions. These include those related to the economically efficient use of cyclical bioresources such as the conservation of national land, environmental preservation, and ensuring landscape stability. In addition, the proper management of forests ensures that they can also play an important role in the management of the national land structure through such means as the prevention of fires and environmental restoration. As such, society has long had a keen awareness of the importance of forests (Takeuchi et al. 2003; Hong et al. 2007a).

The management of forests in Korea began in 1960 with the planting of trees in barren mountains, and was undertaken as part of the difficult task of rebuilding the economy in the aftermath of the Korean War. The *Saemaeul* (new village) Movement of the 1970s changed the appearance of villages by bringing about drastic improvements in roads and land cultivation methods (Hong 1998, 2001).

This paper is a modification of the article 'Village forests of estuary in west sea-forested landscape management for conserving resort area', which was written in Korean with English abstract and published in the Journal of the Island Culture (2007; 29: 441–473).

S.-K. Hong (✉)
Institution for Marine and Island Cultures, Mokpo National University, 61 Dorim-ri, Cheonggye-myeon, Muan-gun, Jeonnam 534-729, Republic of Korea
e-mail: landskhong@gmail.com; skhong@mokpo.ac.kr

J.-E. Kim
Institution for Marine and Island Cultures, Mokpo National University, 61 Dorim-ri, Cheonggye-myeon, Muan-gun, Jeonnam 534-729, Republic of Korea

However, a look back at the past reveals that many things were inadvertently lost during the process that saw Korea become an economic success story. One of the items which suffered the most was the nation's forests. A look at the significant number of trees which have been planted over the past 50 years and at Korea's now green mountains would seem to make the claim that Korea has no forests a nonsensical one. However, one also needs to look at the kinds of trees which were planted in Korean mountains during this period. The majority of the trees which were in fact planted in village forests were black locusts, pitch pines, Japanese green alders, and dawn redwoods (*Metasequoia*), all of which are very familiar to foreign landscapes. The traditional forest spaces which played a large role in the creation of the culture of local villages and served as a communal gathering places for festivals have now disappeared in the name of economic development and modernization. The disappearance of these forests has also entailed the disappearance of a significant aspect of Korean culture and history. Furthermore, the few remaining traditional forest resources have increasingly found themselves under attack as variations in the social environment caused by urbanization, such as changes in the rural lifestyle, the shrinkage of villages amidst an increasingly aging society, and the destruction of the ecosystem caused by changes in land use, have taken root.

Village forests (*Maeul-sup* in Korean, i.e., the natural or anthropogeneous forest type of Korean villages) have long played a key cultural and educational function in promoting a sense of attachment and unity within the community (Lee 2004). Although the modes of expression may have been different, it was nevertheless common practice in fishing, mountain, and agricultural villages to conduct communal festivals in the forests, during which time prayers were made for the continued health and prosperity of their villages. Conducting folk festivals such as the *pungeoje* (豊魚祭) and *dangsanje* (堂山祭) in the forests along coastal areas and estuaries played an important role in the creation of new histories, and in the forging of traditional lifestyles (*see* Chap. 2 in this book). While the culture and history inherent in these forests was passed down from generation to generation, it spawned more than just human culture (Hong 2007).

Various biological organisms which cannot be found in mountainous areas have been discovered in estuaries and coastal areas. These have commonly been referred to as "cultural bio-organisms" (文化生物). As such, village forests can also be regarded as the genesis for various forms of biological culture. However, in addition to their biological value and their value as a source of energy, village forests also played an important aesthetic role in bringing a certain sense of brightness to the daily lives of village residents. Most discussions about the characteristics and functions of forests in Korea generally revolve around the beautiful and scenic forests of Gangwon Province (east coast area). However, the Honam (southwest coast area) of Korea actually boasts a wider variety of traditional forests than any other area on the peninsula. As mentioned above, the forests in coastal areas and along estuaries traditionally served as the site in which the ritual known as *pungeoje* was conducted, and prayers were made for the overall welfare of the fishing village. Therefore, the traditional village forests in coastal and estuary areas are not only of ecological value, but also have a deep historical and cultural significance.

Europe and Japan have recently made efforts to restore traditional village forests as part of efforts to heighten the biodiversity of local communities and provide spaces for residents to gain ecological experience (Fukamachi et al. 2001; Nakagoshi and Hong 2001; Grashof-Bokdam and van Langevelde 2005; Brown et al. 2007). In particular, "*Chinju-no-mori* (鎮守の森), Japanese Forest of the Village Shrine," in which temples and shrines have traditionally been established in Japan, have been employed not only as resources with which to restore the local culture, but also as eco-tourism resources. Such efforts have contributed greatly to the activation of the regional culture. In Korea, villages which can provide an ecological experience have been developed in fishing areas as part of efforts to promote regional activation. In this regard, a number of local autonomies have provided support to these fishing villages, and so has the Ministry of Maritime Affairs and Fisheries (MOMAF). Many people have attempted to experience the various aspects of fishing village culture (such as traditional fishing methods, experiencing tidal flats, and the collection of shellfish). To this end, the traditional village forests found in fishing villages can become another cultural resource through which the culture of fishing villages and folk ecology can be experienced. They can also serve as a major tourism resource through which a better understanding of the history of fishing villages and fishing-related folk customs can be obtained.

7.1.1 Types of Village Forest

The Korea Forest Research Institute (1995) report classified Korea's village forests from ten standpoints: religious significance, education, sanitation, landscape, transportation, agricultural use, timber, military, communal society, and environmental security. In other words, the Japanese colonial government regarded Korean forests as an instrument with which to learn about Korean villages (*Maeul*), ascertain the cultural awareness of residents, and determine the background to the daily life practices of residents. The various types of village forest designated by the Cultural Heritage Administration of Korea include the *seonghwangnim* (城隍林) or *dangsup* (堂林, sacred forests), *hoannim* (護岸林, a forest established along a riverbank or coastline in order to prevent floods), *eoburim* (魚付林, fish shelter forests), *bangpungnim* (防風林, windbreak forests), *bohaerim* (補害林, a forest which under the tenets of geomancy is regarded as helping to overcome the geographical shortcomings of a village), *gunsarim* (軍事林, forests used for military purposes), and *yeoksarim* (歷史林, a forest to which historical stories and legends are attached). However, when this classification system is compared with the six classes of natural monuments in existence in Korea, nine types of village forest can in fact be identified: the *seonghwannim*, *biborim* (裨補林, forests which under the tenets of "*Feng-shui*" (風水) theory are regarded as helping to overcome the geographical shortcomings of a village; Hong 2007; Hong et al. 2007b), *yeoksarim* (historical forests), *jogyeongrim* (造景林, artificial man-made forests), *gyeonggwannim* (景觀林, landscape forests), *gyotongnim* (交通林, forests along transportation arteries), *boannim* (保安林, protected forests), *gunsarim* (軍事林, forests used for military purposes), and *saengsannim* (生産林, production forests).

The Cultural Heritage Administration of Korea's investigation of village forests nationwide was based on the six categories introduced above (Table 7.1).

Trees, along with rocks, water, and animals, have been worshiped as one of the nature gods in traditional folk beliefs. As it was believed that a divine spirit resided in sacred trees, these were also regarded as objects of fear. To this end, people commonly referred to a tree in which a divine spirit was believed to live as a *sinmok* (神木, god's tree) or *sinsu* (神樹, sacred tree). Meanwhile, as Shamans regarded trees as a place where heaven and earth encountered god and man, the *dangnamu* (divine tree) was perceived as lying at the center of the universe. Villagers believed that the reckless felling or destruction of a *dangnamu* would invoke the anger of the tree gods (*moksin,* 木神), who would respond by wreaking disease and disaster on their village. One finds many examples of stone mounds or stone stairs being built around a *dangnamu*. However, belief in the power of the *dangnamu* gradually decreased over time, until it became viewed by some as little more than a superstition. Many of these trees were subsequently cut or destroyed altogether as part of efforts to do away with such superstitions. However, this practice was not unanimously approved by all village residents. Despite all of these changes, many still continue to believe in the power of the *dangnamu*, and its worship continues to be a constant in the daily lives of a great number of Koreans.

7.1.2 The Ecological Characteristics of Traditional Forests

Although we tend to focus on the scenic beauty of forests, our ancestors also placed great importance on their functional aspects. The *dangmok* (divine tree) or *jeongjamok* (亭子木, tree located at the village pavilion) found in a village can be regarded as a symbol of the ecological functions of the entire forest or trees (Korea Forest Research Institute 1995). Although village forests can be observed and classified from various standpoints, including those of culture, ecology, or historical remains, Table 7.1 has become the general classification system used by researchers who study village forests, including those working for the Cultural Heritage Administration of Korea.

7.1.2.1 Dangsannim (Sacred Forest)

One of the key elements of Korean folk beliefs, the *dangsan* (堂山, sacred mountain), was referred to in various ways across different regions. For instance, while it was referred to as *seonangdang* (village shrine), *sansindang* (山神堂, mountain god shrine), *seonghwangdang* (仙皇堂, shrine for a tutelary deity), or *sanjedang* (山祭堂, altar for the mountain spirit) in the Gyeonggi, Gangwon, and Chungcheong provinces, it was called *dangsan* in the provinces of Yeongnam and Honam (Fig. 7.1). Meanwhile, the denizens of Jeju Island referred to it as *bonhyangdang* (本鄉堂, *bonhyang* or shrine to the village guardian deity) or *sancheondang* (山天堂, mountain god shrine). On occasion, it has also been called *sallyeonggak* (山靈閣, shrine to the mountain spirit),

Table 7.1 Examples of traditional village forests

Type	Function
Seonghwangnim	• A forest which had to be preserved based on the belief that it protected the village and ensured the welfare of villagers • The conduct of *dangje* (堂祭, village rituals) involved the installation of *seonangdang* (altar for tutelary deity) and *dangjip* (shrine house) within the forest • Major types of tree found therein: a community of trees that included the zelkova tree (*Zelkova serrata*) and the hackberry tree
Hoannim	• Served as an embankment to prevent water from overflowing during the flood season • *Hoannim* were developed in villages established alongside rivers • Major types of tree found therein: a forest belt that included willows and alder trees
Eoburim	• Artificially or naturally created in coastal areas to protect the village from the ravages of the sea winds and to secure the habitats of fish • *Pungeoje* (豊魚祭) was conducted • Served as windbreak forests along coastal areas • Major types of tree found therein: forest belt which included pine trees and/or zelkova trees
Bangpungnim	• Mainly established in coastal areas or windy places to prevent damage by strong winds • Much like the *hoannim*, forest belts were formed • Served as a village fence called *woosil* which protected the village from winds (*see* Chap. 2 in this book) • Major types of tree found therein: forests of black pine trees situated along the beach
Bohaerim	• Have the significance *of Feng-shui* theory • This type of forest was designed to supplement the "Feng-shui" geomantic shortcomings of the village • Can be regarded as the equivalent of the modern ecologically restored forest or environmentally protected forest
Yeoksarim	• Forest with which a specific historical story or legend is associated • Old and large trees situated in the heart of the village
Summary	• Several types of village forest can be associated with fishing, mountains, and agricultural villages. However, viewed from the modern ecological standpoint, each of these types of forest shares the commonality of having more than one ecological function. For example, the *bangpungnim* (windbreak forest) can play the role of an *eoburim*, as well as that of the *seonghwangnim*, in which prayers for the welfare of the village were traditionally rendered • There are no set types of tree found in specific types of forest. For instance, because of differing climates, the zelkova tree has been used as a *dangmok* (divine tree) in the central and northern provinces; while hackberry trees have been used as a *dangmok* in the southern provinces • However, regardless of the types of forest or tree present, the point of importance remains that these forests served as the venue through which the traditional lifestyle and culture of the common people, based on beliefs in the afterlife and Shamanism, were conveyed

Fig. 7.1 *Dangsan* tree (*Zelkova serrata*). This is the tutelary deity tree of the village at Beopseongpo of Yeonggwang-gun, Jeonnam Province. Village people are praying for peace and prosperity

sansingak (山神閣, shrine to the mountain god), *sanjudang* (山主堂), *halmeonidang* (village grandmother shrine), and *halabeojidang* (village grandfather shrine). Village rituals were generally conducted at an altar or shrine set up in a forest that featured large old trees (老巨樹, *nogeosu*). The village residents prayed for advantageous communal outcomes such as good harvests, an abundance of fish, and rain, as well as to ensure that no misfortune or contagious diseases befell the village. People also came to this site to pray for more personal wishes such as the prosperity of one's descendants, good luck, the fulfillment of wishes, and the birth of sons. The continued health of one's livestock was another common theme. The papers on which such wishes were written were burnt in order to ensure that they made their way to the heavens. However, such village rituals and festivals were prohibited during the Japanese colonial era on the grounds that they amounted to little more than empty superstitions. They vanished once again during the rapid industrial development process that began during the 1960s, to the point that it is now difficult to find any *dangsup* (sacred forests) (Fig. 7.2) today. Recently, the Korea Forest Research Institute (1995) and the Cultural Heritage Administration of Korea (unpublished data) began to collect material related to the *nogeosu* (old and large trees) as part of efforts to preserve village forests. In this regard, these trees have been designated as *bohosu* (nurse trees). However, very few remains of *dangsanje* (shamanist ritual for a good harvest and the prosperity of the village) can be found today.

Fig. 7.2 *Dangsup* (sacred forest) and *dang* (sacred place) (Suseongdang at Jukmak-dong, Byunsan, Jeonbuk Province)

7.1.2.2 Bangpungnim (Windbreak Forest)

Composed mainly of evergreen broad-leaved trees, a *bangpungnim* (防風林) is a type of forest which protects farmland and villages from strong winds (Fig. 7.3). These can be divided into farmland and coastal windbreak forests. In general, the windbreak forests are usually effective in cities, coastal areas, and farming areas, but not in mountainous ones. The windbreak effect tends to be proportionate to the width and density of the relevant forest zone. The influence of an ideal windbreak forest, that is one in which the branches, leaves, and trunks of trees account for 60% of the overall forest area and where the trees are evenly distributed, can be felt for up to five times the height of the trees in the case of high-altitude winds, and at a width of 35 times the height of the trees in the case of low-altitude winds. More specifically, the wind velocity is decreased by as much as 55–65% at a distance of three to five times the height of the trees, and by 20–40% at a distance of ten times the height of the trees. Furthermore, the windbreak forest can decrease the damage emanating from the soil particles, dust, salt, snow, and fog carried by the wind (Forman 1995). In addition to windbreak forests, coastal areas like Muan County, which feature farm and coastal lands, boast artificially made forests designed to attract fish fauna, thereby improving the residents' daily lives by increasing fish yields. These forests, known as *eoburim* (魚付林, fish shelter forest), stand as proof positive of our ancestors' wisdom.

There is a village forest called "*woosil*" in the village of Daeri on Anjwa Island in Shinan County (*see* Chap. 2 in this book). This forest is mainly composed of a belt of hackberry trees (*Celtis sinensis*) and zelkova trees (*Zelkova serrata*). The forest was created to supplement the geographical weaknesses of the area, and also

Fig. 7.3 (**a**) *Bangpungnim* (Yesong-ri evergreen forest, Bogil-myeon, Wando Island, Jeonnam Province, Natural Monument No. 40). *Bangpungnim* have important roles of wind-break forests for fishing village and also *eoburim* (fish shelter forests located near the coastline or on the edge of an island to protect the village from ocean winds and facilitate the gathering of fish, *see* Chap. 2). (**b**). *Bangpungnim* (Hadong pine forest, Kyungnam Province, Natural Monument No. 445). It also has a role as *Bangjorim* to protect river flooding

acts as a border between villages (Hong and Kim 2007b). Thoroughly protected by the residents and featuring relatively good locational conditions, the forest boasts a sound tree form. This particular forest serves as both a windbreak forest and a rest area for farmers. The forests along estuaries also play the role of windbreak forests.

In this regard, the great majority of the beautiful pine tree forests situated along beaches boast functions which are similar to those of windbreak forests. The black pine (*Pinus thunbergii*) forests around Jeungdo Beach, Dongho Beach, and Muan-gun Beach have leaves that are tougher than those of regular pine trees. As these black pine trees grow along the beach, they are also called *haesong* (海松, coastal pine). What's more, this particular type of tree is also called heuksong (黑松) because of the fact that its bark is darker than that of the regular pine tree. Owing to their strong ability to counter sea winds and salt, these black pine trees are routinely planted as part of efforts to create forests for windbreak or tidewater control (*bangjorim*, 防潮林) (Hong et al. 2006; Hong and Kim 2007b).

7.1.2.3 Hoannim

The term *hoannim* (護岸林) refers to a forest established along a riverbank or coastline to prevent floods and protect farmlands and villages (Fig. 7.4) (Kim and Hong 2009). These forests feature various types of tree. Because the *hoannim* protected villages from damage from floods in the agrarian society, they were perceived as the god who protected villages, and they were cast in the role of *dangsannim* (shaman forest). Such forests have also shown themselves to be quite efficient as windbreak forests. While the *hoannim* traditionally provided a resting place for farmers, they have recently also been used as a site for recreational activities during the summer season, as well as a good place, in their capacity as landscape forests, to spend one's

Fig. 7.4 *Hoannim* (Damyang-gun, Jeonnam Province, Natural Monument No. 366). *Gwanbangjerim*, forest to protect river banks

leisure time. As such, the *hoannim* can be regarded as a multipurpose type of forest intricately connected to human everyday life (Lee 2007).

7.1.2.4 Pungchirim

Pungchirim (風致林, ornamental forest) is the name given to a wooded area which has been established for no other purpose than to allow people to enjoy the beauty of the forest (Fig. 7.5). Consisting for the most part of natural forests, these forests serve

Fig. 7.5 *Pungchirim* has the integrative meaning of several types of village forest, such as *Bangpungnim* and *Hoannim*. *Pungchirim* also has an incidental function of landscape beauty (*top*: coastal pine forest in East Coast Sea, *bottom*: coastal pine forest in West Sea)

as places to temporarily escape the stress of an industrial society and engage in recreational and leisure activities. These forests can be perceived as a natural healthy space where people can lose their weariness and be filled with the vitality needed to face the future, all whilst engaging in activities amidst a beautiful landscape in which clean air, a pleasant environment, and soothing sounds such as those of birds and water coexist in perfect harmony. In recent times, traditional village forests such as *bangpungnim, eoburim*, and *hoannim* have had multifunctional roles in educational services and eco-tourism, as well as an ecological function.

7.1.2.5 Haksulrim

Haksulrim (學術林, academic forest) is the name given to a forest which is deemed to have academic research value in terms of ecology-related studies, such as those pertaining to the maintenance and conservation of ecosystems amidst changes in the global environment, the furthering of academic research, or in conjunction with the development of forestry technology (Fig. 7.6). Examples include forest genetic resource reserves, evergreen broad-leaved forests, primary forests, rare plant species, and indicator plants.

Fig. 7.6 Several types of *Haksulrim* exist on islands (**a**) Evergreen forest on Kamakseom Island at Kangjin-gun, Jeonnam Province, (**b**) Evergreen forest on Judo Island at Wando-gun, Jeonnam Province, (**c, d**) Evergreen forest on Bongrae-myeon, Goheung-gun and a sacred house in the forest, Jeonnam Province

7.2 Conservation of Traditional Village Forests

Ecology-related studies dealing with such issues as environmental pollution at the global level, global environmental conservation efforts, and the sustainable management of energy resources have now become more commonplace (Takeuchi et al. 2003; Hong and Lee 2007). In this regard, various studies have shown that Korea's urban forests and adjacent landscapes have been damaged by environmental pollution stemming from the continued drive for rapid economic growth. Furthermore, secondary vegetation, which has traditionally managed to escape human intervention, has now been totally disregarded in circumstances in which traditional management methods have collapsed under the pressure of rapid economic growth. The excessive development of land has resulted in the fragmentation of forests which long served as habitats and ecological spaces for various animals and plant species (Hong et al. 2000; Nakagoshi and Hong 2001). The destruction of ecosystems has not only caused a decrease in species diversity, but also hindered human's ability to enjoy a more enriched and diverse life. Simultaneously, it has also led to a loss of the indigenous culture that linked together humans and the forest environment (Duelli 1997; Brandt et al. 1999; Kim et al. 2006).

Nevertheless, no ecological landscape studies on secondary vegetation, which is so closely related to humans and human life, have been conducted in Korea to date. As part of its efforts to re-establish the traditional cultural relationship between humans and forests, Japan is now actively conducting ecological landscape studies of the fishing villages and forests which were ignored after the fuel revolution of the 1960s (Fukamachi et al. 2001; Nakagoshi and Hong 2001, *see* Chaps. 12 and 15 in this book). Such studies are designed to restore, from an ecological standpoint, the discontinued connections between urban areas and forests. The expansion of alternative energy resources in Korea developed during the drive for rapid economic growth that began in the 1970s, and has resulted in bringing about a general drop in the dependence on forests for energy (Hong 1998, 2003). The drop in the reckless destruction of forests has led to a heightening of the density of the trees in forests, as well as to a diversification of the species found therein. In fact we have now reached the point where such forests can be expected to serve as green spaces possessing the requisite ecological functions (Fig. 7.7).

However, the destruction of secondary vegetation caused by long-term disregard or excessive human intervention has led to the emergence of complicated problems for the types of vegetation that serve as an ecological path connecting the relevant landscape to the agricultural one, such as tidal flats and natural vegetation, as well as for the various regional biota that exist within the ecosystem (Jeanneret et al. 2003; Hong and Kim 2007a, b). Furthermore, fishing villages and estuary areas can expect to suffer economic losses as nearby secondary vegetation is destroyed because of the overall negligence of generally sustainable biota, the destruction of landscapes amidst the reckless deterioration of farmland and forests occasioned by the drive to reorganize the industrial structure and develop residential area and roads, the loss of wild-life habitats stemming from such development, and a decline

Fig. 7.7 The driving forces for the conservation of village forests. Forest management for the beautification of the ancestors' graveyard (**a**), indigenous knowledge about living harmoniously with nature (**b**), and respect and worship for the gods of trees in Korean intrinsic consciousness (**c**)

in the ability to preserve the coastal bioculture created by Korea's traditional fishing culture. In addition, the neglect of agricultural lands resulting from widespread migration to urban areas, and the establishment of large-scale forests containing an inordinate number of foreign species, also constitute undesirable changes in the cultural landscape of rural Korea. As such, Korea, whose rapid economic development has made it possible for it to join the ranks of the advanced countries, must now focus on improving its environmental structure as part of wider efforts to deconstruct the development-oriented economic structure of the past, pursue a social and economic structure that can coexist with the natural environment, and develop a high-quality living environment for its citizens.

Landscape change can be defined as alterations in the types and locations of landscape elements, or their replacement with other elements, occasioned by human activities and natural disturbances in the land mosaic (Forman 1995; Hong 2003; Kim et al. 2006). The characteristics of the relevant ecosystem are altered during these changes, and this not only in terms of landscape heterogeneity, diversity, and types of element, but also with regards to the spatial arrangement of such elements. In this regard, the establishment of a successful landscape conservation and restoration plan as part of efforts to ensure sustainable landscapes is to a large extent predicated on a thorough consideration of the spatial characteristics of landscape elements (Antrop 1997; Berkes et al. 2000), as well as of the ecological characteristics of each landscape element which can emerge during the process of the changes being wrought to the landscape concerned (Volker 1997).

Acknowledgement This research was supported by the National Research Foundation of Korea Grant funded by the Korean Government (NRF-2009-361-A00007).

References

Antrop M (1997) The concept of traditional landscapes as a base for landscape evaluation and planning: the example of Flanders region. Landsc Urban Plan 38:105–117

Berkes F, Colding J, Folke C (2000) Rediscover of traditional ecological knowledge as adaptive management. Ecol Appl 10(5):1251–1262

Brandt J, Primdahl J, Reenberg A (1999) Rural land-use and landscape dynamics – analysis of 'driving forces' in space and time. In: Krönert R, Baudry J, Bowler IR, Reenberg A (eds) Land-use changes and their environmental impact in rural areas in Europe. Parthenon Publishing, Paris, pp 81–102

Brown RD, Lafortezza R, Corry RC, Leal DB, Sanesi G (2007) Cultural patterns as a component of environmental planning and design. In: Hong SK, Nakagoshi N, Fu B, Moriomoto Y (eds) Landscape ecological applications in man-influenced areas: linking man and nature systems. Springer, Dordrecht, pp 395–415

Choung HL, Hong SK (2006) Distribution patterns, floristic differentiation and succession of *Pinus densiflora* forest in South Korea: a perspective at nation-wide scale. Phytocoenologia 36:213–229

Duelli P (1997) Biodiversity evaluation in agricultural landscape: an approach at two different scales. Agric Ecosyst Environ 62:81–91

Forman RTT (1995) Land mosaics: the ecology of landscapes and regions. Cambridge University Press, Cambridge, 632 pp

Fukamachi K, Oku H, Nakashizuka T (2001) The change of a Satoyama landscape and its causality in Kamiseya, Kyoto Prefecture, Japan between 1907 and 1995. Landsc Ecol 16:703–717

Grashof-Bokdam CJ, van Langevelde F (2005) Green veining: landscape determinants of biodiversity in European agricultural landscapes. Landsc Ecol 20:417–439

Hong SK (1998) Changes landscape patterns and vegetation process in the far-Eastern cultural landscapes: human activity on pine-dominated secondary vegetation in Korea and Japan. Phytocoenologia 28:45–66

Hong SK (2001) Factors affecting landscape changes in central Korea: cultural disturbance on the forested landscape systems. In: Zonneveld IS, van der Dick Z (eds) Landscape ecology applied in land evaluation, development and conservation. ITC Publisher, Enschede, pp 131–147

Hong SK (2003) Cause and consequence of landscape fragmentation and changing disturbance by socio-economic development in mountain landscape system of South Korea. J Environ Sci 11:181–187

Hong SK (2007) Linking man and nature landscape systems: landscaping blue-green network. In: Hong SK, Nakagoshi N, Fu B, Moriomoto Y (eds) Landscape ecological applications in man-influenced areas: linking man and nature systems. Springer, Dordrecht, pp 505–523

Hong SK, Kim JE (2007a) Circulation and network of forest-stream-coastal ecosystems. Isl Cult 30:267–286 (in Korean with English abstract)

Hong SK, Kim JE (2007b) Village forest of estuary in West Sea-forested landscape management for conserving resort area. Isl Cult 29:441–473 (in Korean with English abstract)

Hong SK, Lee AJ (2007) Global environmental changes in terrestrial ecosystems. International issues and strategic solutions: introduction. Ecol Res 21:783–787

Hong SK, Rim YD, Nakagoshi N, Chang NK (2000) Recent spatio-temporal changes of landscape structure, heterogeneity and diversity of rural landscape: implements for landscape conservation and restoration. Korean J Ecol 23(5):359–368 (in Korean with English abstract)

Hong SK, Park JW, Yang HS (2006) Ecological characteristics of black pine forest as ecotourism resource-Jeungdo, Shinan-gun, Jeonnam. Isl Cult 28:223–244 (in Korean with English abstract)

Hong SK, Nakagoshi N, Fu B, Morimoto Y (2007a) Landscape ecological applications in man-influenced areas: linking man and nature systems. Springer, Dordrecht, 535

Hong SK, Song IJ, Wu J (2007b) Feng-Shui theory in urban landscape planning. Urban Ecosyst 10:221–237. doi:10.1007/s11252-006-3263-2

Jeanneret Ph, Schüpbach B, Luka H (2003) Quantifying the impact of landscape and habitat features on biodiversity in cultivated landscapes. Agric Ecosys Environ 98:311–320

Kim JE, Hong SK (2009) Landscape ecological analysis of coastal sand dune ecosystem in Korea. J Korean Soc Environ Restor Tech 12:21–32

Kim JE, Hong SK, Nakagoshi N (2006) Changes in patch mosaics and vegetation structure of rural forested landscape under shifting human impacts in South Korea. Landsc Ecol Eng 2:177–195

Korea Forest Research Institute (1995) Traditional environmental forest in Korea. Korea Forest Service, Seoul

Lee D (2004) Ecological knowledge embedded in traditional Korean landscape. Seoul National University Press, 120 pp (in Korean)

Lee KS (2007) Bee-Bo forest: traditional landscape ecological forest in Korea. In: Hong SK, Nakagoshi N, Fu B, Moriomoto Y (eds) Landscape ecological applications in man-influenced areas: linking man and nature systems. Springer, Dordrecht, pp 389–394

Nakagoshi N, Hong S-K (2001) Vegetation and landscape ecology of East Asian 'Satoyama'. Glob Environ Res 5:171–181

Nassauer JI (2005) Using cultural knowledge to make new landscape patterns. In: Wiens J, Moss M (eds) Issues and perspectives in landscape ecology. Cambridge University Press, Cambridge, pp 274–280

Takeuchi K, Brown RD, Washitani I, Tsunekawa A, Yokohari M (eds) (2003) Satoyama: the traditional rural landscape of Japan. Springer, Tokyo, 229

Volker K (1997) Local commitment for sustainable rural landscape development. Agric Ecosys Environ 63:107–120

Chapter 8
"Cho-bun", An Anthropogeneous Landscape in Haui Island, Southwestern Korea

Jong-O Park

8.1 Introduction

A funeral ceremony, one of the rites of passage, can be defined as a ritual way to treat a dead body. Funeral ceremonies are performed in various ways, including interment (土葬), i.e., to bury the dead in the earth, aerial burial (風葬), to put the dead in a secluded place without a burial in order for the dead bodies to be worn away naturally by the rain and the wind, cremation (火葬), to incinerate the dead and to hold a funeral with the remaining ashes, and water burial (水葬), to put the dead into the water. These funeral ceremonies are different depending on the period, region, and social characteristics. However, respect for the ashes of the dead has been a universal phenomenon all over the world (Na and Kurimoto 2002).

The ashes are also respected in Korea, which can be seen in the system of a double funeral (二重葬制). This system is a funeral ceremony where they do not bury (埋葬) the dead body in the earth immediately after death, but let the flesh of the dead body decay completely (肉脫) after the first funeral (一次葬), which is a preliminary burial (假葬). In the second funeral (二次葬), which is the main burial (本葬), they clean the remaining bones and then bury them in the earth (Park 2005).

It seems reasonable to suppose that the tradition of the double funeral system in Korea dates back a very long time, for there are many references to it in the old books.

"*East Okjeo History* (東沃沮傳: 新死者皆假埋之 才使覆形 皮肉盡 乃取骨置槨中 擧家皆共一槨. Upon death, the corpse shall be buried provisionally. That is, it shall be buried just under the surface of the earth, to the extent that it not be revealed, and only the bones shall be collected and put in a coffin after the skin and the flesh is completely decayed)."

J.-O Park (✉)
Institution for Marine and Island Cultures, Mokpo National University, 61 Dorim-ri, Cheonggye-myeon, Muan-gun, Jeonnam 534-729, Republic of Korea
e-mail: kfolk@hanmail.net

S.-K. Hong, et al. (eds.), *Landscape Ecology in Asian Cultures*, Ecological Research Monographs, DOI 10.1007/978-4-431-87799-8_8, © Springer 2011

"*Wi Dynasty* (*魏志*) of *Three Kingdoms* (*三國志*). *Korea History* (高麗傳; 死者殯於屋內 經三年 擇吉日而葬. Upon death, the corpse shall be placed at home for about 3 years, after which the funeral shall be held on a good and fortunate day)."

"*Su Dynasty* (隋書) and *Joseon Dynasty History* (附朝鮮國記. In general, burial shall be done after spring or summer harvest. If the death falls in summer, the corpse shall temporarily be placed in a small mud hut constructed on columns, whose roof is made out of cut rice plants, and be brought back to the home to be put into a coffin with clothes and some stuffs)."

An Account of the Shipwreck of a Dutch Vessel on the Coast of the Isle of Quelpaert together with the Description of the Kingdom of Corea by Hamel (Kim 1996).

Even today, we can observe this double funeral system in the *Cho-bun* (草墳), where the dead body is left somewhere, not in the earth, for some time until the flesh completely decays and the remaining bones are buried.

The National Folk Museum of Korea has closely examined the well-organized Korean practice of Cho-bun as it existed in 2003 (Jong and Choi 2003). These authors have also written papers about the existing tombs of the Cho-bun system on Song-i Island, and the earlier Cho-bun practice on Jin-do Island (Park 2006). Various studies about Cho-bun have been implemented so far, such as case studies of the existence (National Museum of Korea 1993, Choi 1970, Park 1977, Jeong 1985, Kang 1992), constructions about the origin of Korean culture (Lee 1969a, b; Han and Jeon 1969), the custom of the funeral (Jeong 1986). There are many reasons for the Cho-bun in Korea, for example, the absence of the chief mourner, the belief that it is lacking in respect to bury the dead body immediately after death, the death was caused by an infectious disease or had taken place away from home, a failure to obtain the burial site due to lack of money, or it was the death of a child.

The Cho-bun covered by this chapter was made on Haui Island in 2005. The island is located in Dadohae Maritime National Park in the southwest of Korea, 59.7 km from Mokpo. The island is well known as the home of Kim Dae-Jung (金大中, 1924–2009), the 15th President of the Republic of Korea, who received the Nobel Peace Prize in 2000.

The name "Haui Island" means a lotus flower on water, and the island consists of nine inhabited and 47 uninhabited islands. Only one Cho-bun tomb remains, and here we can examine this funeral practice, the process of making a Cho-bun tomb, and the meaning of the practice on Haui Island.

8.2 Funeral and Cho-bun Practice of Haui Island

All people experience many important events from their birth to their death. These events include birth, becoming an adult, marriage, having a baby, and death. Even after death, the Eastern view is that people should go through a special event in order to be revered as ancestors by their offspring. These events act as a qualification to the members of a society so that they can be accepted as legitimate members and have the rights of formal members. These events we call rites of life (一生儀禮) or rites of passage (通過儀禮).

Among these rites, the death rite is a way to treat death, and is usually called a funeral ceremony. Unavoidable death is feared by almost all people. In particular, a tragic death makes a more intense impression on people, and results in more sadness, than a peaceful death. Death rites are divided into seeing the death in order to accept the death as a reality, funeral-related rites, and rites to return to daily life after the death.

The Cho-bun practices of death rites, which were established long ago and are still being performed, are described below in order of the time when they occur. This survey was performed in June 2008, and the informers are Lee Yun-Im (female, born in 1928, and now living in Haui Island), Hong Yeong-Bok (male, born in 1929, and now living in Haui Island), and Kim Hae-Ja (female, born in 1930, and now living in Haui Island).

8.2.1 Seeing the Death (Imjong)

Seeing the death is literally to see the moment when a parent dies, which is a very important funeral rite (喪禮) in Korea. People on Haui Island say that when a person is on the verge of death, a kind of fire ball goes out of their house. The fire ball, the so-called "*hon bul* (soul fire)," indicates the moment of death, and the ball is said to have a long tail when the dying person is a man, while the ball is round when it is a woman. They believe that they should bury the dead person in the direction in which the "*hon bul*" flies.

When a person seems to be about to die, they are moved to the main living room, where they lie with their head directed to the east. Sometimes the dying person is redressed in a new suit of clothes. After this procedure, they inform other family members of the impending death. When a person dies at a place other than their home, they should not be moved into the main living room, but should stay in front of the door of the room.

All the family members should confirm the death, after which they should all cry loudly. We call this loud crying "*gok*." If the dead person's daughter could not see the death, she enters her home from the village, crying loudly, with her hair loose, and carrying her shoes. This peculiar behavior means that she was away from home and has come in great haste.

8.2.2 Calling the Ghost and Setting the Table for the Dead Person (Sajasang)

When the family have confirmed the death, they take off the outer garment which was put on the dead person the moment they died, and throw it onto the roof of the house. This means that there is a death in the house, and is called "calling the ghost." Before they throw the garment onto the roof, they utter the age, the family name, and the first name of the dead person. This means that they hope the ghost of the dead person will return and the dead person will revive. The thrown garment will later be recovered and put into the coffin with the dead body. The people in the village know about the death by the garment thrown onto the roof.

A dining table for the dead person is set in front of the gate of the house. This is for the messenger who is believed to take the dead person's soul to the underworld. On the table are placed three bowls of rice and three pairs of straw sandals, because it is thought three messengers come to take the dead person's soul away. Setting the table is a way of asking the messengers to lead the soul safely.

8.2.3 *Seup* and *Yeom*

After the death is confirmed, the deceased is redressed and the body is placed so that it is directed toward the east. When doing this, the hands are tied at the side of the body and the feet are bound with hemp cloth to straighten the corpse before it gets stiff. Then the corpse is put on the wooden kitchen door and is taken away. The corpse gradually gets stiff over time, but there are some exceptional cases where it does not get stiff. The people call this phenomenon "fortunate suspended," which the islanders consider is an omen for another death in the same family.

The corpse on the kitchen door is hidden by a folding screen, in front of which they set a simple table for the deceased. On the table are simple foods, including fruits and burning incense. If they have a picture of the deceased, it is also put on the table. Every meal time, they put rice, soup, and side dishes on the table as if they were looking after a living person.

Seup is used to wash the body before it is dressed in a new suit of clothes, while *yeom* is to put the corpse into a coffin. They have special ways to carry out these two behaviors, "*dareum-handa*," in the island. When doing *seup*, they use a very weak solution of wormwood or tasteless water. Doing *seup* is not an actual wash but the mere pretense of a wash. There used to be *seup* professionals in the past, but the deceased offspring typically do the *seup* today. After washing the body, they redress the deceased in a shroud, which is a garment for the deceased. This will have been prepared in advance by the deceased or their offspring when they were alive, or hastily by villagers upon the death. Today, they often buy and use a ready-made shroud if it has not been made in advance. In redressing the corpse, they cut the fingernails and toenails of the deceased and put them into a small pouch, which will be put in the coffin at the positions corresponding to the hands and feet.

Yeom is divided into a *soryeom* (small *yeom*) to wrap and tie the dead body after *seup*, and *daeryeom* (big *yeom*) to put the dead body into a coffin. In general, both of these are performed at the same time nowadays. The coffin is often made in advance out of a paulownia tree or a pine tree; otherwise, they use a ready-made coffin. In the coffin, the head of the deceased is put onto a pouch stuffed with yellow earth as a pillow. Some rice is put into the mouth of the dead person, which is to keep the deceased from hunger until they arrive at the underworld. The other space in the coffin is filled with some clothes the deceased wore during their lifetime, or cut rice plants which are to prevent the dead body from shaking in the coffin. Nowadays they use rolls of toilet paper instead of the plants.

After the big *yeom*, they nail the cover on the coffin as long as there are none of the offspring or relatives who have not yet seen the deceased. In the past, they used to use a bamboo blind or a rice plant straw mat to wrap the corpse when they could not afford a coffin.

8.2.4 Funeral Ceremony and Outing (Chulsang) Preparation

The chief mourner (喪主), usually the oldest son of the dead person, is the host of the funeral ceremony. In the absence of the oldest son, the second son (or in his absence the third son) is the chief mourner. If all the sons are dead, the oldest grandson becomes the chief mourner. The preparations on behalf of the chief mourner are in the hands of a senior member of the clan, who leads the general affairs of the funeral, including the preparation of food, the delivery of obituary notices, etc. The obituary notice (訃告狀) is a notice of the death of a person. This is not taken into the house where it is delivered, but is inserted into the main gate or the fence of the house in order that it may disappear by itself.

If the shroud or garments which the offspring or relatives of the deceased put on them throughout the funeral period (喪中) have not been prepared, they make them out of hemp cloth after the death. Members of the family prepare the large amount of food which is necessary for the funeral period. The chief mourner wears a "*gul geon*" on a hempen hood on their head, while the other family members wear only hempen hoods. In addition, the chief mourner ties their head and abdomen with straw strings which are wound to the left. Neighbors or members of "*sangpogye*" help prepare the food. *Sangpogye* is an organization where members collect money to help with future funerals of the members. A "*sangyeo*" is a funeral bier to carry the dead body to the burial site, which looks like a *palankeen*. The *sangyeo* is made by the members of "*sangdugye*" and is decorated with white papers dyed and shaped like flowers. The frame of the *sangyeo* is made by the villagers out of wood before the death. Before using the *sangyeo*, they keep it safe in the *sangyeo* house. When the *sangyeo* goes out, the chief mourner or the family of the deceased make canes out of bamboo, which are called "*sangjeong* rods."

They practice shouldering the *sangyeo*, called "*bamdarae*," on the last day before it goes out of the house of the deceased. The bereaved family treats them with rice cake soup or hot chicken porridge. In the past, they used to sing songs and enjoy themselves, sometimes even with *gisaengs* (singing and dancing girls) when the funeral took place in a rich family. The *bamdarae* was a kind of feast for the whole village, and they took it for granted that it was a time of singing and enjoyment in spite of the death. This was to soothe the bereaved family and help their return to normal daily life.

8.2.5 Balin (Carrying a Coffin Out of the House)

Balin is the procedure of carrying the coffin out of the house to the burial site, and is usually performed on the third day after the death. When starting to carry the coffin out of the room, they take it around the four corners of the room, and step on and break a gourd in front of the room, which is to expel misfortune in the family.

After taking the coffin out of the house, they perform a religious service called "*balinje*." Upon finishing *balinje*, the *sangyeoggun* who shoulder the coffin leave the village by way of the places the deceased used to frequent (Fig. 8.1). Just before leaving the village, they perform another religious service called "*georitje*." When they finally get out of the village, they place the coffin in a standing position for the deceased to make a last bow to the village and the villagers.

8.2.6 Jangji (Burial Site)

Jangji (葬地) means the burial site where they put the coffin in the earth. The site is prepared in advance by the members of the *sangdugye*. Rich families consider *pungsujiri* (風水地理, Feng-Shui Theory, *see* Hong 2007; Hong et al. 2007), a theory of divination based on topography, when determining the burial site, while ordinary families determine their *seonsan* (先山, ancestral burial grounds) or their own grounds as the burial site.

Fig. 8.1 The bier leaving the village in Haui Island (Apr 2009)

After determining the burial site, they perform a brief religious service called *sansinje* (山神祭) before digging. This service is to inform the god in charge of the mountain about the digging. When the *sangyeo* arrives, they separate the coffin from the *sangyeo* and put the coffin in the earth. Upon putting the coffin into the hole in the ground, they wipe it with *gongpo* (a kind of cloth) and cover it with the funeral banner. They burn the *gongpo* with the *sangyeo* later. On Haui Island, they put only the dead body in the earth without the coffin. Before putting in the corpse, they put some pieces of wood, called *chilseongpan*, in the hole.

After they have finished putting the coffin in the hole, they write the name of the deceased and their official post on a sheet of paper, which is called *jibang*, which they later take home along the same route that they carry the *sangyeo*. When the hole has been filled with earth to the level of the surrounding ground, they perform a religious service called *pyeongtoje*. In some places the villagers cover the coffin with soil in the shape of a hemisphere, or burn the *sangyeo* or the garments of the deceased before leaving the burial site.

8.2.7 *Banhonje and Talsang*

Those people who did not go to the burial site remained at home make a place to preserve the *jibang* (紙榜, an ancestral tablet made of paper), which is called *jecheong* (祭廳, an altar set up for a burial rite during interment). After the *pyeongtoje* (平土祭), the chief mourner returns home with the *jibang*, places it at the *jecheong*, and performs a religious service called *banhonje* (返魂祭). The next day, the chief mourner performs a religious service called *wuje* (虞祭), and on the third day after the burial, they perform *samwuje* (三虞祭). From this time on, they perform religious services on the 1st, the 15th, and the last days of every month until *talsang* (脫喪). A 3-year talsang is rarely observed now, while a 1-year *talsang* or *talsang* after 49 *je* is generally observed.

8.2.8 *Cho-bun*

The *Cho-bun* (草墳) can be called a grass guest (草殯), an outer guest (外殯), *goreumjang*, *guto*, or *choibin*. The name "*chobun*" (草墳) in Chinese is believed to originate from the cover made of straw thatch which is usually used in the first funeral. The *Cho-bun* is different from ordinary burials in that the corpse is not put in the earth immediately, but is provisionally put in a *Cho-bun* site. In making the *Cho-bun* site, first stones are heaped up on the ground to a suitable height, the coffin covered with straw thatch is place on this, and then a straw-made *yongmareum* is put on the top of it. It looks like a thatched cottage from a distance (Fig. 8.2). Lastly, the *yongmareum* is tied with ropes to prevent it from blowing away in the wind.

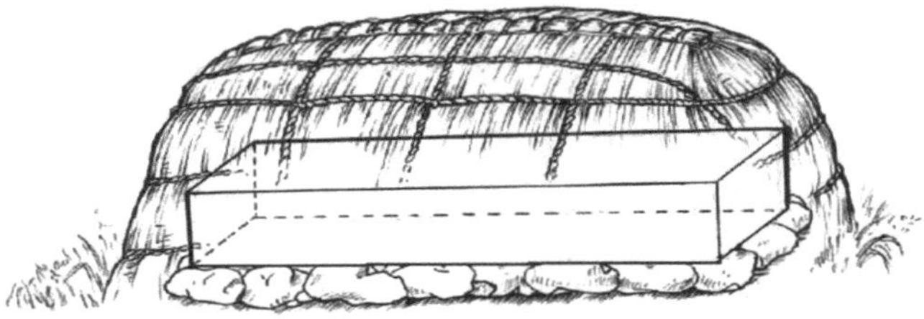

Fig. 8.2 Design for a *Cho-bun*

Every year, the straw thatch and the *yongmareum* are renewed, which is the same behavior as for other burials.

In general, 3 years are sufficient for the flesh of the dead body to decay completely, and only the bones of the dead body remain. The bones are then collected to bury in the earth. Nowadays, however, some 10 years must elapse before the bones are buried, because the flesh of modern people does not decay as quickly as that of people in the past. An exorcism is often performed when the bones are transferred. The method of burying the bones in the *Cho-bun* system is the same as in other ordinary burial systems.

8.3 Change and Preservation of the *Cho-bun* System in Haui Island

In the past, the *Cho-bun* system was generally observed on Haui Island, and they transferred the bones 3 years after the death. The *Cho-bun* system was adopted if the dying person wanted that system, or if the relatives did not want to dig the ground in order to ensure good fortune for the dead or for the living offspring. When the *Cho-bun* system was adopted, they generally changed the straw thatch and *marams* after the rice harvest in the autumn. After the flesh of the dead body had completely decayed (this state is called "*yuktal*" (肉脫, escape of flesh)), they decided on a new burial site for the bones.

There remains one *Cho-bun* tomb in Nanggoji on Haui Island (Fig. 8.3). The tomb is 90 cm high, 90 cm wide, and 230 cm long, and like other grass tombs consists of straw thatch and a *yongmareum*, which is fixed with ropes so that it does not blow about. The shape of this tomb is the same as those in other regions, but the method of making the tomb is different. Unlike the traditional method on the island, this one was made in ground which had been dug over.

In general, the ground is made even by filling any sloping areas with stones in order to place the coffin horizontally, but this tomb was made even by digging out the higher places. This method could save a lot of labor when making the tomb.

This *Cho-bun* tomb was made because it was believed that digging the ground would bring bad fortune. They planned to make a formal tomb later. As in this case,

Fig. 8.3 The *Cho-bun* on Haui Island (June 2008)

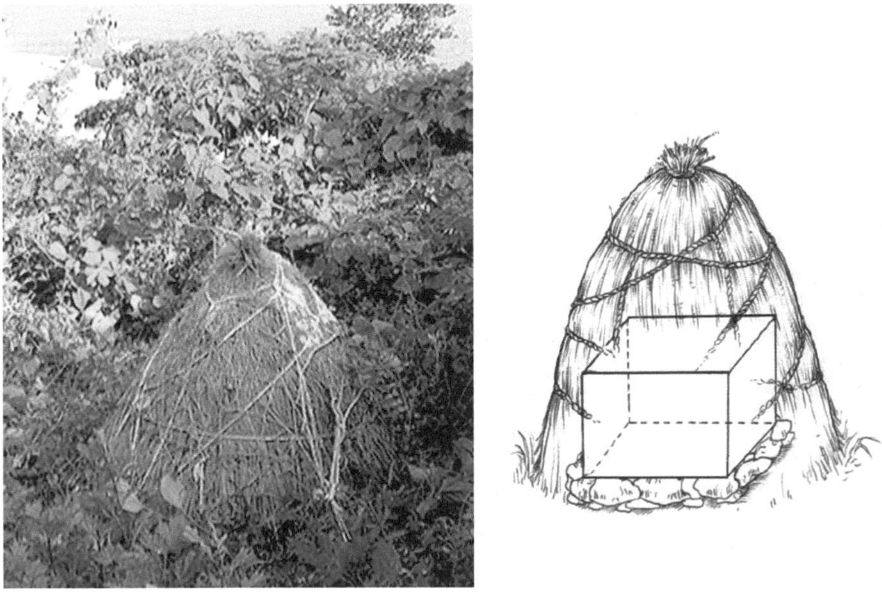

Fig. 8.4 The sitting *Cho-bun* tomb in Song-i Island (*left*) and the inner scene (*right*)

the most common reason for the *Cho-bun* tombs which remain in Korea, or which have been made recently, is related to the fortunes of the offspring.

For example, there is still a so-called "sitting *Cho-bun* tomb" in Song-i Island (Fig. 8.4). In this tomb, the bones are arranged as if the dead person were sitting up. That is why the tomb's lower part is round and the upper part is conical. It is about 1 m high and is covered with mesh as well as straw ropes with stones tied at the ends to protect the tomb against wind or animals. This tomb was meant to "take the bones in custody," and was made after the first burial, with the bones contained in a box.

This is the reason for the down-sizing of the tomb (Park 2005). Most of the tombs are made for personal reasons, that is, the fortunes of the deceased or the offspring.

The tomb on Haui Island was also made for good fortune. Fortunes are unchangeable, and human beings are helpless in front of them. If making a tomb may bring bad luck to the living offspring, then the relatives will avoid making it. Instead, they make a *Cho-bun*, where they do not have to dig the ground. In a sense, this practice was related to realities such as lack of land or money. They would wait until they could afford the formal tomb.

We found that the *Cho-bun* tomb remaining on Haui Island showed many traditional aspects, although the method and purpose of making the tomb were somewhat different from those of traditional tombs.

8.4 Conclusion

The *Cho-bun* in Korea means that the corpse is not put in the earth immediately death, but the relatives wait until the flesh has decayed so that they may collect the bones and make a formal tomb for them. This practice was discouraged by the Sanitary Act during the Japanese Colonization Period, and cremation was then encouraged. However, it has been performed on several islands to the south and west of Korea. Even during the period of the Saemaeul Movement (the new village movement) in the 1970s, it was banned by law.

Korea has some problems with burials, including damage to forests and loss of land at a rate of some 9917355 km^2 a year which is caused by the expansion of burial areas. The responses to these problems include cremation and the preservation of ashes in charnel houses, or natural burials. A charnel house is for preserving the ashes after cremating the dead body. However, a charnel house is made of stones, which raises another environmental problem. A natural burial means to scatter the ashes around trees or on a lawn after cremation, and this has emerged as a new environmentally friendly type of burial (CNBNEWS, "European Style Natural Burial in Gyeonggi-do!," 2008.3.18, http://www.cnbnews.com).

However, change will not be easy because of the traditional belief in cherishing the bones of the dead. In a way, Korean burial systems can be said to be environmentally friendly. A lot of the land used for the burials will eventually return for other uses. The real problem is not the burial system itself, but the stone decoration around the tombs.

The *Cho-bun* practice has evolved along with the environment and the religion. The practice survives to date, in spite of some changes in the shape and the purpose of the tomb, the down-sizing, and the convenience of the offspring. The practice will continue for the near future because of the tendency to respect the bones and the economic conditions, despite the fading-out of the superstitions related to the practice.

Acknowledgement This work was supported by the National Research Foundation of Korea Grant funded by the Korean Government (NRF-2009-361-A00007).

References

Choi DW (1970) A study on the grass guest of Dadohae area. Reports of survey on the formation of scholarship fund (16). Ministry of Cultural and Education

Han SB, Jeon KS (1969) Double funeral system and the mentality of human being. Korean cultural anthropology, vol 2. Korean Society for Cultural Anthropology

Hong SK (2007) Linking man and nature landscape systems: landscaping blue-green network. In: Hong SK et al (eds) Landscape ecological applications in man-influenced areas: linking man and nature systems. Springer, Dordrecht, pp 505–523

Hong SK, Song IJ, Wu J (2007) Feng-Shui theory in urban landscape planning. Urban Ecosyst 10:221–237. doi:10.1007/s11252-006-3263-2

Jeong JS (1985) Burial system. Folklore in Wi Island. The National Folk Museum of Korea

Jeong JS (1986) A study on Korean ancient burial system. Korean Folklore study, vol 19. The Korean Folklore Society

Jeong J Su, Choi SK(2003) The Korean Cho-bun, The National Folk Museum of Korea

Kang NJ (1992) Folklore of the South Sea of Korea. Dungji Publishing, Seoul

Kim TJ (1996) An account of the shipwreck of a Dutch vessel on the coast of the Isle of Quelpaert together with the description of the Kingdom of Corea (original book of Handrik H). Chonnam National University Press, Chonnam

Lee KK (1969a) A study on the Cho-bun of Cho Island – Cho Island's burial system. National Culture Study, vol 13. Institute of National Culture Studies of Korea University

Lee KK (1969b) Origin of Korean culture in the light of ethnology. Korean cultural anthropology, vol 2. Korean Society for Cultural Anthropology

Na KS, Kurimoto Y (2002) A comparative study of bone washing rituals in Okinawa and Konan, The Ryukyuan culture, vol 24. The Institute of Ryukyuan Culture Okinawa International University

National Museum of Korea (1993) Korean Islands of the West Sea. Eulyu Publishing, Seoul, 1957 (National Culture, republished in 1993)

Park SH (1977) On the Cho-bun of Jeonlabuk-do. Summaries of the National gathering. The Korean Folklore Society

Park JO (2005) A study on the Cho-bun in Song-i Island. The Korean Folklore, vol 41. The Korean Folklore Society

Park JO (2006) The Cho-bun in Jin-do Island. The Folklore Study of South Provinces, The South Provinces Folklore Society

Chapter 9
Human Impact on Coastal Sand Dune Ecosystems in Korea

Jae-Eun Kim and Sun-Kee Hong

9.1 Introduction

Coastal sand dunes are a type of ecosystem that is found almost all over the world. They can be regarded as an ecotone, or transition area, where the characteristics of maritime areas are combined with the geographical and ecological features of inland areas. The length and width of sand dunes tend to vary in accordance with the causes of their formation. Generally, sand dunes are formed when the scale of sand sedimentation is larger than that of sand erosion (Greipsson 2002). Sand dune resources generally originate from the erosion of terrestrial ecosystems, with rivers transporting sediments to coastal areas. In fact sand dunes are composed of a mixture of substances emanating from both land and sea. They comprise complicated ecosystems characterized by a distribution of vegetation that tends to vary with topographical characteristics. Sand dunes thus exhibit the characteristics of ecotones, and act as transition areas between terrestrial and marine ecosystems. As a result, researchers have tended to focus not only on the biological and ecological aspects of sand dunes, but also on their standing as a landscape (Roy and Tomar 2000; Walker et al. 2003; Beever et al. 2006; Carboni et al. 2009; Hong et al. 2010). However, the study of sand dune ecosystems in Korea is still primarily focused on species diversity. In addition, the importance of sand dune ecosystems has increased as their value from the standpoint of the function of the ecosystem and visual beauty has been highlighted (Taveira Pinto 2004).

Despite the growing importance attached to the biotopes that represent the habitats of species, very few spatial studies have been conducted on species diversity at the landscape level, or on habitat diversity in Korea (Hong et al. 2004). A limited

This paper is a modification of the article ‘Landscape ecological analysis of coastal sand dune ecosystem in Korea’, which was written in Korean with English abstract and published in the Journal of the Korea Society of Environmental Restoration Technology (2009; 12(3): 21–32).

J.-E. Kim(✉) and S.-K. Hong
Institution for Marine and Island Cultures, Mokpo National University, 61 Dorim-ri, Cheonggye-myeon, Muan-gun, Jeonnam 534-729, Republic of Korea
e-mail: ecokimje@yahoo.co.kr

S.-K. Hong, et al. (eds.), *Landscape Ecology in Asian Cultures*, Ecological Research Monographs, DOI 10.1007/978-4-431-87799-8_9, © Springer 2011

number of studies on village landscapes and the ecosystems of certain island and coastal areas have been carried out to date. Landscape ecological analysis can be used to examine the relation between sand dune ecosystems and their surrounding environment. Landscape indices provide a way to evaluate the natural and ecological attributes of ecosystems at the habitat level of biodiversity.

Sand dunes protect the coastal ecosystem by regulating the amount of sediment that flows between land and sea. The vegetation community, which is composed of herbs, shrubs, and tree layers distributed within the sand dune landscape, helps to protect both the coastline and agricultural land located near the sand dune from sudden storms and tidal waves (Kutiel et al. 1999). Forming a topographically and geographically unique physical structure, sand dunes feature a wide range of ecological spaces. As a result of this fact, terrestrial, marine, and coastal species can simultaneously exist in sand dune areas. As sand dunes are unique biological environments which serve as habitats for endemic species, they represent important sites when it comes to the conservation of biodiversity (Lee and Chon 1984). As mentioned above, sand dunes constitute a transition zone between marine and terrestrial areas that are formed or destroyed based on the physical and biological interactions they have with the surrounding environment. In this regard, the recent physical disturbances of sand dune areas have had more to do with human activities, such as the development of coastal areas, than with any natural phenomenon, especially in Korea.

Studies on the effect of ecological processes on biodiversity can be regarded as an integral element of the basic process of heightening the effectiveness of conservation biology. To this end, studies on the importance of biodiversity, as well as assessments thereof, have been conducted since the early twentieth century. Meanwhile, studies on the importance of biodiversity and the scale of related ecological processes have been conducted for about 20 years (Wiens 1989; Levin 1992; Wu and Loucks 1995). The examination of biodiversity at the landscape level has led to the emergence of new biological definitions. Monitoring biodiversity in sand dunes at the landscape level, as well as assessing ecosystems, is now becoming customary throughout the world (Roy and Tomar 2000; Nigel et al. 2001; Beever et al. 2006). To date, research on the conservation and management of coastal sand dunes has been based on the results of such studies (Carboni et al. 2009).

Many methods exist to assess biodiversity at the landscape level. The majority of the methods used to date have been related to landscape ecological studies, which have sought to analyze the structure and functions of landscapes, the serial changes over time, and the interactions between the structural arrangement of landscapes and their ecosystem functions (Forman 1995; Roy and Tomar 2000; Beever et al. 2006). Assessments of landscape ecological biodiversity generally revolve around the analysis of spatial characteristics such as the size and dispersion, abundance, shape, fragmentation, and matrix of habitat patches. In other words, such studies are designed to facilitate the analysis of species and their habitats, or so-called ecological spaces, and shed some light on their characteristics. Various methods and case studies can be found in conjunction with this particular field of study (Turner et al. 2001).

The formation and disappearance of coastal sand dunes in Korea is closely related to various human activities, including land use. The type of development and density of land use can both directly and indirectly influence sand dune ecosystems. Therefore, studies on sand dune ecosystems and sand dune biodiversity should inevitably analyze the interactions between existing vegetation (species), land use and vegetation patches, and landscape composition (Hong and Kim 2008; Carranza et al. 2007; Hong et al. 2007; Carboni et al. 2009). Based on landscape ecological and landscape-scale analyses, this study examines the landscape structure of coastal sand dune ecosystems in Korea.

9.2 Research Contents and Methodology

9.2.1 Study Areas

This study involved a survey of sand dune areas in the West, South, and East Sea coastal areas of Korea. More specifically, seven study areas were selected, with a large variation in the distribution of species (Fig. 9.1). Thereafter, a geographic information system (GIS) was employed to map out the sand dune ecosystems and identify the uses made of the surrounding land (Acosta et al. 2005). An overview and pictures of each study area can be found in Fig. 9.2.

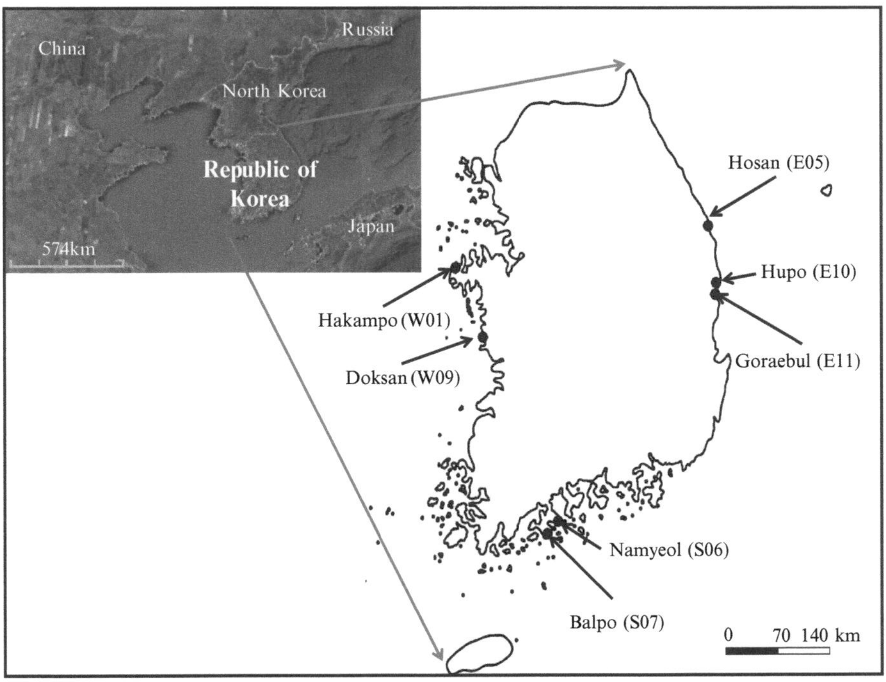

Fig. 9.1 Location of study areas

W01	Hakampo Sand Dune	Banggal-li, Wonbuk-myeon, Taean-gun, Chungnam Province
Overview of the study area: The vegetation of this particular coastal sand dune has been relatively well preserved. One can see hedges and fences in some places which have been set up to prevent the erosion of the sand dune. Good swimming facilities can be found at the beach.		

W09	Doksan Sand Dune	Doksan-li, Ungcheon-eup, Boryeong City, Chungnam Province
Overview of the study area: The construction of roads and artificial flowerbeds has resulted in this area being devoid of the characteristics generally associated with coastal sand dunes. The only flowers which have been planted in these flowerbeds are sweetbriers.		

S06	Namyeol Sand Dune	Namyeol-li, Yeongnam-myeon, Goheung-gun, Jeonnam Province
Overview of the study area: This is a rather wide coastal sand dune area in which the characteristics of coastal sand dunes have been relatively well preserved.		

Fig. 9.2 Overview of study areas

9.2.2 Land-Use Map and Landscape Analysis

A land-use map was prepared in accordance with the digital terrain model (National Geographic Information Institute, 1:5 000 scale). A field survey was also conducted to compare the boundaries and verify the main landscape components. Landscape indices have been widely used as part of landscape ecological analyses in order to investigate and study landscape structures (Forman and Godron 1986; Haines-Young et al. 1996; Kim et al. 2006; Hong et al. 2007; Hong and Kim 2008). As such, after the landscape map had been prepared based on a GIS, the landscape

However, the use of the area as a swimming beach and placing of general facilities and parking lots behind the black pine tree forest raises a very real fear that this particular sand dune will be damaged in the future.

S07	Balpo Sand Dune	Balpo-li, Dohwa-myeon, Goheung-gun, Jeonnam Province

Overview of the study area: the long-standing use of the area as a beach has resulted in practically no coastal vegetation remaining, with the notable exception being the area where the stair-style cement embankment is located.

E05	Hosan Sand Dune	Hosan-li, Wondeok-eup, Samcheok City, Gangwon Province

Overview of the study area: This sand dune has greatly eroded over the last few years. The installation of coastal protection facilities to decrease the damage caused by erosion and the overflowing of waves has in turn resulted in damaging the plants distributed in the sand dune area. As very little sand dune vegetation now remains, the characteristics of coastal sand dune can be said to have all but disappeared.

Fig. 9.2 (continued)

structures of the seven study areas were analyzed using landscape indices available within the software package FRAGSTATS (McGarigal and Marks 1995).

The landscape indices to analyze the landscape structure (McGarigal and Marks 1995) were, in accordance with the scale of analysis, used to assess landscape components (or classes) and the overall landscape. An area index, a patch index, an edge index, and a shape index were analyzed at both levels. The area index, which is the most basic data that can be obtained using GIS, is used to calculate the areas of the individual patches which make up a landscape as well as the overall landscape. Patch indices, such as the number of patches (NP), mean patch size (MPS), patch size standard deviation (PSSD), and patch size coefficient of variation

E10	Hupo Sand Dune	Sanryul-li, Hupo-myeon, Uljin-gun, Gyeongbuk Province

Overview of the study area: This particular coastal sand dune area stands in a state of great disrepair. Various facilities such as the greenhouses and fishing facilities have been arranged in a ramshackle manner.

E11	Gorabul Sand Dune	Byeonggok-li, Byeonggok-myeon, Yeongdeok-gun, Gyeongbuk Province

Overview of the study area: The presence of a military base in the area has meant that civilians' access to this sand dune has been comparatively limited. As a result, this area boasts the largest habitats of coastal vegetation of all the sand dunes investigated in the East Sea area. However, the cultivation of *Glehnia littoralis* in this area raises the possibility that an artificial change in natural conditions could occur.

Fig. 9.2 (continued)

(PSCV), make it possible to ascertain the size of the overall landscape patch by finding the area of each individual patch. Edge indices such as total edge (TE) and edge density (ED) serve as important indicators of species that prefer the edges of a landscape (Acosta et al. 2000). Meanwhile, shape indices, such as mean shape index (MSI), area-weighted mean shape index (AWMSI), mean patch fractal dimension (MPFD), and area-weighted mean patch fractal dimension (AWMPFD), are used to identify individual patch shape complexity. For their part, MPS and MSI are also used to identify habitat fragmentation and spatial heterogeneity. In other words, a landscape which boasts a greater number of patches, but an area that is identical to that of another landscape, can be understood to feature smaller patches. Thus, these indices can be used to explain the fragmentation of habitats from a landscape ecology standpoint.

At the landscape level, also, Shannon's diversity index (SHDI) and Shannon's evenness index (SHEI) were employed to quantify landscape composition. SHDI generally refers to the number of patch types. SHEI refers to the distribution of different patch types. Using both indices can help to explain how various patches are distributed within the overall landscape, which supports a diversity of habitat types as well as of species.

9.3 Results and Discussions

Figure 9.3 shows the land-use maps of the seven study areas. These areas include two areas on the West (Yellow) Sea coast (W01, Hakampo sand dune located in Banggal-li, Wonbuk-myeon, Taean-gun, Chungnam Province; W09, Doksan sand dune located in Doksan-li, Ungcheon-eup, Boryeong City, Chungnam Province), two areas on the South Sea coast (S06, Namyeol sand dune, located in Namyeol-li, Yeongnam-myeon, Goheung-gun, Jeonnam Province; S07, Balpo sand dune, located in Balpo-li, Dohwa-myeon, Goheung-gun, Jeonnam Province), and three areas on the East Sea coast (E05, Hosan sand dune located in Hosan-li, Wondeok-eup, Samcheok City, Gangwon Province; E10, Hupo sand dune located in Sanryul-li, Hupo-myeon, Uljin-gun, Gyeongbuk Province; E11, Goraebul sand dune located in Byeonggok-li, Byeonggok-myeon, Yeongdeok-gun, Gyeongbuk Province).

Table 9.1 shows the results of the class-level spatial analyses conducted using the land-use maps of the seven study areas. A total of nine landscape components were identified in these areas. In terms of the total extent of each study area, the absolute area of the land-use map, which included the surrounding landscape, was determined based on the size and length of the relevant sand dunes. Therefore, the area indices should only be compared after they have been transferred into a comparative value. The highest value assigned under the MSI, calculated using the values of the area and the circumference, was found to have been given to roads. This high value assigned to roads can be understood to be the result of the fact that such structures are generally long and constitute a narrow belt-type patch, a characteristic that causes them to influence the shape index of other landscape patches. To this end, the MSI value for roads was omitted from the spatial analysis aimed at calculating the shape index (Moser et al. 2002). The value of sand dunes became much greater once these roads were excluded from the patch index. This can be taken to mean that the shapes of the patches within these sand dunes were in fact more complex.

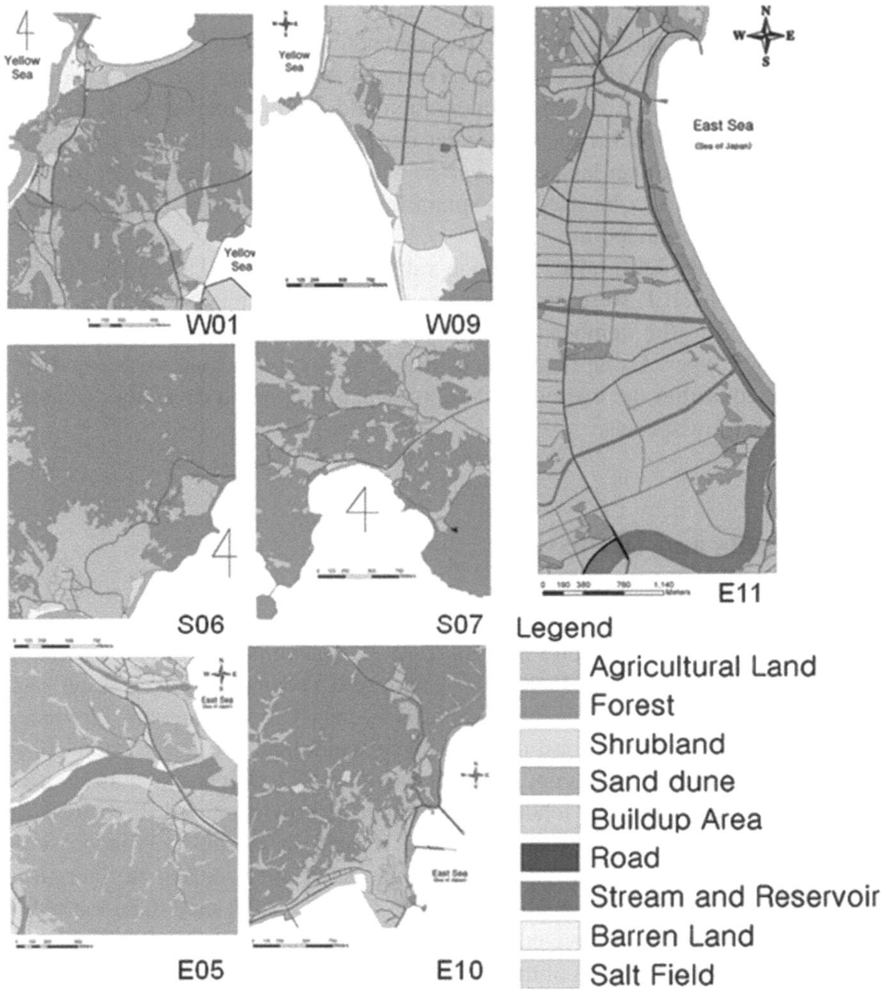

Fig. 9.3 Land-use maps of study areas

Figure 9.4 shows the absolute value of the area index when presented as comparative ratios (ratio of the area and the number of patches). A look at the ratio of patch areas reveals that, with the notable exceptions of the Doksan sand dune (W09) on the west coast and the Goraebul sand dune (E11) on the east coast, forest was the main component of the background landscape of the study areas, followed by agricultural land. Meanwhile, agricultural land was found to be the main component of the background landscape of both areas. In the future, a greater number of areas should be studied in order to understand how different surrounding environments influence sand dune ecosystems.

Table 9.2 shows the results of analyzing the sand dune ecosystems at the landscape level. As PSCV and PSSD include secondary analysis results that the primary

Table 9.1 Landscape indices of the landscape structure of sand dune ecosystems at the class level

Class	CA (ha)	NP	MPS	PSCV	PSSD	TE	ED	MSI	AWMSI	MPFD	AWMPFD
W01											
Agricultural land	850	183	4643	219	10171	55308	0.011	1.460	1.904	1.450	1.397
Forest	3287	21	156531	181	283308	64286	0.012	2.384	3.634	1.400	1.383
Built-up area	329	103	3197	349	11149	25205	0.005	1.476	2.102	1.462	1.410
Road	103	4	25771	170	43713	39242	0.008	11.108	33.513	1.819	1.831
Stream and reservoir	3	1	2790	15	2349	730	0.000	3.896	3.896	1.662	1.662
Sand dune	160	4	40048	73	29357	8018	0.002	3.235	2.932	1.476	1.428
Shrubland	73	8	9160	99	9085	5565	0.001	2.495	2.178	1.495	1.404
Barren land	112	8	14040	143	20100	7641	0.001	2.727	3.044	1.525	1.448
Salt field	241	2	120575	32	38588	3326	0.001	1.337	1.400	1.265	1.270
W09											
Agricultural land	1923966	41	46926	103	48397	41943	0.012	1.448	1.637	1.333	1.306
Forest	310315	8	38789	96	37428	8694	0.002	1.657	1.787	1.350	1.335
Built-up area	558248	56	9969	599	59717	13244	0.004	1.262	1.424	1.427	1.274
Road	86780	3	28927	139	40296	45976	0.013	17.199	43.290	1.876	1.888
Stream and reservoir	19475	8	2434	74	1813	3810	0.001	2.954	2.732	1.618	1.572
Sand dune	80768	4	20192	69	13990	6740	0.002	3.187	3.900	1.493	1.510
Shrubland	299712	6	49952	81	40370	7801	0.002	2.171	1.681	1.410	1.312
Barren land	327526	9	36392	194	70682	9356	0.003	2.797	2.002	1.619	1.335
S06											
Agricultural land	1370	103	13301	406	54031	41813	0.007	1.415	2.310	1.443	1.353
Forest	4010	5	802084	176	1413173	43764	0.008	3.348	4.558	1.426	1.376
Built-up area	86	15	5722	109	6247	5083	0.001	1.384	1.566	1.403	1.369
Road	34	5	6888	129	8883	13650	0.002	8.382	13.778	1.830	1.803
Sand dune	64	2	32168	4	1176	2713	0.000	2.128	2.139	1.387	1.388

(continued)

Table 9.1 (continued)

Class	CA (ha)	NP	MPS	PSCV	PSSD	TE	ED	MSI	AWMSI	MPFD	AWMPFD
Shrubland	36	3	12 043	61	7 322	1 845	0.000	1.594	1.761	1.380	1.379
Barren land	23	2	11 638	65	7 508	1 523	0.000	2.266	2.009	1.461	1.411
S07											
Agricultural land	1 343	125	10 741	282	30 259	52 531	0.012	1.440	2.088	1.432	1.365
Forest	2 987	12	248 905	76	188 736	53 109	0.012	2.516	2.873	1.355	1.360
Built-up area	51	19	2 697	117	3 152	4 673	0.001	1.471	1.705	1.468	1.421
Road	52	5	10 441	162	16 952	22 980	0.005	8.727	23.730	1.796	1.838
Sand dune	13	1	13 023	87	2 654	1 378	0.000	3.405	3.405	1.526	1.526
Shrubland	25	6	4 200	70	2 955	2 973	0.001	2.126	2.603	1.532	1.496
Barren land	9	2	4 374	54	2 375	2 504	0.001	5.102	5.960	1.691	1.706
E05											
Agricultural land	1 042	155	6 723	265	17 846	51 930	0.009	1.497	1.715	1.466	1.355
Forest	3 287	18	182 591	282	514 990	46 584	0.008	2.169	4.164	1.434	1.379
Built-up area	314	88	3 569	131	4 659	23 777	0.004	1.413	1.613	1.437	1.391
Road	182	16	11 358	287	32 650	46 837	0.008	6.914	17.883	1.776	1.744
Stream and reservoir	529	14	37 756	204	76 839	13 514	0.002	2.077	1.897	1.441	1.325
Sand dune	114	4	28 467	168	47 828	4 811	0.001	2.241	3.561	1.528	1.441
Shrubland	253	41	6 170	166	10 249	28 218	0.005	2.842	3.301	1.590	1.506
Barren land	59	2	29 744	76	22 602	2 773	0.000	2.107	2.750	1.385	1.421
E10											
Agricultural land	730	143	5 107	211	10 762	50 391	0.010	1.519	2.217	1.443	1.414
Forest	3 365	8	420 653	200	842 179	59 877	0.012	3.018	6.641	1.425	1.434
Built-up area	535	46	11 625	342	39 707	27 843	0.006	1.728	3.798	1.443	1.443
Road	70	4	17 536	160	28 109	24 020	0.005	8.221	24.089	1.748	1.799
Sand dune	87	5	17 471	73	12 695	7 247	0.002	3.216	3.491	1.521	1.500
Shrubland	14	2	7 215	3	219	681	0.000	1.130	1.130	1.313	1.312

Barren land	2	1	1 645	13	276	297	0.000	2.069	2.069	1.538	1.538
E11											
Agricultural land	6 827	136	50 202	176	88 245	139 875	0.014	1.602	1.925	1.380	1.314
Forest	1 029	24	42 863	150	64 471	37 547	0.004	2.180	3.095	1.416	1.407
Built-up area	484	73	6 630	154	10 235	25 961	0.003	1.417	1.641	1.415	1.363
Road	27	12	2 230	122	2 728	12 284	0.001	5.671	8.172	1.794	1.796
Stream and reservoir	791	13	60 859	119	72 233	27 625	0.003	3.057	2.511	1.450	1.366
Sand dune	511	5	102 208	98	100 147	16 745	0.002	2.997	3.584	1.423	1.422

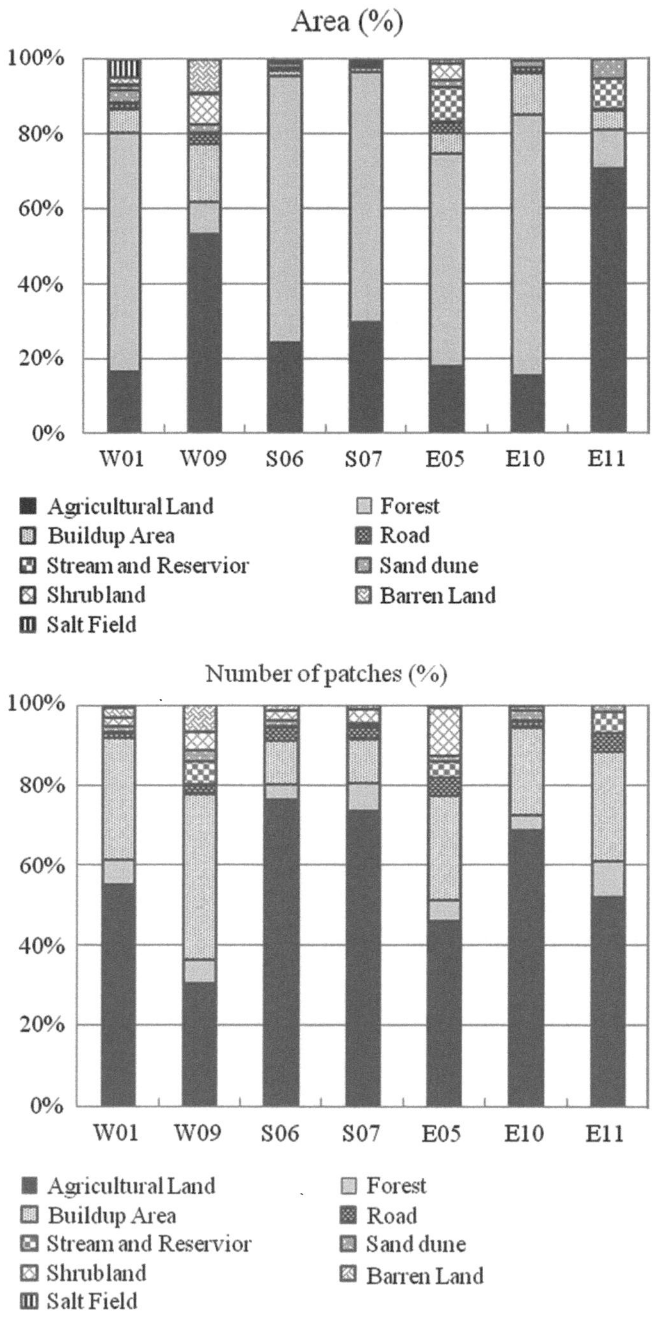

Fig. 9.4 Indices of the area, number, and average shape of the patches that make up the landscape structure

Table 9.2 Landscape indices of sand dune ecosystems at the landscape scale

Study area	CA (ha)	NP	MPS	PSCV	PSSD	TE	ED	MSI	AWMSI	MPFD	AWMPFD
W01	5 159	334	15 446	527	81 445	209 319	0.041	1.722	3.689	1.458	1.394
W09	3606 789	135	26 717	206	55 107	137 563	0.038	1.996	2.712	1.429	1.327
S06	5 624	135	41 663	753	313 774	110 390	0.020	1.769	3.965	1.450	1.373
S07	4 480	170	26 352	317	83 469	140 147	0.031	1.813	2.873	1.448	1.370
E05	5 779	338	17 099	744	127 300	218 444	0.038	1.967	3.743	1.486	1.389
E10	4 804	209	22 985	801	184 120	170 356	0.035	1.790	5.831	1.449	1.438
E11	9 669	263	36 765	200	73 468	260 036	0.027	1.887	2.188	1.416	1.338

MPS does not, they can be regarded as being more useful in the comparative analysis at the landscape level (McGarigal and Marks 1995). The PSCV and PSSD revealed that the Namyeol sand dune, (S06) located on the south coast, had the largest values. This can be taken to mean that the size of the patches in this area was larger than that of other sand dunes.

While four kinds of patch shape indices exist, the use of AWMSI and AWMPFD are more strongly recommended when one or two patches are found to dominate the others within a landscape (McGarigal and Marks 1995). In this respect, the Hupo sand dune (E10) showed the highest values, while the Doksan sand dune (W09) on the west coast and Goraebul sand dune (E11) on the east coast had the lowest values. Viewing the notion of habitats from the standpoint of landscape ecology, it is generally believed that an area features stronger natural attributes when the value of the shape index is high. However, as the inclusion of roads as one of the landscape components results in higher values being recorded on the shape index, the claim that the Hupo sand dune (E10) features stronger natural attributes cannot be substantiated.

Figure 9.5 shows the landscape diversity and landscape evenness indices for the study areas. The Doksan sand dune (W09) on the west coast and the Hosan sand dune (E05) on the east coast were found to exhibit a high degree of landscape diversity, implying that these particular two sand dunes are composed of heterogeneous patches. Meanwhile, the two sand dunes located on the south coast showed the most homogenous landscape patterns. Landscape ecologists argue that such differences in the shape of heterogeneous and homogenous landscapes are in fact the result of the degree of disturbance, i.e., disturbance intensity, size, frequency, severity, interval, rotation period, and so on (Forman 1995; Turner et al. 2001). Homogenous landscapes emerge when the disturbance is maximized in a certain area. However, as time passes, the landscape can gradually become more heterogeneous again as a result of multiple factors ranging

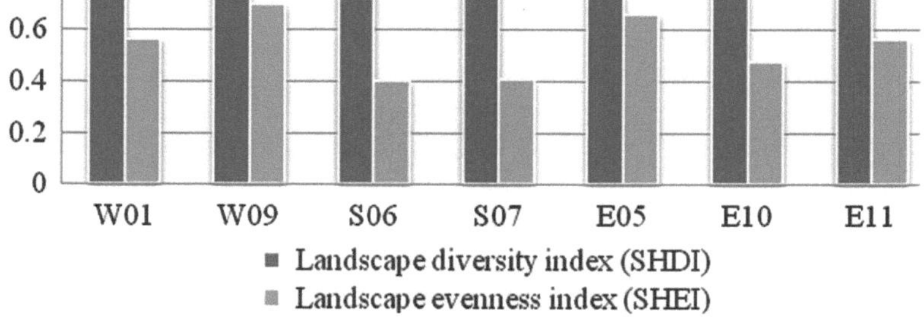

Fig. 9.5 Shannon's landscape diversity and landscape evenness indices for the study areas

from human activities to natural phenomena such as vegetation succession. While the background to such a theory has been rooted in studies of forest ecosystems (examples include the study of Yellowstone National Park in the United States, Romme and Knight 1982; Turner et al. 1993), it remains unclear at present whether this theory can also be applied to ecosystems such as coastal sand dunes.

9.4 Conclusion

The formation of the sand dune ecosystems located along the west, south, and east coasts of Korea has had much to do with physical forces such as the waves and winds found on each coast. While sand dune ecosystems are formed through complex physical actions, the maintenance and preservation of sand dunes is greatly influenced by the development of species that adopt these areas as their habitats. However, the large-scale changes that have occurred in sand dune ecosystems and their landscapes in the modern era have not been brought about by such natural phenomenon, but rather by human activities. In particular, changes in the use of the land surrounding these sand dunes have had a great influence on the possibility of preserving these sand dune ecosystems (Table 9.3). The growing influence of widespread human activities in such areas can be investigated and understood through multi-disciplinary approaches such as landscape ecological analyses (Williams et al. 2001).

Studies of habitats using GIS have become customary in the field of landscape ecology. By studying the temporal and spatial structures, functions, and changes that

Table 9.3 Non-biological elements influencing the distribution of vegetation communities in coastal sand dune ecosystems (Greipsson 2002)[a]

Element	Areas of coastal sand dune ecosystems				
	Shore	Fore dunes	Main dunes	Dune slacks	Back dunes
Exposure to wind	++	++	++	–	+
Increase of sand	+	++	+	+–	–
Erosion of sand	++	++	+–	–	–
Salt concentration of sand	++	+	–	+–	–
Salt spray	++	++	+	+–	–
Underground water	–	–	–	++	+
Light intensity	++	++	+	+	–
Evaporation	++	++	+–	+–	–
Dryness	+–	++	++	+–	–
Heat stress	+–	++	++	–	–
Soil moisture	–	++	++	–	–
Soil N	–	–	+–	+	++
Soil S	–	–	+–	++	++
Fire	–	+–	+	–	++

[a] Stress level: ++, strong stress; +, medium stress; +, weak stress; –, no stress

have taken place within a landscape, the quality, measures for management, and future of a habitat can be predicted (Yoccoz et al. 2001). However, when using and analyzing landscape indices there is a need to be aware of the limitations associated with such indices, and to ensure that they are used correctly (Li and Wu 2004). Rather than dealing with sand dune ecosystems only, this study analyzed the internal and external factors which influence sand dune ecosystems at the landscape level. An analysis of the selected areas revealed that the main background landscape components influencing the sand dune ecosystems in Korea were forest and agricultural areas. Therefore, the preservation and restoration of sand dune ecosystems will require strategies to manage such background landscapes. In this regard, analysis of the interactions between landscape components, and of the main factors which cause changes in sand dune ecosystems, can be used to establish the long-term plans needed to preserve and maintain sand dune ecosystems, especially in Korea.

Acknowledgment This research was supported by the National Research Foundation of Korea Grant funded by the Korean Government (NRF-2009-361-A00007).

References

Acosta A, Blasi C, Stanisci A (2000) Spatial connectivity and boundary patterns in coastal dune vegetation in the Circeo National Park, Central Italy. J Veg Sci 11:149–154

Acosta A, Carranza ML, Izzi CF (2005) Combining land cover mapping of coastal dunes with vegetation analysis. Appl Veg Sci 8:133–138

Beever EA, Swihart RK, Bestelmeyer BT (2006) Linking the concept of scale to studies of biological diversity: evolving approaches and tools. Divers Distrib 12:229–235

Carboni M, Carranza ML, Acosta A (2009) Assessing conservation status on coastal dunes: a multiscale approach. Landsc Urban Plan 91(1):17–25

Carranza ML, Feola S, Acosta A, Stanisci A (2007) Using between patch boundaries for conservation status assessment on coastal dune ecosystems. In: Bunce RGH, Jongman RHG, Lojas L, Weel S (eds) 25 years landscape ecology: scientific principles in practice. IALE Publication Series 4, Wageningen, The Netherlands

Forman RTT (1995) Land mosaics: the ecology of landscapes and regions. Cambridge University Press, New York, 632p

Forman RTT, Godron M (1986) Landscape ecology. John Wiley & Sons Inc., New York, 619p

Greipsson S (2002) Coastal dunes. In: Perrow MR, Davy AJ (eds) Handbook of ecological restoration, vol 2. Cambridge University Press, Cambridge, pp 214–237

Haines-Young R, Green DR, Conusins S (1996) Landscape ecology and GIS. Taylor & Francis, London 288p

Hong SK, Kim JE (2008) Applying geographical information system and landscape indices to landscape ecological analysis of maritime villages in Korea: case study on fishery villages in Tae-An Peninsula, Chungnam. Island Cult 31:278–294 (in Korean with English abstract)

Hong SK, Kim S, Cho KH, Kim JE, Kang S, Lee DW (2004) Ecotope mapping for landscape ecological assessment of habitat and ecosystem. Ecol Res 19:131–139

Hong SK, Nakagoshi N, Fu B, Morimoto Y (2007) Landscape ecological applications in man-influenced areas: linking man and nature systems. Springer, Dordrecht, 535p

Hong SK, Koh CH, Harris RR, Kim JE, Lee JS, Ihm BS (2010) Land use in Korean tidal wetlands: impacts and management strategies. Environ Manage 45(5):1014–1026

Kim JE, Hong SK, Nakagoshi N (2006) Changes in patch mosaics and vegetation structure of rural forested landscapes under shifting human impacts in South Korea. Landsc Ecol Eng 2:117–195

Kutiel P, Zhevelev H, Harrison R (1999) The effect of recreational impacts on soil and vegetation of stabilized coastal dunes in the Sharon Park, Israel. Ocean Coast Manag 42:1041–1060

Lee WT, Chon SK (1984) Ecological studies on the coastal plants in Korea: on the sand dune vegetation of the western coast. Korean J Ecol 7(2):74–84 (in Korean with English abstract)

Levin SA (1992) The problem of pattern and scale in ecology. Ecology 73:1943–1967

Li H, Wu J (2004) Use and misuse of landscape indices. Landsc Ecol 19:389–399

McGarigal K, Marks B (1995) FRAGSTATS: spatial pattern analysis program for quantifying landscape structure. General Technical Report PNW-GTR-351, US Department of Agriculture, Forest Service, Portland, Oregon: Pacific Northwest Research Station

Moser D, Zechmeister H, Plutzar C, Sauberer N, Wrbka T, Grabherr G (2002) Landscape patch shape complexity as an effective measure for plant species richness in rural landscapes. Landsc Ecol 17(7):657–669

Nigel G, Yoccoz JDN, Boulinier T (2001) Monitoring of biological diversity in space and time. Trends Ecol Evol 16(8):446–453

Romme WH, Knight DH (1982) Landscape diversity: the concept applied to Yellowstone Park. Bioscience 32:664–670

Roy PS, Tomar S (2000) Biodiversity characterization at landscape level using geospatial modelling technique. Biol Conserv 95:95–109

Taveira Pinto F (2004) The practice of coastal zone management in Portugal. J Coast Conserv 10:147–158

Turner MG, Romme WH, Gardner RH, O'Neill RV, Kratz TK (1993) A revised concept of landscape equilibrium: disturbance and stability on scaled landscape. Landsc Ecol 8:213–227

Turner MG, Gardner RH, O'Neill RV (2001) Landscape ecology in theory and practice: pattern and process. Springer, New York, 401p

Walker S, Bastow WJ, Steel JB, Rapson GL, Smith B, King WM, Cottam YH (2003) Properties of ecotones: evidence from five ecotones objectively determined from a coastal vegetation gradient. J Veg Sci 14:579–590

Wiens JA (1989) Spatial scaling in ecology. Funct Ecol 3:385–397

Williams AT, Alveirinho-Dias J, Novo FG, Garcia-Mora MR, Curr R, Pereirae A (2001) Integrated coastal dune management: checklists. Cont Shelf Res 21:1937–1960

Wu J, Loucks OL (1995) From balance-of-nature to hierarchical patch dynamics: a paradigm shift in ecology. Q Rev Biol 70:439–466

Yoccoz NG, Nichols JD, Boulinier T (2001) Monitoring of biological diversity in space and time. Trends Ecol Evol 16(8):446–453

Chapter 10
Have Ecological Human Rights Been Globally Lost? A Conflict of Ecological Spatial Requirements and Cultural Landscape Opportunities in Modern *Homo sapiens*

Anastassia M. Makarieva, Victor G. Gorshkov, and Bai-Lian Li

10.1 Introduction

The mission of science, as of any other meaningful enterprise, is to contribute to the well-being of humanity. For a very long period of human history, this task has been predominantly understood in the utilitarian terms of finding the most efficient ways of exploiting the biosphere, to spend less time and labor to obtain more and better products (food, clothing, housing, transport, medicines). At present, amidst acute global and regional environmental and ecological problems, it has become clear that a conceptually novel scientific goal can be put forward: to understand what the inherent environmental requirements (rights) of our species are and how (whether) they are (can be) satisfied in the cultural landscape that substitutes for a natural environment for the majority of people in modern society. Here we outline the ecological problem of space use by large social mammals from that perspective.

All living organisms must consume energy from the environment and spend it within their bodies to sustain life processes. For a majority of species, their mass-specific metabolic rate falls within 1–10 $W kg^{-1}$ (Makarieva et al. 2005a, 2008), with the metabolic rate of *Homo sapiens* falling close to the lower limit of this range. In physiological rest, the human body consumes about 10^2 W. Because the area that a human body projects on the Earth's surface is of the order of 0.5 m^2, the individual energetic requirements of a human being could in principle be covered by solar radiation, the power of which constitutes about 2×10^2 W m^{-2} as a global average (Mitchell 1989). The area of the Earth's surface is 5×10^{14} m^2. The laws of physics do not prohibit a technology that would convert solar power to the power of an edible biomass with an efficiency close

B.-L, Li (✉)
Ecological Complexity and Modeling Laboratory, Department of Botany and Plant Sciences, University of California, Riverside, CA 92521-0124, USA
e-mail: bai-lian.li@ucr.edu

A.M. Makarieva and V.G. Gorshkov
Theoretical Physics Division, Petersburg Nuclear Physics Institute, Russian Academy of Sciences, 188300, Gatchina, St. Petersburg, Russia

S.-K. Hong, et al. (eds.), *Landscape Ecology in Asian Cultures*, Ecological Research Monographs, DOI 10.1007/978-4-431-87799-8_10, © Springer 2011

to unity. It follows that in principle it is possible to put two human beings per square meter over the entire planetary surface and to satisfy their energetic food requirements. This would correspond to a global population of 10^{15} individuals, which is over a hundred thousand times larger than the modern population. This simple estimate shows that there are no tangible *physical* limits to population growth.

At the same time, the biology of every species is correlated with the properties of the ecosystem to which the species belongs. With the characteristic time of the existence of a species being of the order of several million years (Stanley 1979), while the time-scale of the turnover of organic matter in most ecosystems does not exceed a hundred years (Whittaker and Likens 1975), it is obvious that the ecological and biological principles of ecosystem organization should include mechanisms which ensure ecosystem stability. These principles will impose *ecological* limits to the population density of any particular species, i.e., limits that are compatible with ecological and environmental sustainability. These principles should also be encoded in the biological design of the species.

The productivity of the biosphere does not exceed $p_{max} = 2\ W m^{-2}$ (Gorshkov 1995), while a medium-sized mammal, as discussed above, can consume about $j \sim 10^2\ W m^{-2}$ on the area occupied by its body on the Earth's surface. This means that the presence of the larger organisms in the ecological community, with $j > p_{max}$, introduces fluctuations in the plant biomass, because in any local area, such organisms must consume biomass at a rate greatly exceeding its production, thus disturbing the standing biomass store. To guard the ecosystem against the disturbing impact of large animals, their population density, and hence energy consumption per unit area of the ecosystem, declines with the increasing body size of the species in natural ecosystems (Makarieva et al. 2004). As a consequence, the individual territory controlled by the animal (home range) increases with the increasing body size of the species (Makarieva et al. 2005b and references therein). As we discuss below, this has profound implications for the physiological health and general well-being of the animals, with humans being no exception.

10.2 Quantifying the Natural Space Requirements of *Homo sapiens*

Our species originally belonged to a tropical ecosystem. It is a locomotive species. The encoded territorial requirements of human beings can be estimated from their physiological abilities related to movement. *Homo sapiens* is a relatively poor long-distance runner. The human world record speed for long-distance running is 22 $km h^{-1}$, while animals of the same size can reach about 60 $km h^{-1}$ (Fig. 10.1). It is noteworthy that our species is the most inefficient sprinter (world record 36 $km h^{-1}$ versus over 100 $km h^{-1}$ in similarly sized animals). However, humans are among the best walkers in the animal world. Man can sustain a walking speed of about 4 $km h^{-1}$ for a long period of time. The world record speed in race-walking for 20 km is 14 $km h^{-1}$. Military troops can undertake daily marches of over 40 km.

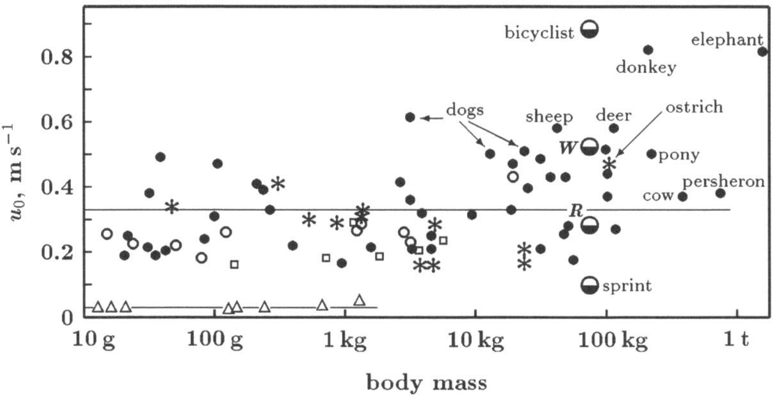

Fig. 10.1 Daily available speed u_0 for vertebrates of various body sizes. The daily available speed is the average speed that can be supported by the animal's field metabolic rate. *Horizontal lines* in the figure denote averages for homeothermic (*upper line*) and poikilothermic (*lower line*) vertebrates. *Triangles*, *asterisks*, *open dots*, *squares*, and *black dots* stand for reptiles, birds, marsupials, echidnas and hedgehogs, and placental mammals, respectively. For *Homo sapiens*, four regimes (sprint, R running, W walking, and cycling) are shown. Note that u_0 does not coincide with the actual speed of movement. For example, by making very fast leaps (sprint), humans are able to cover a much smaller daily distance at the same energetic cost than, for example, while walking. After Gorshkov et al. (2000)

On average, humans in natural environments cover about 10 km per day (Rappaport 1971; Winterhalder et al. 1988). Taking the width of the band scanned for food to be about 1 m, and a time-scale of vegetation regeneration of 1 year (=365 days), for the area of individual territory needed for humans, we obtain $H = 10 \times 10^3$ m day$^{-1} \times 365$ days $\times$ 1 m, which is approximately 4 km^2. The linear size of an individual territory is therefore of the order of 2 km.

The ecological meaningfulness of this estimate is sustained by the fact that it can be obtained in a different way by considering the well-established dependence of home range on body size in mammals (Kelt and Van Vuren 2001) (Fig. 10.2). For herbivorous mammals, the area H (ha) of individual territory grows as $\log_{10} H = -2.64 + 1.08 \log_{10} M$, where M is body mass (g). For a human body mass of $M = 80$ kg, we obtain $H = 450$ ha = 4.5 km^2. This is a minimum estimate, because omnivorous animals like *H. sapiens* have larger home ranges than herbivores of equal size (Kelt and Van Vuren 2001).

Assuming that the population density of humans, which is compatible with ecological stability and is encoded into the genome of our species, is of the order of 0.2 individuals km^{-2}, it is clear that in modern civilizations this spatial requirement, or the ecological right for space, is dramatically violated. The global mean population density (calculated for the total surface area of the planet, including the oceans) is 13 individuals km^{-2}, or 45 individuals km^{-2} if the oceans are excluded (Table 10.1). This corresponds to a per capita area of 0.022 km^2, which is hundreds of times less than the ecologically sustainable area. In the urban landscapes of the world's largest cities, the per capita available area A can be many thousands of

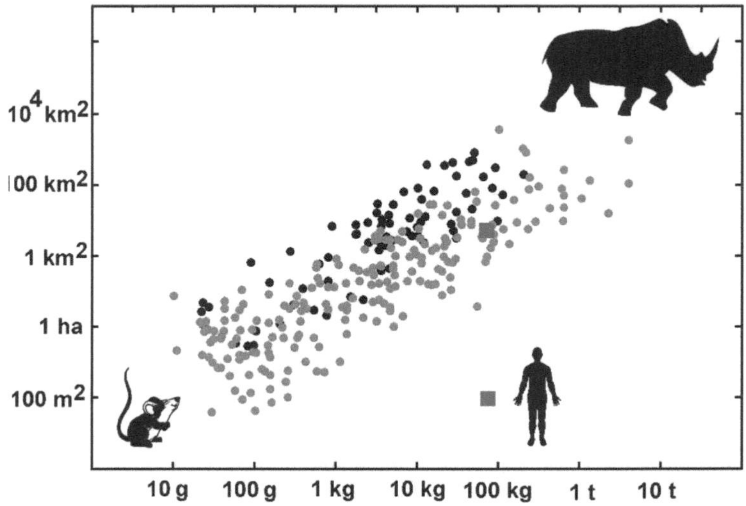

Fig. 10.2 The dependence of individual territory on body mass in mammals. *Green circles* denote herbivores, and *black circles* denote carnivores and omnivores according to the data of Kelt and Van Vuren (2001). The *upper square* corresponds to a natural home range $H=4$ km^2 per individual for a herbivorous mammal of equal size to *Homo sapiens*. The *lower square* corresponds to the per capita area available in modern urban landscapes (see Table 10.1)

Table 10.1 Population density and per capita available area A at different spatial scales in modern civilizations

Region	Population density, individuals km^{-2}	Per capita area A, km^2 per individual	H/A, $H=4$ km^2 per individual
Earth (incl. water surface)	13	0.078	51
Earth (excl. water surface)	45	0.022	182
Eastern Asia	129	0.0078	513
South–Central Asia	153	0.0065	615
South–Eastern Asia	124	0.0081	494
Western Asia	44	0.023	174
Russia	8	0.13	31
Western Europe	168	0.0060	667
Central America	58	0.017	235
South America	21	0.048	83
USA	31	0.032	125
Canada	3	0.33	12
Australia	3	0.33	12
Urban landscapes (cities)			
Moscow	15000	0.000066	60606
New York	10500	0.000095	42105
São Paulo	7300	0.00013	30769
Hong Kong	6400	0.00016	25000
Sydney	2000	0.00050	8000

The last column shows the ratio between the area, H, of the natural individual territory of *Homo sapiens* and the per capita available area, A, in the regions considered. (Data are for 2005; taken from the UN World Population Prospects, http://esa.un.org/unpp, retrieved November 14, 2008).

times smaller than the natural individual area *H*, which is of the order of a hundred square meters (Table 10.1). This is equivalent in size to the individual territory of a medium-sized rodent in natural ecosystems (Fig. 10.2).

10.3 Possible Consequences of the Violated Ecological Right for Space in Humans

Since the evolutionary survival of species is tightly dependent on the stability of the ecosystem to which the species belongs, it would not be surprising to find that the mechanisms ensuring ecosystem stability are encoded into the biological properties of the species. In other words, since the overpopulation of large animals poses a threat to the stability of the biomass store due to the high individual energy consumption rate of large animals, one can expect that the behavioral mechanisms preventing overpopulation should be clearly manifested in, and have profound implications for, the biology of species.

Extensive ecological literature exists on the subject of how animals defend their individual territories from intrusion (e.g., Mitani and Rodman 1979; Baker 1983; Adams 2001), thus sustaining optimal population density. There are many fewer studies that have addressed the biological effects of territory deprivation in natural animal species. However, the available evidence testifies that overcrowding in many species results in a considerable decrease in fertility (e.g., Sibly and Hone 2003). The pervasive importance of territory for normal biological performance can explain why it is often impossible to obtain viable progeny of large animals in captivity (e.g., in zoos), where animals are forced to occupy tiny areas compared with their natural territory requirements. For example, analyses of zoo environments, undertaken to investigate the causes of high mortality and suboptimal, unsustainable reproduction in captive populations of black rhinoceros (*Diceros bicornis*), revealed that the breeding success of captive animals was positively correlated with the size of their enclosure (Carlstead et al. 1999). Perhaps even more vivid is the fact that rhinoceros kept in open areas (where they could see, albeit not move across, a large territory) reproduced better than those enclosed by high walls. This means that the territorial requirements of the species are encoded on the physiological level; even a visual signal that a large territory is potentially available is essential for facilitating the reproduction process. Remarkably, an identical pattern was found in tiny jerboas (rodent family Dipodoidae), where, unless provided with a very large enclosure imitating their natural conditions, the animals could not reproduce, females did not care for their progeny (if they were captured when pregnant), the offspring refused to eat and did not gain weight, and pregnancy occurred very rarely and often led to female mortality (Fokin 1978). Data for captive animals show unequivocally that it is not food shortage (possibly related to overcrowding under natural conditions) that leads to a reduction in fertility, but rather that a signal of territory deprivation per se is received by the animal at the physiological level and translated to the observed inhibition of reproduction.

Until very recently, the effects of population density on fertility in humans remained practically unstudied. Recent work by Lutz and colleagues (Lutz and Qiang 2002; Lutz et al. 2006) revealed a significant correlation between population density and fertility in different countries, with fertility invariably reducing with growing population density. The researchers were able to uncouple population density effects from other potentially influential parameters, and concluded that population density is a key factor in the determination of human fertility. It can be proposed that the finding of the critical importance of individual territory for biological performance could be fruitfully studied from several other aspects, including a possible correlation with aggression, suicide rates, and other destructive behavioral manifestations.

10.4 Discussion: How Was the Major Ecological Right of Humans Lost?

Two inherent human features directly contributed to the loss of the major ecological human right. First, *H. sapiens* is a species with a genetically encoded ability/need to accumulate cultural information and pass it on to the next generation. Cultural heritage includes humanitarian and scientific information consisting of the cumulative memory store of the population. No other biological species ever existed on Earth which possessed this ability to accumulate cultural information *across* generations. Progress in the accumulation of scientific and technical information made our species the most competitive one in the biosphere, and also put our species in opposition to the rest of the biosphere. {Notably, this very property has brought humanity to the verge of ecological and environmental collapse. Natural biological and ecological laws and negative feedback [including emotion-(instinct-) based behavioral reactions] that ensure the stable existence of all other biological species do not function efficiently in the human population. The only way out of the crisis is therefore to use the accumulated cultural information effectively, to solve the survival problem scientifically, and then to implement the solution on a global scale by appealing to the reason of the peoples of the Earth.}

Second, *H. sapiens* is a social species. Sociability arises as a consequence of highly organized behavior and the corresponding complexity of the genetic program of our species (Gorshkov and Makar'eva 2001). The higher the level of organization of a species, the greater the rate of its spontaneous degradation. To compensate for this degradation, it is necessary to intensify the processes of competitive interaction among individuals in the population, identify individuals with a degraded (eroded) genetic program, and effectively exclude them from the population. These processes become most intense when a social structure is formed, i.e., when the population consists of individuals who continuously interact with each other and form a hierarchical structure where the status of each individual is determined.

Each social group possesses its own territory, which is equal in size to the product of the number of group members and the size of each individual territory.

The necessity of continuous interactions between group members ensures that the natural group size will not exceed a hundred members (e.g. Hayden 1981). (The footprint of this species-specific property can be traced in the sizes of the world's professional elites, where all members know each other and each other's status. Political and financial elites in various states are all approximately the same size.) Group territory is defended in competition with neighboring groups. In a natural environment with low a population density, this interaction did not give rise to any conflicts. Until very recently, such a lifestyle could still be observed in the aboriginal tribes of Papua New Guinea, Amazonian Indians, and tribes in the far north of Eurasia.

Many species are social, but only *H. sapiens* accumulates cultural knowledge. Cultural information changes the character of competitive interaction between social groups. When some new cultural information was occasionally discovered that imparted a competitive advantage to the corresponding social group, that group was able to out-compete the neighboring groups, invade and appropriate their territories, and subdue the population. Unlike in the rest of the animal world, that was done in spite of the genetic equivalence (equal orderliness of the genetic program) of the dominant and subordinate populations, and solely at the expense of cultural advancements occasionally acquired by the invaders. The increased size of the social group further contributed to its military power and facilitated further invasions. As a result, the social group was transformed into a hierarchical state, where the majority was compulsorily subdued to the elite minority. As a logical consequence, female and male social equivalence was gradually lost; in practice, female individuals functioned as biological machines producing military manpower for the state. Another essential feature of such radically transformed social groups was the inheritance of political and financial privileges. Corresponding moral, ethical, religious, and juridical norms gradually evolved that secured the status quo.

Importantly, in a very large social formation (state), the majority of the population is deprived of the opportunity to have a competitive interaction with the elite and to find out the true biological status of individuals that form the ruling elite. This undermines the fundamental biological and ecological meaning of social group formation, which consists in the identification of the most competitive and excellently performing individuals on the basis of competitive interactions between *all* members of the social group. The majority of the population was deprived of their individual territories and lost the ability to control their own population density. Their reproduction was further encouraged by the elite. Since they did not participate in competitive interactions with the majority of the population, the elite did not actually consider the population to be individuals genetically equivalent to themselves. Accordingly, the population was deprived of all major political and social rights, and in many cases was socially brought down to the level of domestic animals. Because they were not considered to be equal (or belonging to the same "species"), the presence of a dense population in the territory controlled by the elite did not impose any social or physiological stress on the elite.

Generally, the violation of the inherent hierarchical structure in a developing civilization inevitably leads to the genetic degradation of *H. sapiens* as a biological species, which is manifested as the growing percentage of individuals born with mental or physical disorders.

The situation could be globally changed if the genetically encoded ecological requirements of our species became the target of focused studies. We have outlined a few perspectives along which the research could proceed. Currently, the Universal Declaration of Human Rights does not contain any reference to the ecological rights of our species. This situation is likely to change in the coming decades, when the necessity of re-evaluating our understanding of *H. sapiens* and our environmental interactions with the rest of the biosphere will be widely recognized by the world (scientific) community.

Acknowledgments BLL thanks the US National Science Foundation and UC Agricultural Experiment Station for their support.

References

Adams ES (2001) Approaches to the study of territory size and shape. Annu Rev Ecol Syst 32:277–303

Baker PR (1983) Insect territoriality. Annu Rev Entomol 28:65–89

Carlstead K, Fraser J, Bennett C et al (1999) Black rhinoceros (*Diceros bicornis*) in U.S. Zoos: II. Behavior, breeding success, and mortality in relation to housing facilities. Zoo Biol 18:35–52

Fokin IM (1978) The Dipodoidae. Leningrad University Press, Leningrad

Gorshkov VG (1995) Physical and biological bases of life stability: man, biota, environment. Springer, Berlin

Gorshkov VG, Makar'eva AM (2001) On the possibility of physical self-organization of biological and ecological systems. Dokl Biol Sci 378:258–261

Gorshkov VG, Gorshkov VV, Makarieva AM (2000) Biotic regulation of the environment: key issue of global change. Springer-Praxis, London

Hayden B (1981) Subsistence and ecological adaptations of modem hunter/gatherers. In: Harding RSO, Teleki G (eds) Omnivorous primates: gathering and hunting in human evolution. Columbia University Press, New York, pp 344–421

Kelt DA, Van Vuren DH (2001) The ecology and macroecology of mammalian home range area. Am Nat 157:637–645

Lutz W, Qiang R (2002) Determinants of human population growth. Philos Trans R Soc Lond B Biol Sci 357:1197–1210

Lutz W, Testa MR, Penn DJ (2006) Population density is a key factor in declining human fertility. Popul Environ 28:69–81

Makarieva AM, Gorshkov VG, Li B-L (2004) Body size, energy consumption and allometric scaling: a new dimension in the diversity-stability debate. Ecol Complex 1:139–175

Makarieva AM, Gorshkov VG, Li B-L (2005a) Energetics of the smallest: do bacteria breathe at the same rate as whales? Proc R Soc Lond B Biol Sci 272:2219–2224

Makarieva AM, Gorshkov VG, Li B-L (2005b) Why do population density and inverse home range scale differently with body size? Implications for ecosystem stability. Ecol Complex 2:259–271

Makarieva AM, Gorshkov VG, Li B-L et al (2008) Mean mass-specific metabolic rates are strikingly similar across life's major domains: evidence for life's metabolic optimum. Proc Natl Acad Sci USA 105:16994–16999

Mitani JC, Rodman PS (1979) Territoriality: the relation of ranging pattern and home range size to defendability, with an analysis of territoriality among primate species. Behav Ecol Sociobiol 5:241–251

Mitchell J (1989) The "greenhouse" effect and climate change. Rev Geophys 27:115–139
Rappaport RA (1971) The flow of energy in an agricultural society. Sci Am 225:117–132
Sibly RM, Hone J (2003) Population growth rate and its determinants: an overview. In: Sibly RM, Hone J, Clutton-Brock TH (eds) Wildlife population growth rates. Cambridge University Press, New York, pp 11–40
Stanley SM (1979) Macroevolution: pattern and process. Freeman, San Francisco
Whittaker RH, Likens GE (1975) The biosphere and man. In: Lieth H, Whittaker RH (eds) Primary productivity of the biosphere. Springer, Berlin, pp 305–328
Winterhalder B, Baillargeon W, Cappelletto F et al (1988) The population ecology of hunter-gatherers and their prey. J Anthropol Archaeol 7:289–328

Part II
Measuring and Managing Patterns and Processes of Cultural Landscapes

Chapter 11
A Methodological Framework to Quantify Anthropogenic Effects on Landscape Patterns

Jan Bogaert, Yao Sabas S. Barima, Jian Ji, Hong Jiang, Issouf Bamba, Léon Iyongo Waya Mongo, Adi Mama, Edgar Nyssen, Farid Dahdouh-Guebas, and Nico Koedam

11.1 Introduction: Spatial Pattern and Its Effects on the Ecological Processes in Landscapes

Landscape ecology is defined as the study of ecological patterns and processes in their spatial context (Antrop 2001). Landscape ecology is motivated by a need to understand the development and dynamics of pattern in ecological phenomena, the role of disturbance in ecosystems, and characteristic spatial and temporal scales of ecological events (Urban et al. 1987). Many other definitions of landscape ecology can be found in the literature (Turner et al. 2001; Wu and Hobbs 2007). The same observation can be made for the landscape concept (Farina 2000a). Application of the principles of hierarchy theory (Allen and Starr 1982; Urban et al. 1987; Forman 1995; Burel and Baudry 2003; Bogaert and Mahamane 2005) to the biosphere leads

J. Bogaert(✉), Y.S.S. Barima, I. Bamba, L. Iyongo Waya Mongo, and A. Mama
Service d'Ecologie du Paysage et Systèmes de Production Végétale, Université Libre de Bruxelles, CP 169, 50 Avenue F.D. Roosevelt, 1050 Bruxelles, Belgium
e-mail: jan.bogaert@ulb.ac.be

J. Ji and H. Jiang
Nanjing University, 22 Hankou Road, Nanjing, Jiangsu 210093, China

J. Ji
Chengdu University of Technology, 1 Dongsan Road, Erxianqiao, Chengdu
Sichuan 610059, China

E. Nyssen
Vrije Universiteit Brussel, Vakgroep Elektronica en Informatica, Pleinlaan 2, 1050 Brussel, Belgium

F. Dahdouh-Guebas
Laboratoire de Complexité et Dynamique des Systèmes Tropicaux, Université Libre de Bruxelles, CP 169, 50 Avenue F.D. Roosevelt, 1050 Bruxelles, Belgium

F. Dahdouh-Guebas and N. Koedam
Onderzoeksgroep Algemene Plantkunde en Natuurbeheer, Vrije Universiteit Brussel, Pleinlaan 2, 1050 Brussel, Belgium

S.-K. Hong, et al. (eds.), *Landscape Ecology in Asian Cultures*, Ecological Research Monographs, DOI 10.1007/978-4-431-87799-8_11, © Springer 2011

to a definition of a landscape as the level of spatial or biological organization situated between the ecosystem level (lower level of organization than the landscape) and the regional level (higher level of organization than the landscape). A landscape can consequently be defined as a cluster of interacting and contiguous ecosystems; several contiguous landscapes will constitute a region.

To describe an individual landscape, the patch–corridor–matrix model is generally applied and three types of elements are identified (Forman and Godron 1986; Forman 1995); every point of the landscape is within either a patch, a corridor or the background matrix. In a landscape, different types or classes of land cover can be distinguished. Every class is consequently composed of one or more patches or landscape elements; a patch is the elementary unit of a landscape and corresponds to a homogeneous area. The landscape matrix is the patch type dominating the landscape, and is characterized by the lowest degree of fragmentation and the largest area (Bogaert and Mahamane 2005). Patches characterized by an elongated or linear shape, often present as a kind of network structure, are denoted as corridors. The distinction between patches and corridors is functional, and is not applied a priori to pattern analysis; its utility is often debated (Hobbs 2002). A landscape can therefore also be defined as a mosaic of patches, the latter being considered the components of pattern (Urban et al. 1987). The universality of the patch–corridor–matrix model has been questioned, particularly in relation to the influence of the matrix, which may not always be hostile to all elements of the biota (McIntyre and Hobbs 1999 in Hobbs 2002), as is indirectly assumed by the model.

The importance of the study of landscape spatial patterns is justified by the pattern/process paradigm (Turner 1989; Coulson et al. 1999), which states that there is a direct relationship between the ecological functioning of the systems in a landscape and the presence and spatial distribution of the landscape elements. The quantification of landscape pattern is an area of broad practical interest, driven by the premise that ecological processes are linked to, and can be predicted from, some broad-scale spatial pattern (Krummel et al. 1987; Baskent and Jordan 1995; Gustafson 1998; Noon and Dale 2002; Bogaert and Hong 2004).

To describe landscape pattern, two distinct concepts are often referred to. Firstly, landscape pattern is determined by the presence of particular land cover types, and by their proportional presence. This component of landscape pattern is denoted as landscape composition. The dominance of a land cover in a given landscape is consequently an example of the interpretation of its composition. Secondly, the spatial arrangement of the landscape elements determines the spatial pattern. This component is generally referred to as landscape configuration, and describes the way in which a patch type is spatially arranged. Classes can be composed of one single patch, or can be present as a group of patches of different sizes and shapes scattered across the landscape. In Table 11.1, some examples are given of frequently used landscape metrics for composition and configuration. It should be noted that many more metrics are reported in the literature for configuration than for composition, the former pattern component having a higher potential complexity. Some metrics can be used for both composition and configuration, such as the Shannon and Simpson metrics. When these indices are used as

Table 11.1 Examples of landscape metrics for composition and configuration

Composition	Configuration
Landscape dominance index	Area to perimeter shape index
Gini evenness index	Average nearest neighbor distance
Number of patch types	Average patch size
Proportion of the dominant class	Bogaert fragmentation index*
Ratio between patch type areas	Contagion index
Shannon diversity index	Contiguity fragmentation index
Shannon evenness index	Largest patch index
Simpson diversity index	Number of patches (per type)
Simpson evenness index	Monmonier fragmentation index*
	Patch (type) fractal dimension
	Patch perimeter length
	Proximity index
	Shannon evenness index
	Simpson evenness index
	Twist number statistic

For most metrics, a detailed description and formula can be found in the FRAGSTATS documentation (McGarigal et al. 2002), in the r.le documentation (Baker and Cai 1992) or in Colson et al. (2005). Metrics indicated by an "*asterisk*" are named after the first author of the publication in which the metric was presented, in order to avoid confusion between similar names.

composition metrics, the patch-type area will be used as input and the dominance of one of the patch types is measured; as a configuration metric, patch size will be used, and the heterogeneity of the patch sizes within one type or across the types is assessed. Between certain metrics, mathematical relationships can be evidenced (Bogaert et al. 2002a, b), as shown later in this chapter.

Using the aforementioned definition of spatial pattern as a combination of composition and configuration, the pattern/process paradigm can be extended to a triangular relationship between landscape composition, landscape configuration and ecological function (Fig. 11.1). The concepts of composition and configuration are also used to interpret landscape patterns; the notion of (spatial) heterogeneity was developed for this purpose (Burel and Baudry 2003; Bogaert 2005; Bogaert and Mahamane 2005). A landscape will be considered more heterogeneous when its configuration is characterized by complex patch shapes, and by a large variability of patch sizes and shapes. Heterogeneous landscapes show many patch types and an equal partition of the landscape area over the types. Homogeneous landscapes, as opposed to heterogeneous ones, will be dominated by simple patch shapes, and by small numbers of patches of comparable size. They will also contain a smaller number of patch types, and one of the types will dominate the landscape by a proportional area exceeding all other types. Human activity is expected to homogenize patterns through land use (Urban et al. 1987).

In order to quantify landscape pattern, a large number of landscape metrics have been presented and tested since the end of the 1980s, and several software packages have been developed to facilitate their calculation. Many indices have

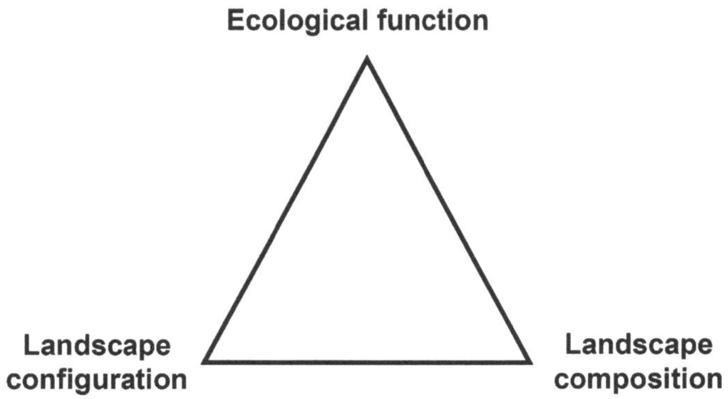

Fig. 11.1 Relationship between landscape pattern and ecological processes in the landscape (pattern/process paradigm). Landscape spatial pattern is accepted to be a combination of landscape composition and configuration, which both influence and are influenced by the ecological processes in the landscape. Any change of one of the three components will influence the remaining two. After Noon and Dale (2002) and Bogaert and Mahamane (2005), modified

been shown to be correlated (Bogaert et al. 2002a, b), which introduces redundant information in pattern analysis and complicates interpretation of the pattern (Bogaert and Hong 2004). Li and Wu (2004) identified three major problems associated with landscape metrics: conceptual flaws in landscape pattern analysis, inherent limitations of landscape indices, and improper use of landscape indices. A possible solution for this enigma is to return to the basics of pattern measurement and to avoid any correlation of metrics before starting the analysis, in order to limit the number of metrics and to concentrate the spatial information in a limited series of well-selected metrics (Li and Reynolds 1994; Giles and Trani 1999; Bogaert and Hong 2004). Bogaert and Mahamane (2005) proposed a conceptual framework for this selection. A subdivision of pattern complexity in four components was presented: patch size, patch shape, the number of patches and patch spatial dispersion. For every component, a sufficient number metrics are available in literature. This type of approach is recommended as a first, exploratory analysis of pattern before launching into a more profound analysis involving a higher number of complex metrics and focusing on detailed aspects of landscape structure. Nevertheless, the debate on the use and interpretation of landscape metrics still continues.

In this chapter, methods and strategies to analyze spatial pattern and landscape dynamics are discussed. Emphasis is put on pattern features due to human intervention and on anthropogenic disturbance. "Anthropization" is the process of landscape change as a consequence of anthropogenic effects. According to the pattern/process paradigm, human intervention on pattern will influence the ecological functioning of landscape ecosystems. The detection of anthropogenic effects and its quantification should therefore be considered as key activities of landscape ecologists.

Human-influenced landscapes will be characterized by fragmented natural land covers, a high frequency of edge habitat, simple patch geometries, and dominant anthropogenic patch types; their dynamics will show landscape transformations leading to a disintegration of natural land cover and to a reinforcement of anthropogenic types.

11.2 Fragmentation, Diversity and Entropy: Related Concepts

Fragmented landscapes are characterized by patchiness, large distances between patches (patch isolation), and patch size distributions skewed towards smaller patch sizes. Many metrics are available for fragmentation measurement (Colson et al. 2005), although the performance of many of the metrics has been questioned (Bogaert et al. 2002a). In this section, an unexpected relationship between fragmentation metrics currently agreed upon and widely used diversity metrics is put forward.

To measure diversity, three types of indices are generally used: richness metrics, which quantify the number of categories, species or patch types; evenness metrics, which quantify the number of elements in each category, such as the number of individuals of each species or the area of every patch type; and diversity metrics, which combine both richness and evenness.

The degree of fragmentation F_j of a patch type j is often quantified using the number of patches (n_j) and the total class area (a_j) (Monmonier 1974; Johnsson 1995; Bogaert et al. 2002b).

$$F_j = \frac{n_j - 1}{a_j - 1} \tag{11.1}$$

where n_j is defined using four-connectivity or eight-connectivity and is unit-less, and a_j is measured as the number of pixels in the class being considered. Note that $a_j \geq n_j$, and hence $0 \leq F_j \leq 1$. Equation (11.1) describes how a given number of pixels is spread over a given number of patches. The use of eight-connectivity, in which eight neighboring pixels are considered to define patches, will have a tendency to underestimate the degree of fragmentation, since patch fusion will be observed for patches considered to be separated by four-connectivity (Fig. 11.2).

A second parameter influencing (11.1) is the spatial resolution (g). The spatial aggregation of data will influence the fragmentation assessment (Fig. 11.3) as well as the use of a coarser resolution to represent fine-resolution data without information loss (Fig. 11.4). For coarse-resolution data, lower values of n_j and a_j will be observed, since every pixel represents a larger area on the ground, and since small patches and types tend to disappear (Bogaert and Hong 2004; Hu et al. 2008) as a consequence of spatial aggregation by the majority rule. For fine-resolution data, n_j and a_j tend to take high values; while n_j will remain directly related to the fragmentation of the landscape itself, a_j will influence the F_j ratio independently. For very fine resolution data ($g \rightarrow 0$), $a_j \approx (a_j - 1)$ and $a_j >> n_j$ hence $F_j \rightarrow 0$ for all observations, i.e.,

$$\lim_{g \rightarrow 0} F_j = \frac{n_j - 1}{a_j - 1} = 0 \qquad (11.2)$$

This shortcoming of (11.1) has already been observed by Kabulu Djibu et al. (2008). Application of the fragmentation index for vector data is complicated because of the determination of a_j. To estimate a_j, the class area is often divided by the size of the smallest patch observed. It should be noted that for $a_j \approx (a_j - 1)$ and $n_j \approx (n_j - 1)$, which corresponds to a patch type characterized by many patches and by a large class area, $F_j \approx n_j (a_j)^{-1}$, i.e., the inverse of the average patch area.

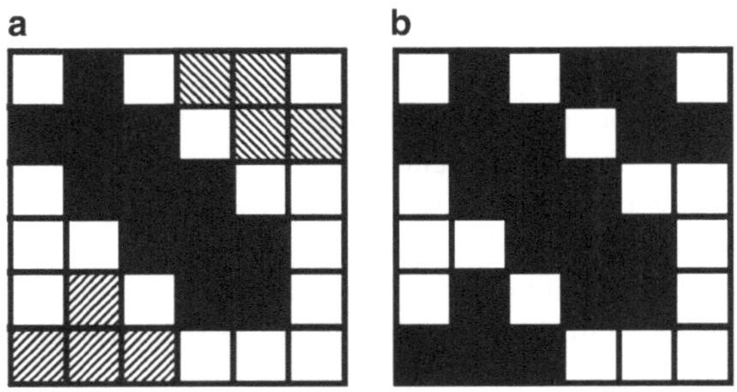

Fig. 11.2 Influence of patch definition on fragmentation measurement. (**a**) Patch definition based on four-connectivity: number of patches equals 3; class area equals 20 pixels; $F = 0.11$. (**b**) Patch definition based on eight-connectivity: number of patches equals 1; class area equals 20 pixels; $F = 0$. Pixels characterized by the same fill patterns belong to the same patch

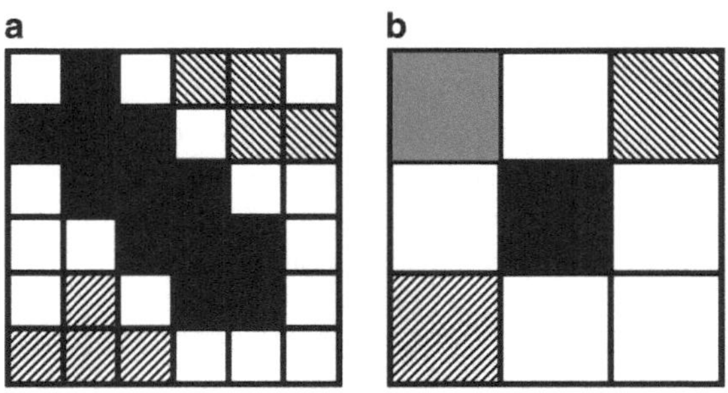

Fig. 11.3 Influence of spatial aggregation on fragmentation measurement. Both maps represent the same landscape, but the map in (b) is the result of spatial aggregation of (a) based on the majority rule (Hu et al. 2008) and using a 2×2 window. Patch definition is based on four-connectivity. (**a**) Fine-resolution landscape: number of patches equals 3; class area equals 20 pixels; $F = 0.11$. (**b**) Coarse-resolution image: number of patches equals 4; class area equals four pixels; $F = 1$. Pixels characterized by the same fill patterns belong to the same patch. As a consequence of spatial aggregation, the number of patches has increased

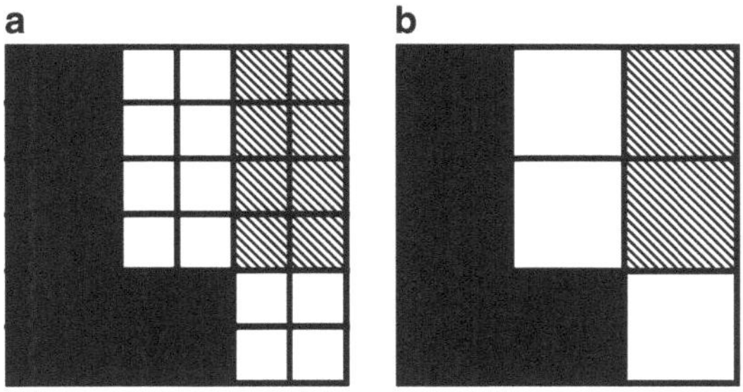

Fig. 11.4 Influence of spatial resolution on fragmentation measurement. Both maps represent the same landscape, but the map in (b) is a representation of (a) using a coarser spatial resolution. No information is lost when the coarser resolution is used. Patch definition is based on four-connectivity. (**a**) Number of patches equals 2; class area equals 24 pixels; $F=0.04$. (**b**) Number of patches equals 2; class area equals six pixels; $F=0.20$. Pixels characterized by the same fill patterns belong to the same patch

It is appealing to compare F to other widely used metrics in ecology. To measure the diversity of a community, the richness metrics of Margalef (R_{Mg}) and Menhinick (R_{Mn}) are often used (Magurran 2004):

$$R_{\mathrm{Mg}} = \frac{S-1}{\ln(N)} \quad (11.3)$$

$$R_{\mathrm{Mn}} = \frac{S}{\sqrt{N}} \quad (11.4)$$

where S is the number of species in the community and N is the number of individuals. Ease of calculation is one great advantage of the Margalef and Menhinick indices (Magurran 2004). Division by N is included to allow comparisons between communities with different numbers of individuals, i.e., it constitutes a correction for sample size (Magurran 2004), since $S \leq N$. Both indices describe how a given number of individuals are spread over a given number of species. For use in landscape ecology, the number of species is substituted by the number of patches, and the number of individuals by the number of pixels composing the class j, i.e.,

$$R_{\mathrm{Mg},j} = \frac{n_j - 1}{\ln(a_j)} \quad (11.5)$$

$$R_{\mathrm{Mn},j} = \frac{n_j}{\sqrt{a_j}} \quad (11.6)$$

and the correspondences between (11.1), (11.5) and (11.6) are obvious. The relation between F_j, $R_{\mathrm{Mg},j}$ and $R_{\mathrm{Mn},j}$ is given by

$$F_j = R_{\mathrm{Mg},j} \frac{\ln(a_j)}{a_j - 1} = \frac{\left(R_{\mathrm{Mn},j}\sqrt{a_j}\right) - 1}{a_j - 1} \tag{11.7}$$

For large class areas, i.e., $a_j \approx (a_j - 1)$ and $(a_j)^{-1} \approx 0$, (11.7) can be simplified to

$$F_j \approx R_{\mathrm{Mg},j} \frac{\ln(a_j)}{a_j} \approx \frac{R_{\mathrm{Mn},j}}{\sqrt{a_j}} \tag{11.8}$$

This strong link between a fragmentation metric and a diversity metric is not unique. In Bogaert et al. (2005), an equality has been reported between the coherence fragmentation metric (Jaeger 2000) and the Simpson diversity metric. Coherence was measured by calculating the diversity, based on the Simpson index, of the proportional patch areas of the same type, and consequently constituted no new metric. Its use as a metric to characterize population data (sensu opposite to sample data) was also questioned. It could be accepted that a landscape represents, by definition, the entire population unless the study zone considered was used as a representative spatial sample of a larger zone.

Owing to their link to the entropy concept (Forman and Godron 1986), the use of diversity metrics to quantify human impact was exemplified in Bogaert et al. (2005). Long-term data analysis of patch size diversity revealed increasing entropy levels associated with anthropogenic fragmentation. These entropy trends defied the laws of thermodynamics and signaled the impact of human action on landscape integrity. According to the second law of thermodynamics, only isolated systems, with no exchange of energy or matter with their surroundings, should become more disordered with time (Odum 1983; Smith and Smith 2000); biological systems should tend towards order, owing to (solar) energy input and the presence of internal feedbacks (Forman and Godron 1986; Smith and Smith 2000). Bogaert et al. (2005) speculated that the observed anthropogenic fragmentation had overruled this former energy input. Anthropogenic pattern changes were assumed to be based primarily on fossil fuel energy that had formerly been extracted from solar energy by plant photosynthesis tens or hundreds of millions of years ago. It was concluded that by consuming this stored energy with land cover changes for agricultural and economic development, leading to landscape fragmentation, the original energy balance had been reversed.

11.3 Edge Effects: Measuring the Impact of Anthropogenic Effects on Landscape Pattern

The quantification of edge effects is an appropriate tool to assess the ecological impact of land cover changes in general, and fragmentation in particular, on natural habitats. Measuring the presence and the extent of edge effects can be considered an elementary way to study anthropogenic effects on landscapes. However, the methods used to perform this type of analysis are not yet free from debate.

11.3.1 Importance and Assessment of Potential Edge Effects

Edge effects are observed when two different land cover types are adjacent and when both types are sufficiently different in structure, i.e., when the edge contrast is high enough (Forman 1995; Farina 2000a), such as for an agricultural field of corn next to a forest. This type of contact between contrasting land cover types is generally the consequence of the substitution of natural land cover by anthropogenic types. Edges and ecotones between habitats are unique as a result of both biotic and abiotic influences (Groom and Schumaker 1993). The effect of this juxtaposition of different land cover types on environmental variables is illustrated in Fig. 11.5. Owing to the proximity of a contrasting land cover, the peripheral contact zones of both patches involved are altered with regard to their microclimates, which can be assessed by means of variables such as wind velocity, air temperature, relative humidity, soil water content and light intensity (Chen 1991; Groom and Schumaker 1993; Alongo Longomba 2007; Litucha Bakokola Makeu 2007; Lokonda O. Wa Kipifo 2007). Consequently, the impact of land cover change will be underestimated if only the area which is converted to an anthropogenic land cover is measured (Chen 1991).

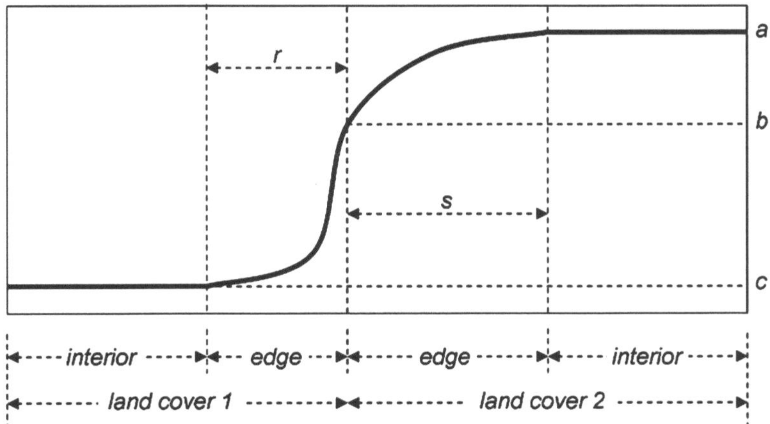

Fig. 11.5 Conceptual framework of the edge effect (Iyongo Waya Mongo 2008). The *black curve* describes the evolution of a theoretical environmental variable across the perimeter separating both land covers. For large patches, a zone can be observed which is remote from the contact area, and for which a constant magnitude of the variable is measured. This zone is the interior (habitat) of the patch. The peripheral zone, in which the magnitude of the variable is influenced by the adjacent land cover, is denoted as the edge of the patch. This edge effect can be quantified using the widths of the edges (*r*, *s*) and the changes in the variable considered (*a*–*b*, *a*–*c*, *b*–*c*). Both land covers are considered to be homogeneous

Because the microclimate changes in the contact zone, faunal and floral differences with more centrally situated parts of the patches will be observed, owing to the strong link between ecological conditions and biodiversity. Both interior habitat and edge habitat are characterized by their flora and fauna. Too much fragmentation of natural habitats will therefore lead to many edge zones and a rarefaction of the interior, which will have important consequences for the conservation of the biodiversity of the interior habitats (Naiman et al. 1988 in Farina 2000b). Knowledge of the magnitude of the edge effect, i.e., its absolute impact on the microclimate, and the distance inside the patch across which it is observed, can therefore be considered crucial to an assessment of the real ecological impact of land cover change on the remaining natural habitats. This distance is often referred to as the "edge influence distance," "edge width" or "penetration distance."

The simplest way to quantify potential edge effects is by measuring the perimeter of the landscape patches, since the edge effect is observed along the perimeter of the patches. This method could only be used as a proxy, since it does not take into account the two-dimensional properties of the edge zone (Fig. 11.5). To correctly assess the zone influenced by the neighboring land cover, the edge width should also be determined. For old-growth Douglas fir fragments, edge width has been shown to vary between 60 and 400+ m depending on the environmental variable considered (Chen 1991). Edge widths observed in situ are also influenced by vegetation structure (density, three-dimensional architecture, leaf type, heterogeneity) and seasonality (absence of leaves or strata during a particular period of the year) (Bogaert et al. 2001a, b). Patch exposition (Forman 1995) and topography can be added as parameters to be considered when determining edge width. Consequently, it can be concluded that it is not feasible to determine edge presence at a landscape level, and that approximate methods should be applied to estimate potential edge effects. Efforts should be made to collect empirical data in situ for a variety of vegetation types, contrast levels and ecological conditions to allow more realistic estimates of edge effects and their consequences to be made (e.g., Alongo Longomba 2007; Litucha Bakokola Makeu 2007; Lokonda O. Wa Kipifo 2007; Iyongo Waya Mongo 2008).

11.3.2 Quantification of Edge Effects for Single Patches

At the patch level, edge presence, measured by the interior-to-edge ratio, will be determined by patch size, patch shape and by the distance of the edge influence (Forman and Godron 1986). For a given edge width, large, isodiametric or compact patches will be characterized by larger interior-to-edge ratios than small, elongated ones. Unfortunately, edge width determination remains subject to debate, since it is very variable in space and in time. Therefore, instead of determining the interior-to-edge ratio as a function of edge width and patch size and shape, an alternative approach has been proposed (Bogaert et al. 2001a) in order to incorporate the edge effect in patch evaluation. The novelty of this approach is that the edge width is considered to be a variable instead of a constant value; its value will be determined by patch size and shape, and is observed when an equal proportion of edge and interior area are present. In this case, the edge width is denoted as the "interior-to-edge breakpoint distance" d_{bp}, since it marks the transition between a patch dominated by the interior habitat and the same patch dominated by edge habitat. To illustrate this concept, a rectangular patch could be considered, with side lengths equal to *a* and *b*. Assuming a perpendicular centripetal disturbance *d*, the area of the interior habitat will be given by $I=(a-2d)\times(b-2d)$ and the area of the edge zone by $E=2d(a+b-2d)$. In the case of $I=E$, *d* will be equal to d_{bp} and will be given by

$$d_{bp} = \frac{(a+b)-\sqrt{a^2+b^2}}{4} \quad (11.9)$$

For a patch with $a=350$ m and $b=420$ m, d_{bp} will equal 55.82 m. In Bogaert et al. (2001a), the ecological interpretation of d_{bp} is explained. For elliptical shapes, d_{bp} will be higher when the shapes are more compact, and when the patch size is larger. Large, isodiametric patches will consequently be characterized by larger edge widths before reaching a situation of an equal presence of interior and edge habitat. Small, elongated patches will be characterized by small values of d_{bp}, which indicates that even limited disturbances will have an influence throughout the patch. Disturbance will immediately have a large impact on these landscape elements. Thus, the d_{bp} quantifies at the same time patch size and shape, and reflects the resistance of patches to matrix influence. The higher the value of d_{bp}, the higher the probability of finding interior habitat conditions inside the patch considered. Quantifying d_{bp} for a patch type can consequently be used as a proxy for edge zone presence.

To estimate or determine d_{bp} for irregular shapes, the following procedure is proposed for landscapes analyzed in raster data format (Wu 2007). For every pixel in the patch, the distance is calculated towards the nearest pixel in the image outside the patch. This distance is traditionally calculated as the Euclidean distance or as the Manhattan or "city block" distance, the latter being equal to the sum of the differences between the row and column numbers of the two pixels considered. Consequently, a cumulative distribution function of the distance values is constructed, which will be convex in the case of a dominance of low distances, and which will be concave in the case of a dominance by high distances. This function also shows for which edge distance half of the patch will belong to the edge zone and half of the patch will belong to the interior habitat. This concept is illustrated in Figs. 11.6 and 11.7 for three landscape patches of the mangrove ecosystem in Galle, Southern Sri Lanka (06°01′N, 80°13′E) taken from land cover maps based on in situ observations, aerial photographs and remote sensing imagery (Dahdouh-Guebas et al. 2000; De Smet 2005). The patches represent mangrove vegetation (*Bruguiera gymnorrhiza*, *Rhizophora apiculata*, *Excoecaria agallocha*), and for this chapter are analyzed with a spatial resolution of 10 m using spreadsheet software. Since Manhattan distances were used, linear interpolation was applied. Patch b is characterized by the smallest value of d_{bp}, owing to its small size and rather elongated shape. Patch a is characterized by the largest value of d_{bp}, owing to its compact shape and large area.

The program PatchCalc has been developed, and made available as free and open software, to calculate the interior-to-edge distance with high precision

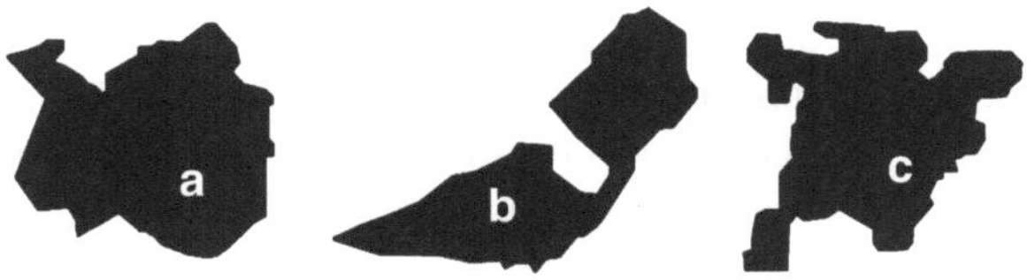

Fig. 11.6 Patch examples used to illustrate the concept of the interior-to-edge breakpoint distance. (**a**) Patch of *Bruguiera gymnorrhiza* in the 2004 Galle mangrove ecosystem; the area of the patch is 3.13 ha. and the breakpoint distance is 25.9 m. (**b**) Patch of *Rhizophora apiculata* in the 2004 Galle mangrove ecosystem; the area of the patch is 1.24 ha, and the breakpoint distance is 12.6 m. (**c**) Patch of *Excoecaria agallocha* vegetation in the 1974 Galle mangrove ecosystem; the area of the patch is 2.66 ha, and the breakpoint distance is 19.1 m. A pixel resolution equal to 10 m has been used to determine the patch area and the breakpoint distance. Distances are measured as Manhattan distances to the nearest pixel outside the patch. The breakpoint distance is determined by linear interpolation. Patches are rescaled for illustration purposes

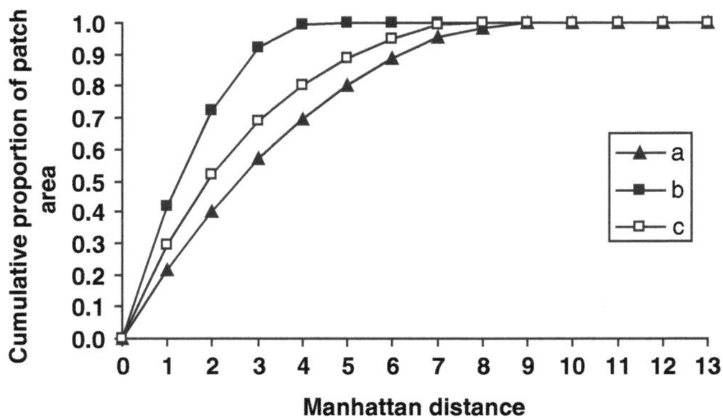

Fig. 11.7 Manhattan distance distribution for the three landscape patches of the mangrove ecosystem in Galle, Southern Sri Lanka, illustrated in Fig. 11.6. For every pixel, the shortest distance to the patch perimeter is determined. For every patch, the distribution of the distance values will be determined by its size and shape; small elongated patches will be characterized by low distance values, while large isodiametric patches will be characterized by greater distance values. The convexity of the curves indicates dominance by short pixel distances. The spatial resolution of the data was equal to 10 m. The Manhattan distance in the graph is expressed as pixel side lengths

(Wu 2007); it is considered to be a more practical solution to the procedure presented in Bogaert et al. (2001a). The procedure to download and implement PatchCalc is explained in the Appendix.

11.4 Fractal Dimension as an Indicator of Anthropogenic Effects on Spatial Pattern

Fractal theory offers methods for describing the inherent irregularity of natural objects, and the fractal dimension can be considered as a measure of landscape complexity, or as an index of the scale-dependency of pattern (Kenkel and Walker 1996).

It is also considered to be an index of landscape "anthropization." Anthropogenic patches such as agricultural fields, gardens or urban zones tend towards more regular and uniform shapes such as rectangles or squares (Krummel et al. 1987; Urban et al. 1987). This phenomenon is illustrated in Fig. 11.8 for Baoxing county (Sichuan basin, PR China), home town of the Giant Panda (*Ailuropoda melanoleuca*) (Ji et al. 2008). This landscape of about 3 250 km², which is mainly influenced by national policy, plays a key role in the protection of the Giant Panda. Based upon analyses of remote sensing imagery, a decrease in forest and grassland areas has been observed between 1975 and 1994. During the following 8 years (1995–2002), forest, especially conifers, and grassland areas increased again, while agricultural land use had strongly decreased, which had positive effects for the Giant Panda population (Ji et al. 2008). Nevertheless, the decreasing trend of the fractal dimension between 1975 and 2002 for the deciduous forests, conifer forests and croplands signals anthropogenic effects on landscape patterns throughout the same period.

Usually, the fractal dimension D describing perimeter "roughness" is determined using a power–law relationship between the perimeter P of a patch and its area A (Lovejoy 1982; Mandelbrot 1983; Burrough 1986; Voss 1988; Cheng 1995; Halley et al. 2004):

$$P = k\sqrt{A}^{D} \tag{11.10}$$

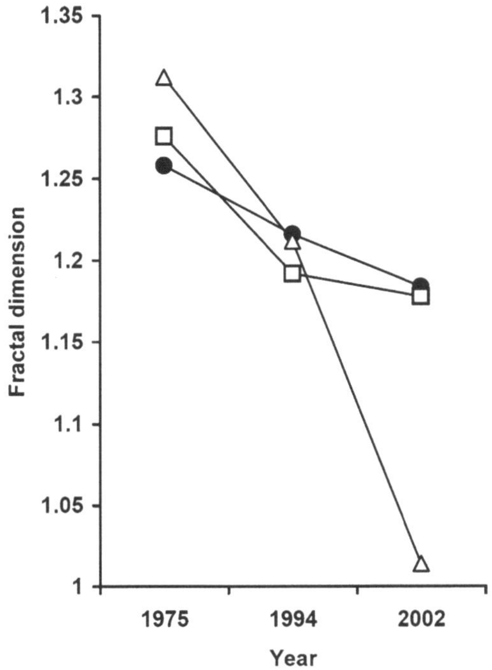

Fig. 11.8 Human impact on landscape pattern quantified by the patch type fractal dimension for Baoxing (PR China). Three patch types are shown: deciduous forest (*filled circle*), conifer forests (*open square*) and agricultural fields (*open triangle*). Decreasing values of the fractal dimension reflect anthropogenic effects on landscape pattern, i.e., a tendency towards more regular shapes. Fractal dimensions are calculated using the log–log regression method

where k is a scaling factor or "prefactor" (Halley et al. 2004) relating A to P, and is assumed to be constant and to reflect patch shape (Kenkel and Walker 1996; De Cola 1989; Cheng 1995; Addison 1997). The exponent D obtained from an analysis of the perimeter–area can be used only as a reliable estimate of the dimension of the perimeter if the dimension of the measured area equals the Euclidean dimension (Cheng 1995; Imre and Bogaert 2004). For a perfectly regular shape such as a square, $D=1$ and $k=4$. For a disk, $D=1$ and $k=2\sqrt{\pi}$.

D cannot be calculated directly from (11.10) since both D and k are unknown for an individual patch. To solve (11.10) for D, the power–law relationship is generally transformed into (Halley et al. 2004)

$$\log(P)=\frac{D}{2}\log(A)+\log(k) \tag{11.11}$$

A log–log area–perimeter plot for a set of patches therefore generates D (slope) and k (intercept), a technique referred to as the "log–log regression method." It is based on the analysis of patches of different sizes at one scale as a "surrogate" for a change in scale (Frohn 1998). Its application requires sufficient data (Krummel et al. 1987), and self-similarity among the patches considered (Voss 1988; De Cola 1989; Imre 2006). This latter condition is often postulated when patches are generated by the same process (Krummel et al. 1987; Imre and Bogaert 2004), a postulate which is more easily acceptable for anthropogenic patches. Although k is generally considered to be a constant (Kenkel and Walker 1996; De Cola 1989; Cheng 1995), different definitions are encountered in the literature. In Turner et al. (2001), k is a constant that takes on different values depending on the way that P is estimated, since different methods or definitions can be applied (e.g., Rosenfeld 1974; Bribiesca 1997). In Olsen et al. (1993) and Ricotta et al. (1997), k is considered to be determined by the cell shape composing the raster grid layers. It should be noted that this latter assumption was not derived from the seminal ideas of Mandelbrot (1983). Although the algebra used to estimate D is simple and straightforward, a variety of alternative formulae have been used, resulting in considerable confusion when attempting to make comparisons between studies (Turner et al. 2001). It is therefore useful to limit the definition of k to (11.10). Since $1\le D\le 2$, the boundary values of k are given by

$$\frac{P}{A}\le k\le\frac{P}{\sqrt{A}} \tag{11.12}$$

Equation (11.12) confirms the connotation of k as a shape descriptor, since two perimeter-to-area ratios are found as limits which are often used to quantify patch shape compactness (Patton 1975; Bogaert et al. 2000). Nevertheless, the definition of k as a constant form factor is erroneous. We show here that k varies with the unit of measurement used to express the P and A of the patches measured.

Consider P' and A', which are defined as $P'=\lambda P$ and $A'=\lambda^2 A$, respectively. If, for example, P and A were expressed in km and km^2, respectively, $\lambda=10$ would signify that P' and A' measure the same patch features in hectometres (10^2 m) and hectares (10^4 m^2), respectively. This arbitrary choice of the units of length and area does not influence the value of D. The cluster of data points used to determine D by linear regression is only translated and not rotated, i.e., the relative position of the patches in the log–log graph is not affected by λ. This can be shown by considering two patches, characterized by areas a_i and a_j and perimeters p_i and p_j for $\lambda=1$, and by areas $\lambda^2 a_i$ and $\lambda^2 a_j$ and perimeters λp_i and λp_j for $\lambda \neq 1$. The slope (s) of the line connecting patches i and j determines the relative position of the patches to each other. It can be shown that

$$s_{\lambda\neq 1} = \frac{\ln(\lambda p_j)-\ln(\lambda p_i)}{\ln(\lambda^2 a_j)-\ln(\lambda^2 a_i)} = \frac{\ln(p_j)-\ln(p_i)}{\ln(a_j)-\ln(a_i)} = s_{\lambda=1} \tag{11.13}$$

The Euclidean and Manhattan translation distances between both clouds are given by $\sqrt{5}\log\lambda$ and $3\log\lambda$, respectively. Since D is not dependent on λ, (11.10) and (11.11) can be rewritten as

$$P' = k'\sqrt{A'}^{D} \Leftrightarrow \log P' = \frac{D}{2}\log A' + \log k' \tag{11.14}$$

where k' is the new shape or scaling factor relating P' to A'. The substitution of P' by λP and of A' by $\lambda^2 A$ and combining this with (11.11) leads to

$$\log k' = \log k + (1-D)\log\lambda \tag{11.15}$$

or

$$k' = k\lambda^{1-D} \tag{11.16}$$

Equation (11.16) shows that the scaling or shape factor is not a constant, and that it changes as a function of λ and D. Equality between k and k' will be observed if $\lambda=1$ or if $D=1$, this latter condition corresponding to "anthropized" landscapes and generating $\lambda^{1-D}=\lambda^0=1$. For complex shapes characterized by $D=2$, i.e., landscape configurations dominated by natural processes, $k'=k/\lambda$. For $2\geq D>1$ and $\lambda>1$, $k'<k$ since $\lambda^{1-D}<1$. When $2\geq D>1$ and $\lambda<1$, $k'>k$ since $\lambda^{1-D}>1$. These theoretical findings were confirmed by a study of a landscape situated in the Bas-Congo province of the Democratic Republic of the Congo (Bamba et al. 2009).

The aforementioned analysis does not reopen the debate on the regression method to determine D or on the interpretation of D itself. Its importance lies in the significance sometimes attached in the literature to k as a shape factor; this latter interpretation of the scaling factor should be avoided, since shape assessment should not be dependent on λ, a principle implicitly adhered to by fractal theory.

11.5 Landscape Transformation Processes and Anthropogenic Landscape Change

The conversion of native landscapes into anthropogenic ones results in widespread changes in landscape spatial patterns (Collinge 1998). Two opposite processes can be observed in today's landscapes that have profound consequences on their structure and functioning (1) intensification of agriculture and urbanization, and (2) the abandonment of land (Farina 2000a). August et al. (2002) discuss a list of possible types of land transformation by human influences, going from forest cutting and wetland drainage to herbicide use and the release of nonnative species. Consequently, the subject of land transformation is significant for all human issues that involve land (Forman 1995). A study of the spatial pattern of landscapes and their dynamics can reveal the impact of human activities on these landscapes.

Regardless of the patch types concerned in land conversion, and considered from a geometric point of view, there appears to be a limited number of common spatial configurations that result from such land transformation processes. Consequently, a list of ten spatial landscape transformation processes has been defined in order to describe these typical conversions (Collinge 1998; Collinge and Forman 1998; Jaeger 2000; Bogaert et al. 2004; Koffi et al. 2007; Bogaert et al. 2008): aggregation, attrition, creation, deformation, dissection, enlargement, fragmentation, perforation, shift and shrinkage. A similar approach to study spatial changes in forest patches was recently proposed by Saura et al. (2008). For a definition and schematic illustration of every process, and for a practical decision tree model enabling the analyst to determine the process characterizing the patch type of interest rapidly, the reader is referred to Bogaert et al. (2004).

The aforementioned processes are class-neutral, i.e., they can theoretically be observed for any class present in a landscape being studied. Nevertheless, certain processes will be more frequently observed for particular classes than for others. For example, fragmentation is typical for natural land covers such as forests, in which human intervention generates a dispersed pattern of the natural resources across an anthropogenic matrix. It is therefore useful in the framework of landscape monitoring to classify the aforementioned spatial transformation processes into two groups in order to detect anthropogenic effects on landscape pattern. The first group of processes can be observed for anthropogenic classes such as cropland, urban zones or industrial plants; the second group of processes describes the transformation of natural land covers (Fig. 11.9).

Aggregation, creation, and enlargement will be seen for anthropogenic land-use and land-cover classes when the landscape structure is modified by human activities, since they are associated with an increase in the total area of these types (Bogaert et al. 2004). This phenomenon, which is often a consequence of population density increase, is often quantified by a measure of disturbance (U):

$$\log k' = \log \tag{11.17}$$

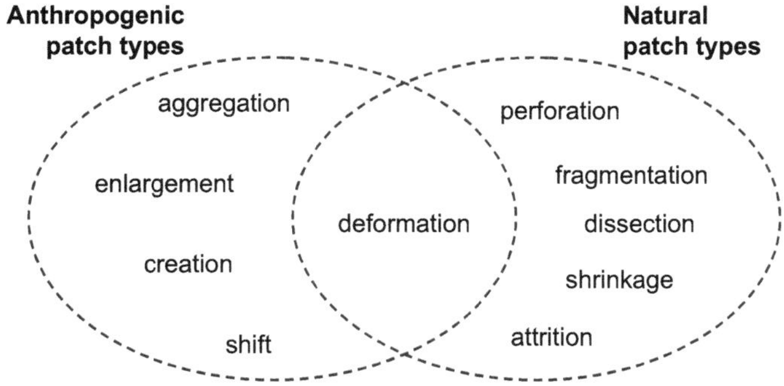

Fig. 11.9 Classification of landscape transformation processes to detect anthropogenic effects on landscape patterns. Patch types associated with anthropogenic activities are characterized by different transformations than patches representing natural land cover in case of landscape "anthropization". Deformation can be observed for both patch types

where D_i is the proportion of areas associated with intensely managed, disturbed or anthropogenic land covers and uses (such as cities, croplands, or road infrastructure), and N_j is the proportion of areas of natural or relatively undisturbed land use, such as forests or wetlands (O'Neill et al. 1988; August et al. 2002). Large agricultural and (sub)urban zones will merge or aggregate, and an anthropogenic landscape matrix will lead to a more uniform landscape. This evolution in landscape structure is often observed when anthropogenic land cover and land-use types, which were initially dominated by natural ones, become dominant themselves. By definition, a dominant land cover is characterized by small patch densities as a consequence of fusions between formerly separated patches. Centers of human activity (cities, industrial zones, agriculture) will expand, which is known as the nuclei type of mosaic sequence (sensu Forman (1995)); this enlargement of man-made patches will often cause patch aggregation. In Robinson et al. (2000), the phenomenon of urban growth is illustrated by means of photographs from Apollo, Skylab and Shuttle-Mir for the San Francisco Bay Area, Mexico City, Vancouver, Dallas, Chicago and Las Vegas in the period 1969–1996. Aggregation and enlargement can also cause the patches to change shape (deformation). During landscape "anthropization," new zones will also be created for human activities. These new patches, often characterized by a simple geometry, can be purely anthropogenic (e.g., urban or industrial land use), or can keep a seminatural trait (e.g., forest plantations (Bogaert et al. 2008)). This creation of anthropogenic patches spatially separated from others of the same type will increase total patch density. Finally, shift is also bound to anthropogenic land covers, since a rearrangement of landscape structure, keeping patch density, total class area and perimeter unchanged, seems unlikely for natural landscape dynamics, which are bound by ecological constraints and in which random processes reduce the probability of a repetition of identical pattern features.

For natural land covers, attrition, dissection, fragmentation, perforation and shrinkage will dominate the pattern changes. Often, a sequential dominance of one of the processes is observed (Forman 1995; Jaeger 2000; Bogaert et al. 2004).

Landscape transformation often starts with perforation or dissection; both processes involve a limited loss of the original habitats. Perforation is also observed in cases of natural disturbance of forests, e.g., gaps created by fallen trees (Salvador-Van Eysenrode et al. 1998, 2000). Dissection is observed when (rail)road networks are created; it therefore seems a key process of landscape "anthropization," since dissection directly increases the accessibility of a landscape. Accessibility constitutes the trigger or a priori condition of landscape dynamics; without accessibility (August et al. 2002), the probability of anthropogenic effects on pattern is expected to decrease sharply. Fragmentation is accepted to be the next step (Forman 1995; Jaeger 2000), which often aggravates the initial mosaic changes initiated by dissection and perforation. The zones of human influence become wider and better connected, and anthropogenic perforations become larger and can eventually merge. A patchy landscape, dominated by a man-made matrix, is now observed. This shift from a natural matrix to an anthropogenic one, which is often a consequence of population density (Bamba et al. 2008; Bogaert et al. 2008), will have profound implications for the connectivity of the natural habitats. The role of population density as a driver for landscape dynamics cannot be underestimated, and is very obvious in cases of landscape degradation (Fig. 11.10). This shift of an intact or variegated landscape towards a fragmented or relictual one (sensu Hobbs 2002) will have profound consequences for species diversity. Species occurrence at any place in the landscape does not only depend on the immediate characteristics at that spot, but is also affected by characteristics of the larger landscape; this means that in assessing a planning connectivity, it is important to consider current and forecasted landscape geometry, since the location with respect to other landscape elements can influence its use (Hilty et al. 2006). Inside the patchy landscape created by dissection

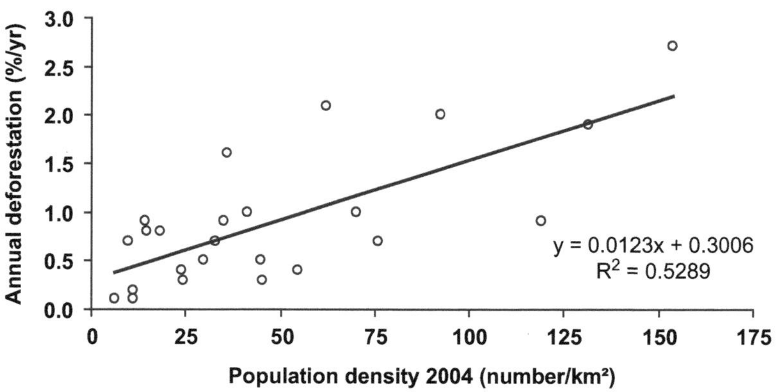

Fig. 11.10 Illustration of the impact of population density on deforestation rate between 1990 and 2005 in African countries with forested areas exceeding 10% in 2005. The following countries are included in the graph: Angola, Benin, Burkina Faso, Cameroon, Congo, Equatorial Guinea, Ethiopia, Ghana, Guinea, Guinea-Bissau, Liberia, Madagascar, Mali, Malawi, Mozambique, Nigeria, Central African Republic, Democratic Republic of the Congo, Senegal, Sierra Leone, Sudan, Tanzania, Uganda and Zambia. Data source: Global Forest Resources Assessment of 2005 of the Food and Agriculture Organization (FAO 2006)

and fragmentation, dynamics can continue, although often at a smaller scale. Changes take place at the patch level, where peripheral perturbations will be observed leading to the edge effect, and where land cover change can take place, which will lead to shrinkage and/or deformation of the patches, and eventually attrition. A similar sequence has been observed for Cadiz Township (Green County, Wisconsin, USA) between 1831 and 1950, where an initial phase of fragmentation was followed by attrition and shrinkage (Bogaert et al. 2004).

The aforementioned transformations of patch types are taking place in a closed system, i.e., an area moves between classes, and when one or several classes are characterized by an increase in their area, other land covers will be characterized by the opposite trend. These exchanges of area between patch types can be quantified using a transition matrix (Forman and Godron 1986; Parks 1991), which, in its turn, can be converted in a probability matrix describing the probability that a unit area of a given land cover is transformed into another land cover (Urban and Wallin 2002). Transition matrices are useful to disentangle overlapping or interfering dynamics in the landscape, and to identify the underlying drivers of landscape change (Bamba et al. 2008; Barima et al. 2009).

11.6 Conclusions

A terrestrial landscape is a mosaic of heterogeneous land forms, vegetation types and land uses (Urban et al. 1987). Natural and anthropogenic disturbances are common to many landscapes (Turner and Dale 1991). Landscapes are consequently dynamic, and are often referred to as shifting mosaics (Gustafson and Diaz 2002). Different modeling approaches have been proposed to understand landscape disturbances, such as Markov chains, cellular automata and probabilistic spread models (Turner and Dale 1991; Urban and Wallin 2002).

Human activity is affecting landscape pattern in different ways (Urban et al. 1987). A primary influence of humans is to rescale patterns in time and space. Secondly, humans rescale natural regions by establishing new boundaries. Thirdly, humans introduce novel perturbations that might differ in spatial or temporal scales from natural regimes. Finally, humans homogenize patterns through land use. Appropriate metrics should be developed and tested to quantify the aforementioned anthropogenic effects on landscapes. In August et al. (2002), typical changes in landscape and biological characteristics during the conversion of natural land to human-dominated landscapes were described. Nineteen landscape and biological indicators were listed which could be monitored to detect anthropogenic effects on landscapes: the presence of native or core habitat, patch size, patch shape, patch density, edge density, connectedness, landscape heterogeneity, roads, human population density, richness of native species, richness of introduced/weedy species, neotropical migratory birds, predators, wide-ranging species, core forest species, total species diversity, native species density, probability of extinction and edge-related mortality. Since some of the characteristics are overlapping, and because some indicators are related through the pattern/process paradigm (Turner 1989; Coulson et al. 1999), it is proposed to summarize these 19 characteristics into a limited series of landscape metrics or analysis types in order to detect anthropogenic effects early enough to authorize counteraction. The following analysis pathways are therefore suggested to monitor landscape "anthropization."

- Assessment of the degree of fragmentation of natural land covers in order to measure landscape heterogeneity and entropy, since these contain two basic components of pattern, patch density and patch type area, and because they are easy to calculate and interpret. The Monmonier index (Monmonier 1974) can be recommended. An alternative procedure using the Brillouin index is described by Bogaert et al. (2005). Average patch size could be used as complementary information.
- Analysis of patch type fractal dimension in order to detect the homogenization and simplification of landscape patterns as a consequence of changes in land cover. The log–log regression method (Krummel et al. 1987) can be considered as reliable.
- Quantification of edge presence in the landscape, via patch perimeter length or the interior-to-edge breakpoint distance (Bogaert et al. 2001a), in order to assess patch-matrix interactions and to link spatial pattern to diversity patterns.
- Analysis of the evolution of the relative presence of anthropogenic land covers versus natural land covers by means of the disturbance index (O'Neill et al. 1988; August et al. 2002), and/or by interpretation of the transition matrix (Forman and Godron 1986; Parks 1991).
- Analysis of human population density in the region of interest and its evolution over time.
- Determination and interpretation of the spatial transformation process characterizing natural land covers and anthropogenic land covers using the decision tree model proposed in Bogaert et al. (2004).

Acknowledgements The authors acknowledge the government of Ivory Coast for the fellowships of I. Bamba and Y.S.S. Barima. BTC/CTB is acknowledged for the fellowship of L. Iyongo Waya Mongo. This publication was made possible by a research grant from the Fund for Scientific Research – Flanders "Dynamiek in diversiteit, functionaliteit en stabiliteit van mangroven, benaderd vanuit een retrospectieve en actuele teledetectie-aanpak m.b.v. nieuwe patroonherkenningstechnieken."

11.7 Appendix

11.7.1 Calculation of the Interior-to-Edge Breakpoint Distance Using PatchCalc

The program PatchCalc has been developed and made available as free and open software (Wu 2007). The basic algorithms used in this package implement the identification of the distinct patches, and the calculations for each patch, on a distance map, i.e., a two-dimensional map (say, a matrix) that holds for each pixel inside the patch the minimum distance to the pixels outside the patch. Two distance measures have been implemented.

- The city block distance (CBD) considers all paths between the pixel being considered and the outside pixels, defined by "stepping" from one pixel to a horizontally or vertically adjacent pixel. The CBD is the minimum number of steps required.
- The Euclidean distance.

Such a distance map allows us to derive most of the relevant information for each patch.

- The area, i.e., the number of pixels.
- The maximum perimeter, i.e., the number of horizontal and vertical edges separating an inner pixel and an adjacent outer pixel.
- The minimum perimeter, i.e., the number of boundary pixels.
- The breakpoint distance, which is obtained from a histogram of the values from the distance map, and is derived by taking the smallest integer distance that makes the interior area less than the edge area (median value derived from the histogram).
- The largest radius, i.e., the maximum of the distance map.
- The core area.

The software package (including all sources) is available free for both UNIX/Linux and Windows systems. It can be retrieved from the following ftp site from the Vrije Universiteit Brussel – Electronics Department, using the ftp command to access both ftp.etro.vub.ac.be and the subdirectory ETRO/EDGARD_NYSSEN, or using any kind of web browser (Firefox, Explorer, etc.) via the following URL:
ftp://ftp.etro.vub.ac.be/ETRO/EDGARD_NYSSEN/
Use the following credentials to access the ftp site:
Username: etroguest
Password: anonymous1!
The name of the file to be downloaded is PatchCalc_software_<version>.zip (where <version> stands for the version number or version date of the package). This zip archive needs to be unpacked (accessible in a straightforward manner under Windows; the files can be extracted under Linux using the unzip command). The main software consists of the PatchCalc program. The PatchCalc program is normally used in conjunction with generic support software for image processing (mainly for file format conversion). For this purpose, we recommend the use of the free ImageMagick® package (ImageMagick® is a registered trademark of ImageMagick Studio LLC). For more details about the installation, see below.

(a) Instructions for Windows users

The main directory of the PatchCalc_software package contains the subdirectory Windows, in turn including the PatchCalc.exe command. Subdirectory code-Windows contains the sources, so that the executable can be rebuilt from scratch if necessary. The ImageMagick® package can be obtained via the World Wide Web: http://www.imagemagick.org/script/index.php. It is recommended that the most recent version of the ImageMagick® package is downloaded and installed. ImageMagick Studio LLC has an ftp site (ftp.imagemagick.org) from which all necessary files can be downloaded, using anonymous ftp or using a web browser (URL: ftp://ftp.imagemagick.org/pub/ImageMagick/):

Username: anonymous
Password: <your e-mail address>
The instruction file and the directory containing the installation scripts are, respectively, ftp://ftp.imagemagick.org/pub/ImageMagick/QuickStart.txt
ftp://ftp.imagemagick.org/pub/ImageMagick/binaries/
The installation of PatchCalc (and ImageMagick®) can be tested using the test data you find in the package.

Use

- Open a command line dialog (launch Start Run and enter cmd in the dialog).
- Go to the folder (say, C:\test) where PatchCalc.exe is stored (use the cd command).
- Copy the image file (say, X.bmp) in that folder.
- Convert the file X.bmp into a gray-formatted image file. ImageMagick® provides as useful tools the commands identify (to figure out the dimensions <width>x<height> of the image) and convert (to perform format conversion).

C:\test> identify X.bmp
C:\test> convert -depth 8 -size <width>x<height> X.bmp X.gray

- Apply PatchCalc.

C:\test> PatchCalc.exe X.bmp X.gray X.txt
Beside text output on X.txt, the program produces a color_0.gray image file showing the distinct patches. This can be converted to any kind of image file using the ImageMagick® command convert, e.g.,

C:\test> convert -depth 8 -size <width>x<height> color_0.gray color_0.jpg

(b) Instructions for Linux users

The PatchCalc program can be compiled and installed under a Linux system through the following procedure.

- The extraction of the files of PatchCalc_software_<version>.zip yields a directory named PatchCalc_software. Go to the Linux source directory
- % cd PatchCalc_software/code_Linux
- This directory contains the source code and makefile. If necessary you can edit this file to change the installation directory (by default, INSTALLDIRECTORY = ../Linux). The comments of the makefile also describe in detail how the standard UNIX/Linux command make can be called to manage the compilation/installation process.
- For compiling and installing PatchCalc simply execute the command.
- % make install

The ImageMagick® package can be obtained via the World Wide Web: http://www.imagemagick.org/. However, as this package is very popular, your brand of Linux almost certainly allows you to fetch and install it using the software installation tools available to you (e.g., synaptic). The installation of PatchCalc (and ImageMagick®) can be tested using the test data you find in the package.

Use

- Open a shell window.
- Go to the folder (say, /home/dominique/software/patchcalc/) where PatchCalc.exe is stored (use the cd command).
- Copy the image file (say, X.bmp) in that folder.
- Convert the file X.bmp into a gray-formatted image file. ImageMagick® provides as useful tools the commands identify (to figure out the dimensions <width>x<height> of the image) and convert (to perform format conversion).

software/patchcalc% identify X.bmp
software/patchcalc% convert -depth 8 -size <width>x<height> X.bmp X.gray

- Apply PatchCalc

software/patchcalc% ./PatchCalc X.bmp X.gray X.txt
Beside text output on X.txt, the program produces a color_0.gray image file showing the distinct patches. This can be converted to any kind of image file, using the ImageMagick® command convert, e.g.,
software/patchcalc% convert -depth 8 -size <width>x<height> color_0.gray color_0.jpg

References

Addison PS (1997) Fractals and chaos. An illustrated course. Institute of Physics Publishing, Bristol and Philadelphia

Allen TFH, Starr TB (1982) Hierarchy: perspectives for ecological complexity. University of Chicago Press, Chicago

Alongo Longomba S (2007) Etude de l'effet des lisières sur l'humidité équivalente et la température du sol d'un écosystème forestier de la cuvette centrale congolaise. Mémoire présenté en vue de l'obtention du DEA en Gestion de la biodiversité, Université de Kisangani

Antrop M (2001) The language of landscape ecologists and planners – a comparative content analysis of concepts used in landscape ecology. Landsc Urban Plan 55:163–173

August P, Iverson L, Nugranad J (2002) Human conversion of terrestrial habitats. In: Gutzwiller KJ (ed) Applying landscape ecology in biological conservation. Springer, New York

Baker WL, Cai Y (1992) The r.le programs for multi-scale analysis of landscape structure using the GRASS geographical information system. Landsc Ecol 7:291–302

Bamba I, Mama A, Neuba DFR et al (2008) Influence des actions anthropiques sur la dynamique spatio-temporelle de l'occupation du sol dans la province du Bas-Congo (R.D. Congo). Sci Nat 5(1):49–60

Bamba I, Iyongo Waya Mongo L, Imre A et al (2009) La variabilité du facteur de graduation utilisé dans la méthode d'estimation de la dimension fractale des mosaïques paysagères. Ann Inst Supér Etu Agron Bengamisa 4:168–176

Barima YSS, Barbier N, Bamba I et al (2009) Dynamique paysagère en milieu de transition forêt-savane ivoirienne. Bois For Trop 299:15–25

Baskent EZ, Jordan GA (1995) Characterizing spatial structure of forest landscapes. Can J For Res 25:1830–1849

Bogaert J (2005) Metriche del paesaggio: definizioni ed utilizzo. Estimo e Territorio 68(9):8–15
Bogaert J, Hong SK (2004) Landscape ecology: monitoring landscape dynamics using spatial pattern metrics. In: Hong SK, Lee JA, Ihm BS et al (eds) Ecological issues in a changing world. Status, response and strategy. Kluwer, Dordrecht
Bogaert J, Mahamane A (2005) Ecologie du paysage: cibler la configuration et l'échelle spatiale. Ann Sci Agron Bénin 7(1):1–15
Bogaert J, Rousseau R, Van Hecke P et al (2000) Alternative area–perimeter ratios for measurement of 2-D shape compactness of habitats. Appl Math Comput 111:71–85
Bogaert J, Salvador-Van Eysenrode D, Impens I et al (2001a) The interior-to-edge breakpoint distance as a guideline for nature conservation policy. Environ Manage 27:493–500
Bogaert J, Salvador-Van Eysenrode D, Van Hecke P et al (2001b) Geometrical considerations for evaluation of reserve design. Web Ecol 2:65–70, Erratum. Web Ecol 2:74
Bogaert J, Myneni RB, Knyazikhin Y (2002a) A mathematical comment on the formulae for the aggregation index and the shape index. Landsc Ecol 17:87–90
Bogaert J, Zhou L, Tucker CJ et al (2002b) Evidence for a persistent and extensive greening trend in Eurasia inferred from satellite vegetation index data. J Geophys Res 107:14. doi:10.1029/2001JD001075
Bogaert J, Ceulemans R, Salvador-Van Eysenrode D (2004) A decision tree algorithm for detection of spatial processes in landscape transformation. Environ Manage 33:62–73
Bogaert J, Farina A, Ceulemans R (2005) Entropy increase of fragmented habitats signals human impact. Ecol Indic 5:207–212
Bogaert J, Bamba I, Koffi KJ et al (2008) Fragmentation of forest landscapes in Central Africa: causes, consequences and management. In: Lafortezza R, Chen J, Sanesi G et al (eds) Pattern and processes in forest landscapes. Multiple use and sustainable management. Springer, New York
Bribiesca E (1997) Measuring 2D shape compactness using the contact perimeter. Comput Math Appl 33:1–9
Burel F, Baudry J (2003) Ecologie du paysage. Concepts, méthodes et applications. Editions Tec&Doc, Paris
Burrough PA (1986) Principles of geographic information systems for land resources assessment. Oxford University Press, Oxford
Chen J (1991) Edge effects: microclimatic pattern and biological responses in old-growth Douglas-fir forests. PhD dissertation, University of Washington
Cheng Q (1995) The perimeter–area fractal model and its application to geology. Math Geol 27:69–82
Collinge SK (1998) Spatial arrangement of habitat patches and corridors: clues from ecological field experiments. Landsc Urban Plan 42:157–168
Collinge SK, Forman RTT (1998) A conceptual model of land conversion processes: predictions and evidence from a microlandscape experiment with grassland insects. Oikos 82:66–84
Colson F, Bogaert J, Ceulemans R (2005) Spatial pattern analysis to address reliability issues in remotely sensed data. In: Veroustraete F, Bartholomé E, Verstraeten WW (eds) Proceedings of the second international spot/vegetation users conference. Office for Official Publications of the European Communities, Luxemburg
Coulson RN, Saarenmaa H, Daugherty WC et al (1999) A knowledge system environment for ecosystem management. In: Klopatek JM, Gardner RH (eds) Landscape ecological analysis. Issues and applications. Springer, Berlin
Dahdouh-Guebas F, Verheyden A, De Genst W et al (2000) Four decade vegetation dynamics in Sri Lankan mangroves as detected from sequential aerial photography: a case study in Galle. Bull Mar Sci 67:741–759
De Cola L (1989) Fractal analysis of a classified Landsat scene. Photogramm Eng Rem S 55:601–610
De Smet K (2005) Spatio-temporal dynamics of mangroves detected using aerial photography, Ikonos satellite imagery and fieldwork in Galle-Unawatuna, Sri Lanka. Thesis submitted in fulfilment of the requirements for the degree of Licentiaat Biologie, Vrije Universiteit Brussel

FAO (2006) Global forest resources assessment of 2005. Progress towards sustainable forest management, vol 147, FAO forestry paper. FAO, Rome
Farina A (2000a) Landscape ecology in action. Kluwer, Dordrecht
Farina A (2000b) Principles and methods in landscape ecology. Kluwer, Dordrecht
Forman RTT (1995) Land mosaics. The ecology of landscapes and regions. Cambridge University Press, Cambridge
Forman RTT, Godron M (1986) Landscape ecology. Wiley, New York
Frohn RC (1998) Remote sensing for landscape ecology. New metric indicators for monitoring, modelling, and assessment of ecosystems. Lewis Publishers, Boca Raton
Giles RH Jr, Trani MK (1999) Key elements of landscape pattern measures. Environ Manage 23:477–481
Groom MJ, Schumaker N (1993) Evaluating landscape change: patterns of worldwide deforestation and local fragmentation. In: Kareiva PM, Kingsolver JG, Huey RB (eds) Biotic interactions and global change. Sinauer, Sunderland
Gustafson EJ (1998) Quantifying landscape spatial pattern: what is the state of the art. Ecosystems 1:143–156
Gustafson EJ, Diaz N (2002) Landscape pattern, timber extraction, and biological conservation. In: Gutzwiller KJ (ed) Applying landscape ecology in biological conservation. Springer, New York
Halley JM, Hartley S, Kallimanis AS et al (2004) Uses and abuses of fractal methodology in ecology. Ecol Lett 7:254–271
Hilty JA, Lidicker WZ Jr, Merenlender AM (2006) Corridor ecology. The science and practice of linking landscapes for biodiversity conservation. Island Press, Washington
Hobbs RJ (2002) Habitat networks and biological conservation. In: Gutzwiller KJ (ed) Applying landscape ecology in biological conservation. Springer, New York
Hu J, Bogaert J, Tan B et al (2008) A rank-based algorithm for aggregating land cover maps. In: Dupont A, Jacobs H (eds) Landscape ecology research trends. Nova Science, New York
Imre A (2006) Artificial fractal dimension obtained by using perimeter area relationship on digitalized images. Appl Math Comput 173:443–449
Imre AR, Bogaert J (2004) The fractal dimension as a measure of the quality of habitats. Acta Biotheor 52:41–56
Iyongo Waya Mongo L (2008) Etude des effets de lisière sur les populations de rongeurs dans la reserve forestière de Masako. Mémoire présenté en vue de l'obtention du DEA en Sciences de la vie – Biologie végétale, Université libre de Bruxelles
Jaeger J (2000) Landscape division, splitting index, and effective mesh size: new measures of landscape fragmentation. Landsc Ecol 15:115–130
Ji J, Jiang H, Yang W (2008) National policy's influence on Baoxing county's landscape patterns and Giant Panda population. In: Chen J, Lui S, Lucas R et al (eds) Proceedings of the international conference "Landscape ecology and forest management", IUFRO8.01.02, Chengdu
Johnsson K (1995) Fragmentation index as a region based GIS operator. Int J Geogr Inf Syst 9:211–220
Kabulu Djibu JP, Bamba I, Munyemba F et al (2008) Analyse de la structure spatiale des forêts au Katanga. Ann Fac Sci Agron Univ Lubumbashi I(2):12–18
Kenkel NC, Walker DJ (1996) Fractals in the biological sciences. Coenoses 11:77–100
Koffi KJ, Deblauwe V, Sibomana S et al (2007) Spatial pattern analysis as a focus of landscape ecology to support evaluation of human impact on landscapes and diversity. In: Hong SK, Nakagoshi N, Fu B et al (eds) Landscape ecological applications in man-influenced areas. Linking man and nature systems. Springer, Dordrecht
Krummel JR, Gardner RH, Sugihara G et al (1987) Landscape patterns in a disturbed environment. Oikos 48:321–324
Li H, Reynolds JF (1994) A simulation experiment to quantify spatial heterogeneity in categorical maps. Ecology 75:2446–2455
Li H, Wu J (2004) Use and misuse of landscape indices. Landsc Ecol 19:389–399

Litucha Bakokola Makeu J (2007) Etude de l'effet de lisière sur le microclimat thermique d'un écosystème forestier de la cuvette centrale congolaise. Mémoire présenté en vue de l'obtention du DEA en Gestion de la biodiversité, Université de Kisangani

Lokonda O. Wa Kipifo M (2007) Etude de l'effet de lisière sur la réaction ou pH du sol dans un paysage fragmenté de la cuvette centrale congolaise. Mémoire présenté en vue de l'obtention du DEA en Gestion de la biodiversité, Université de Kisangani

Lovejoy S (1982) Area–perimeter relation for rain and cloud areas. Science 216:185–187

Magurran AE (2004) Measuring biological diversity. Blackwell, Oxford

Mandelbrot BB (1983) The fractal geometry of nature. Freeman and Company, New York

McGarigal K, Cushman SA, Neel MC et al (2002) FRAGSTATS: spatial pattern analysis program for categorical maps. Computer software program produced at the University of Massachusetts, Amherst. Available at the following web site: http://www.umass.edu/landeco/research/fragstats/fragstats.html

McIntyre S, Hobbs RJ (1999) A framework for conceptualizing human effects on landscapes and its relevance to management and research models. Conserv Biol 13:1282–1292

Monmonier MS (1974) Measures of pattern complexity for choroplethic maps. Am Cartogr 1:159–169

Naiman RJ, Holland MM, Decamps H et al (1988) A new UNESCO program: research and management of land: inland water ecotones. Biol Int 17:107–136

Noon BR, Dale V (2002) Broad-scale ecological science and its application. In: Gutzwiller KJ (ed) Applying landscape ecology in biological conservation. Springer, New York

O'Neill RV, Krummel JR, Gardner RH et al (1988) Indices of landscape pattern. Landsc Ecol 3:153–162

Odum HT (1983) Systems ecology: an introduction. Wiley, New York

Olsen ER, Ramsey RD, Winn DS (1993) A modified fractal dimension as a measure of landscape diversity. Photogramm Eng Rem S 59:1517–1520

Parks PJ (1991) Models of forested and agricultural landscapes: integrating economics. In: Turner MG, Gardner RH (eds) Quantitative methods in landscape ecology. Springer, New York

Patton DR (1975) A diversity index for quantifying habitat "edge". Wildlife Soc B 3:171–173

Ricotta C, Olsen ER, Ramsey RD et al (1997) A generalized non-regression technique for evaluating fractal dimension of raster GIS layers consisting of non square cells. Coenoses 12:23–26

Robinson JA, McRay B, Lulla KP (2000) Twenty-eight years of urban growth in North America quantified by analysis of photographs from Apollo, Skylab and Shuttle-Mir. In: Lulla KP, Dessinov LV (eds) Dynamic earth environments: remote sensing observations from shuttle-*Mir* missions. Wiley, New York

Rosenfeld A (1974) Compact figures in digital pictures. IEEE Trans Syst Man Cyb 4:221–223

Salvador-Van Eysenrode D, Bogaert J, Van Hecke P et al (1998) Influence of tree-fall orientation on canopy gap shape in an Ecuadorian rain forest. J Trop Ecol 14:865–869

Salvador-Van Eysenrode D, Bogaert J, Van Hecke P et al (2000) Forest canopy perforation in time and space in Amazonian Ecuador. Acta Oecol 21:285–291

Saura S, Torras O, Gil-Tena A et al (2008) Shape irregularity as an indicator of forest biodiversity and guidelines for metrics selection. In: Lafortezza R, Chen J, Sanesi G et al (eds) Pattern and processes in forest landscapes. Multiple use and sustainable management. Springer, New York

Smith RL, Smith TM (2000) Elements of ecology. Benjamin Cummings Science Publishing, San Francisco

Turner MG (1989) Landscape ecology: the effect of pattern on process. Annu Rev Ecol Syst 20:171–197

Turner MG, Dale V (1991) Modeling landscape disturbance. In: Turner MG, Gardner RH (eds) Quantitative methods in landscape ecology. Springer, New York

Turner MG, Gardner RH, O'Neill RV (2001) Landscape ecology in theory and practice. Pattern and process. Springer Science+Business Media Inc., New York

Urban DL, Wallin DO (2002) Introduction to Markov models. In: Gergel SE, Turner MG (eds) Learning landscape ecology. A practical guide to concepts and techniques. Springer, New York

Urban DL, O'Neill RV, Shugart HH Jr (1987) Landscape ecology. A hierarchical perspective can help scientists understand spatial patterns. Bioscience 37:119–127

Voss RF (1988) Fractals in nature: from characterization to simulation. In: Peitgen HO, Saupe D (eds) The science of fractal images. Springer, New York

Wu D (2007) Development and implementation of image processing algorithms related to the estimation of the interior-to-edge ratio. Thesis submitted in partial fulfilment of the requirements for the Master's degree of Applied Computer Science, Vrije Universiteit Brussel

Wu J, Hobbs RJ (2007) Landscape ecology: the state-of-the-science. In: Wu J, Hobbs RJ (eds) Key topics in landscape ecology. Cambridge University Press, Cambridge

Chapter 12
Analysis of Factors Affecting the Landscape Dynamics of Islands in Western Japan

Yoko Ohta and Nobukazu Nakagoshi

12.1 Introduction

The results of ecological research depend on the scale of the research (Turner et al. 2001). The scale that is important to humans is not necessarily important to other organisms or to ecological processes. Biological interactions occur at multiple scales. The landscape is a holistic ecosystem that is complicated by human and natural factors (Naveh and Lieberman 1984). A landscape can be analyzed at various scales, and the challenge is to select the best scale to solve regional problems.

The islands of Japan are isolated from economic centers, and in consequence, emigration has been a serious problem since World War II. Some islands have become uninhabited, and some have not been able to maintain their former rural activities. However, other islands have developed successful industries such as tourism, agriculture, and fishing (Suyama 2003).

The conservation and diversity of regional characteristics is critical to regional planning. Individual plans must be established for each region, and landscape ecology can contribute greatly to such planning (Ahern 1999).

In areas where humans have had a great influence on the landscape, socioeconomic information on changes in land use and human activities is an indispensable resource for understanding the changes in the landscape (Hong and Nakagoshi 1995; Nakagoshi and Hong 2001). Here, we analyze factors affecting landscape change on islands at three spatial scales, identify the relationships between them, and compare factors at different scales.

Y. Ohta (✉)
Open Network for Nature Conservation, Shimane 694-0064, Japan
e-mail: qzr11101@nifty.ne.jp

N. Nakagoshi
Graduate School for International Development and Cooperation, Hiroshima University, Higashi-Hiroshima 739-8529, Japan

S.-K. Hong, et al. (eds.), *Landscape Ecology in Asian Cultures*, Ecological Research Monographs, DOI 10.1007/978-4-431-87799-8_12, © Springer 2011

12.2 Seto Inland Sea

The Seto Inland Sea, the largest sea in Japan, lies in the western part of Japan, surrounded by Honshu, Shikoku, and Kyushu (latitude 33°20′–34°50′N, longitude 131°59′–134°29′E; Fig. 12.1). It is surrounded by nine prefectures: Osaka, Hyogo, Okayama, Hiroshima, Yamaguchi, Kagawa, Ehime, Fukuoka, and Oita. It covers about 21 800 km^2 and has a maximum depth of 105 m. The region has a mild and sunny climate; the annual average temperature is 15°C and the annual precipitation is 1 000–1 600 mm (Japan Meteorological Agency 2005). The Sea has 1 050 islands, of which 160 are inhabited (Sugata 1995). The islands range in size from less than 1 ha to 595 km^2 (Awajishima); most are small, ranging from 0.6 to 1 ha.

Because of its comparatively dry climate and its many islands, the Seto Inland Sea is often called the Mediterranean Sea of Japan. However, its density of marine resources (fish and shellfish) is far greater than that in the Mediterranean. Although the Seto Inland Sea has an area less than 0.3% of the Mediterranean, the population and level of industrial production are similar to those on the Mediterranean islands.

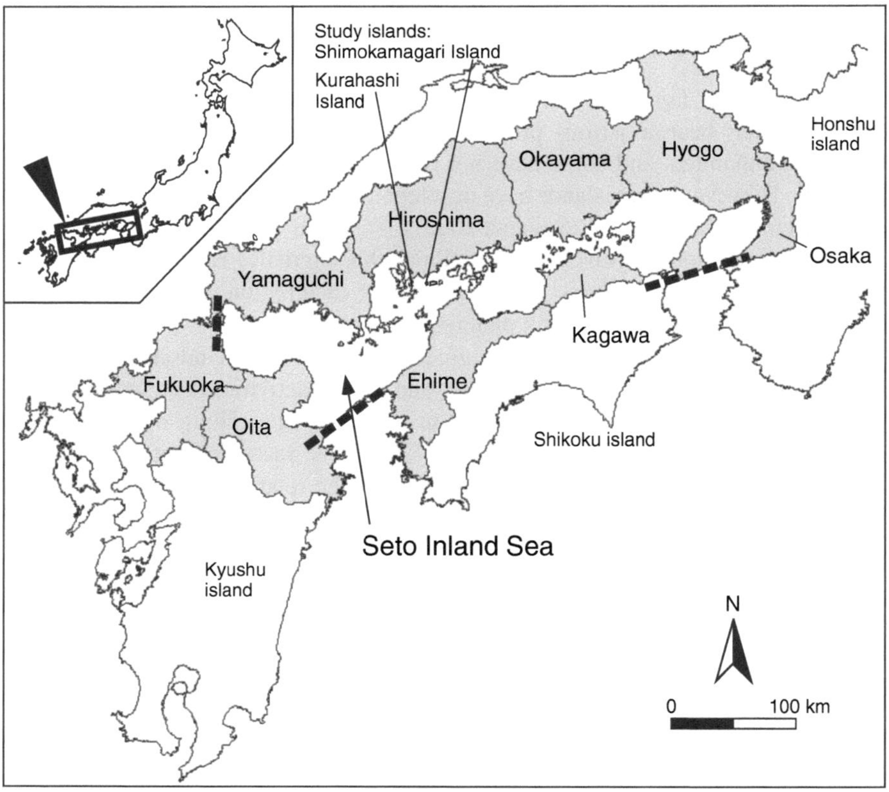

Fig. 12.1 Map of the Seto Inland Sea and survey area

In ancient times, such as the Heian Era, people composed odes to the landscape of the Seto Inland Sea, but they focused only on beautiful or historic sites. In the Middle Ages, foreign missionaries and sailors visited, but they had no interest in the landscape (Nishida 1995). In contrast, Asian visitors, such as Korean sailors, acclaimed the landscapes (Nishida 1996). From the 19th century, after Japan opened its doors to foreigners, many European and American visitors praised the islands spread on calm seas; the seashores, terraced fields, and traditional streets (Nishida 1999). These opinions were new to the Japanese people, who adopted them from the latter period of the Meiji Era (1868–1912). The modern practice of sightseeing altered the act of evaluation from understanding the meaning of a sight to recognizing it visually. Visitors viewed the entire landscape from the perspective of natural science, and valued its authenticity (Nishida 2004).

However, in the 1960s, a period of rapid economic growth, the landscape changed greatly. Modern tourism homogenized it, erasing its historical and cultural values. Many uniform developments sprang up, especially resorts. Many such plans failed, but the value of the landscape declined. Now illegal dumping of waste is a big problem (Seto Inland Sea group of reporters of Chugoku Shimbun 1998). These landscape changes had both physical and spiritual components.

12.3 Landscape Characteristics and Changes at Three Scales

12.3.1 Landscape Changes at the Regional Scale

Vegetation is an important element of the landscape because it represents the environmental conditions and forms the basis of the living world. At the regional scale, we analyzed the distribution of vegetation types. We prepared a map with a 1 km × 1 km grid from a vegetation map prepared during 1993–1998. These maps were published on the Internet by the Environment Ministry of Japan. It revealed many paddy fields and Japanese red pine (*Pinus densiflora*) forests, in addition to urban areas. We also prepared other maps of site conditions – bedrock, soil type, inclination, elevation, and slope direction – to analyze relationships with the vegetation distribution. Conjoint analysis showed the site conditions which contributed to the distribution of vegetation (Table 12.1). Conjoint analysis has been used in marketing studies, and it examines individual preferences for private and public goods which have multiple attributes (Nagashima and Nakagoshi 2001). This analysis is appropriate for evaluating which factor has more influence on determining the vegetation distribution. Natural site conditions, such as bedrock, elevation, and slope direction, are related to it.

Next we classified the islands by economics. Municipalities were classified into six groups according to industry (Ohta and Nakagoshi 2006). Primary industry, especially agriculture and fisheries, and population structure strongly influenced vegetation structure. The post hoc test showed that each group had a close association with

Table 12.1 Contribution of site conditions for the distribution of vegetation

Vegetation types	Multiple coefficient of determination (R^2)	*P* value	Range of regression coefficients				Attributes which made a high contribution to regression			
			Bed rocks	Elevation (m)	Inclination (degrees)	Slope direction	Bed rocks	Elevation (m)	Inclination (degrees)	Slope direction
Evergreen broad-leaved forest	0.54	0.000	4.23	4.54	–	7.27	Sedimentary rocks	<100	–	S
Deciduous broad-leaved forest	0.46	0.000	1.36	2.20	–	1.48	Sedimentary rocks	100–200	–	S
Pine forest	0.62	0.000	2.67	2.95	–	3.38	Felsic plutonic rocks	<100	–	S
Cedar and Japanese cypress plantation	0.51	0.000	2.51	1.64	–	2.68	High pressure metamorphic rocks	600 <	–	S
Pine plantation	0.57	0.000	7.56	4.73	–	3.80	Mafic volcanic rocks	100–200	–	S
Orchard	0.36	0.001	1.80	3.14	2.00	3.13	Felsic plutonic rocks	<100	3–8	S
Paddy field	0.43	0.043	10.08	9.54	–	11.05	Sedimentary rocks	<100	–	S
Urban areas	0.36	0.000	1.83	2.08	2.35	3.30	Sedimentary rocks	<100	< 3	S

a particular vegetation type. The post hoc value indicates the difference in the observed frequency of each group between the actual distribution and the expected value (Nagata 1999). Thus, a positive value of the cell contribution ratio shows that a group prefers the vegetation type, and a negative value shows avoidance. For example, group D (municipalities where fruit production is prosperous) related to pine forests and orchards. However, they had no relation to evergreen broad-leaved forest and paddy fields (Table 12.2).

Figure 12.2 shows the landscape elements and the factors affecting landscape change at the regional scale. The elements were classified into society, the inorganic environment, and vegetation. Society was characterized by industrial characteristics and social structure. The inorganic environment was characterized by sea area, coastal land area, and warm climate. Vegetation was characterized by the dominance of secondary vegetation. Society influenced the vegetation by using land for activities associated with living (home-building, farming, ranching) and by causing forest fires. The characteristics of the inorganic environment directly influenced the establishment of vegetation types, and were related to natural succession. Agricultural land use and forest fires re-set natural succession to the first stage (Kamada et al. 1991), and helped to maintain the traditional landscapes in the region. Society also had serious impacts on the inorganic environment through the use of natural resources, such as mining, stone cutting, sand mining, and landfill, with consequent irreversible land use. Moreover, islands lacking in natural resources

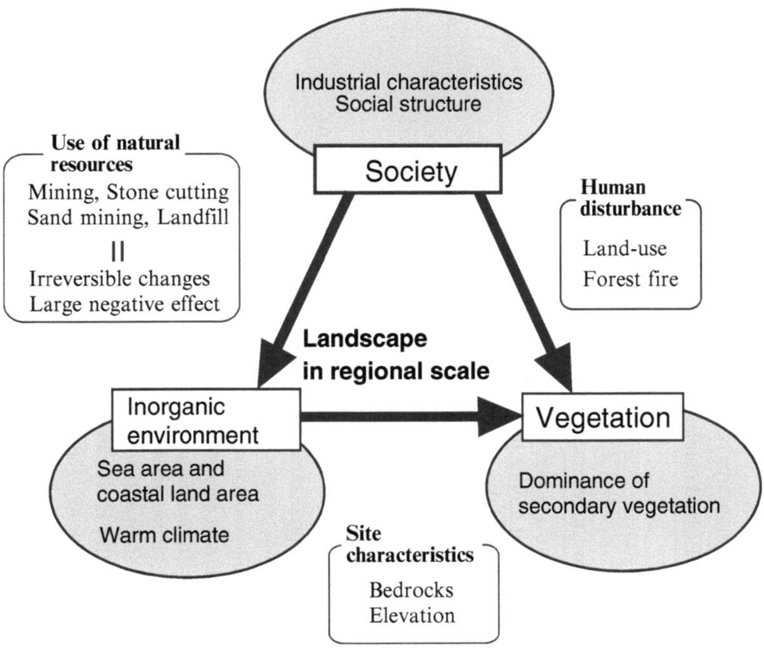

Fig. 12.2 Structure of landscape and factors causing changes at the regional scale

Table 12.2 Relationships between municipal groups and vegetation types

Classification of island municipality		Vegetation types							
Group	Industrial characteristics	Pine forest	Evergreen broad-leaved forest	Deciduous broad-leaved forest	Pine tree plantation	Cedar and Japanese cypress plantation	Orchard	Paddy field	Urban area
A	Manufacturing, mining, and service industries are prosperous. Ratio of elderly people is large	−2.22	−4.57	1.04	2.54	6.32	−2.01	1.59	0.75
B	Farmers have a large income from agriculture	−2.34	−0.75	0.87	−0.91	−3.75	28.97	−7.62	−5.09
C	Agriculture is declining. Fruit production is not prosperous	−9.17	19.61	0.06	6.40	−6.78	−6.70	10.68	−1.61
D	Fruit production is prosperous	5.94	−8.70	−1.87	−3.19	−2.62	4.88	−5.62	0.38
E	Farmers earn more from sideline businesses than from agriculture	4.65	−2.73	0.64	−3.83	5.21	−11.04	−1.55	2.32
F	Fishery is also an important industry. Ratio of elderly people is small	−2.87	−1.82	−0.23	2.46	−2.17	2.70	3.42	0.65

were used for the disposal of industrial wastes or for the construction of atomic power plants. These activities imposed dynamic changes on the landscape.

The changes in the inorganic environment also indirectly caused vegetation changes, and further influenced society by degradation of the environment. In summary, Fig. 12.2 shows that the landscape elements were closely related to each other, and that both society and the inorganic environment influenced the vegetation.

12.3.2 Landscape Changes at the Municipal Scale

We mapped the vegetation of Shimokamagari Island, Hiroshima Prefecture, from just after the end of World War II to recent times. The island's dominant vegetation types throughout this period were mandarin orange (*Citrus unshiu*) orchards and tall (≥8 m) red pine forests (Nakagoshi and Ohta 1992; Ohta and Nakagoshi 2003). Changes in the vegetation of the island were closely related to economic changes (Nakagoshi and Ohta 1991; Fig. 12.3). During the period of rapid economic growth, mandarin oranges were grown commercially in terraced orchards, which were converted from upland fields. Mandarin production was prosperous on the island, and the Japanese traditional *Satoyama* landscape (mixed forests and fields) dominated by red pine forests prevailed (Kamada and Nakagoshi 1996). Natural materials became unnecessary because of fossil fuels and chemical fertilizers, so the use of forest materials largely stopped. As a result, the abandoned forests underwent succession: the red pine forests changed into deciduous broad-leaved forests. During the 1970s, the price of mandarin oranges decreased considerably, and farmers began to abandon the orchards. Kudzu (*Pueraria lobata*) invaded and rapidly occupied the abandoned orchards. Bamboo (*Phyllostachys* sp.) also invaded abandoned fields and forests.

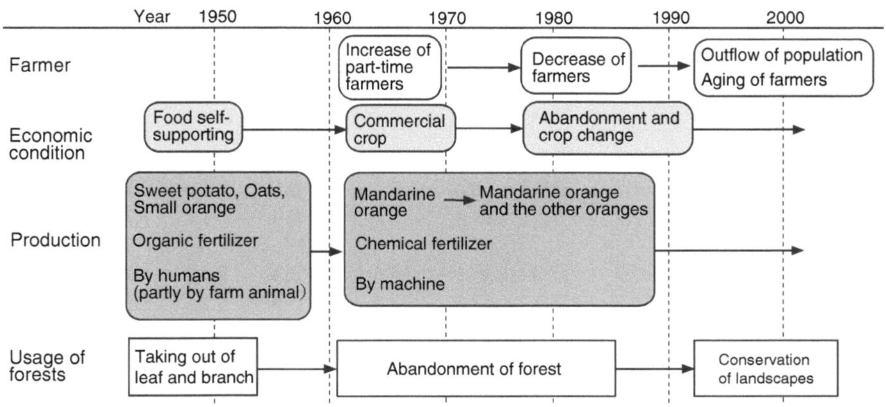

Fig. 12.3 Social changes in Shimokamagari-cho town

Natural site conditions played important roles here. Orchards that were abandoned were inaccessible because they were at a high altitude or had steep slopes. The forest succession was influenced by bedrock, soil, and surface compaction.

We prepared a Markov chain model (Turner and Dale 1991) to predict the area of vegetation in the near future (Nakagoshi and Ohta 2000). We then tried to compare the expected value in 2004 with the actual value in 2002. Table 12.3 shows the results of this comparison and the areal changes in vegetation. The orchards remaining in 2002 were larger than the value expected in 2004. However, tall, medium, and small red pine forests decreased greatly, and these areas were smaller than expected. On the other hand, some vegetation types, such as tall oak forest, bamboo plantation, and upland fields increased, and their areas were larger than expected. Concerning the Kudzu (*P. lobata*) community, the actual area was almost the same as expected. In particular, vegetation types in urban areas, such as grounds and parks, roads and parking areas, and bare land, were also greater than the

Table 12.3 Changes in vegetation at three periods and the difference from prediction on Shimokamagari Island

	Area (ha)				Proportion (%)			
Vegetation type	1976	1990	2002	Expected value in 2004	1976	1990	2002	Expected value in 2004
Orchard	341.3	252.0	198.4	161.7	44.1	32.3	25.0	20.5
Tall pine forest (≦8 m)	263.2	248.9	198.0	246.7	34.0	31.9	25.0	31.3
Medium pine forest (3–8 m)	16.0	11.5	9.7	12.7	2.1	1.5	1.2	1.6
Small pine forest (<3 m)	7.8	0.8	0.1	0.7	1.0	0.1	0.0	0.1
Tall oak forest (≦8 m)	55.9	92.9	127.2	141.7	7.2	11.9	16.0	18.0
Bamboo plantation	7.0	11.9	29.6	16.0	0.9	1.5	3.7	2.0
Pueraria lobata community	6.6	73.0	97.3	96.3	0.9	9.4	12.3	12.2
Erigeron canadensis –E. sumatrensis community	4.5	15.9	6.8	23.3	0.6	2.0	0.9	3.0
Upland fields	2.5	7.8	12.6	5.9	0.3	1.0	1.6	0.7
Paddy fields	4.7	0.9	0.4	0.1	0.6	0.1	0.1	0.0
Grounds and parks	2.7	6.2	13.4	8.2	0.4	0.8	1.7	1.0
Inhabited areas	25.7	27.3	39.1	33.5	3.3	3.5	4.9	4.3
Roads and parking areas	19.3	25.1	43.6	29.5	2.5	3.2	5.5	3.8
Bare land	5.2	2.1	9.2	1.1	0.7	0.3	1.2	0.1
Others	12.1	4.7	7.9	9.4	1.6	0.6	1.0	1.2
Total number of meshes	774.6	781.0	793.4	787.1				

expected areas. They had increased with the construction of bridge and wide roads going round the islands. The Markov chain model used the transition probability matrix during 1976–1990 for the prediction. The difference between the actual area and the expected area indicated that the factors affecting vegetation changes were different from the assumption. It was suggested that human activities were impossible to predict, and often promoted the irreversible changes.

Figure 12.4 shows the landscape elements and the factors affecting landscape change at the municipal scale. The society was characterized by social and economic conditions and bridge construction between the island and the mainland. The inorganic environment was characterized by limited land area and the neighboring sea. Vegetation was characterized by changes to orchards and red pine forests. In the absence of forest fires or mining, disturbances affecting landscape change were comparatively small and were limited to agricultural land use, forest cutting, and the construction of roads. However, the connection of the island to the mainland encouraged the exchange of materials and people, and permitted the invasion of foreign species. The construction of roads, parking lots, and parks greatly altered the inorganic environment. Thus, society influenced the vegetation through land use and invasive species. In the absence of large and irreversible disturbance to the inorganic environment, vegetation types favored specific site conditions.

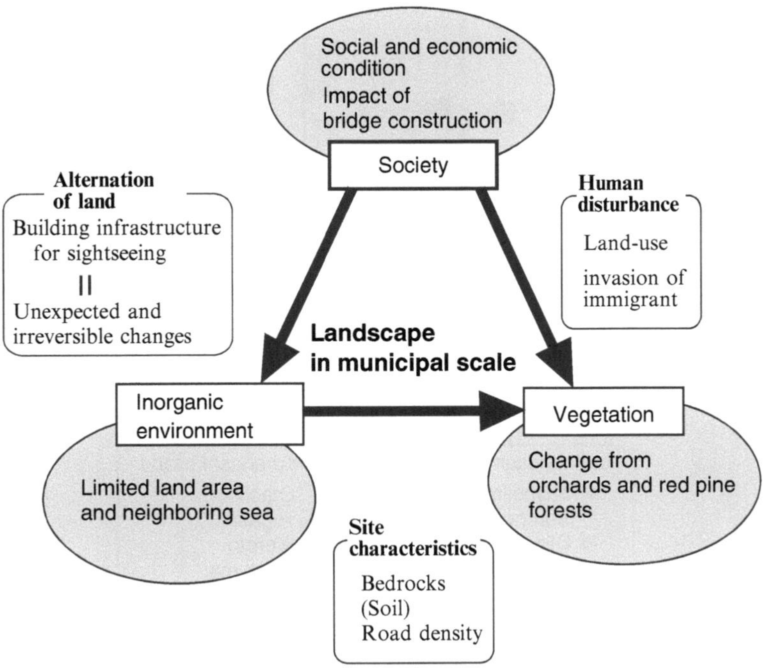

Fig. 12.4 Structure of the landscape and factors causing changes at the municipal scale

12.3.3 Landscape Changes at the District Scale

The smallest unit of a community is a settlement. In Japan, the district is the smallest administrative unit, which is on a par with a settlement. An island district has its own characteristics. For example, people feel more affinity with a settlement on the island opposite than with one on the opposite side of the same island (Nan-nan-sya, 2007). The results of analyses at this scale were reflected in the characteristics of industrial structure and the way the vegetation change differed from district to district (Nakagoshi and Ohta 2001).

The similarities in agricultural conditions in seven districts on two islands are shown in Fig. 12.5. This graph was the result of multidimensional scaling with the statistical data of 1970, 1985, and 2000. Multidimensional scaling (MDS) is a set of related statistical techniques for exploring similarities or dissimilarities in data (Jongman et al. 1995). We extracted statistical data from the population censuses and the world agricultural and forestry industries censuses. The horizontal axis shows the decrease in farmers and agricultural populations, and the vertical

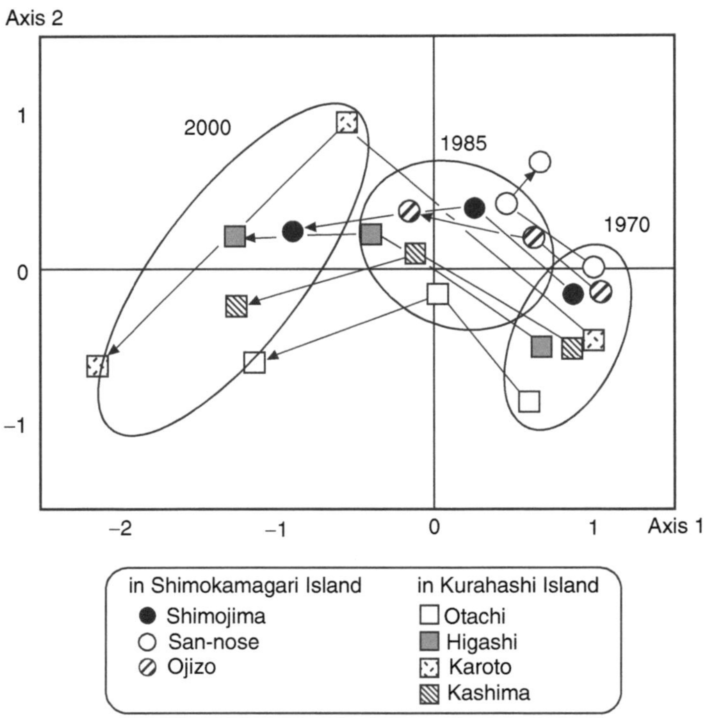

Fig. 12.5 Changes in agricultural conditions in districts on Shimokamagari and Kurahashi Islands

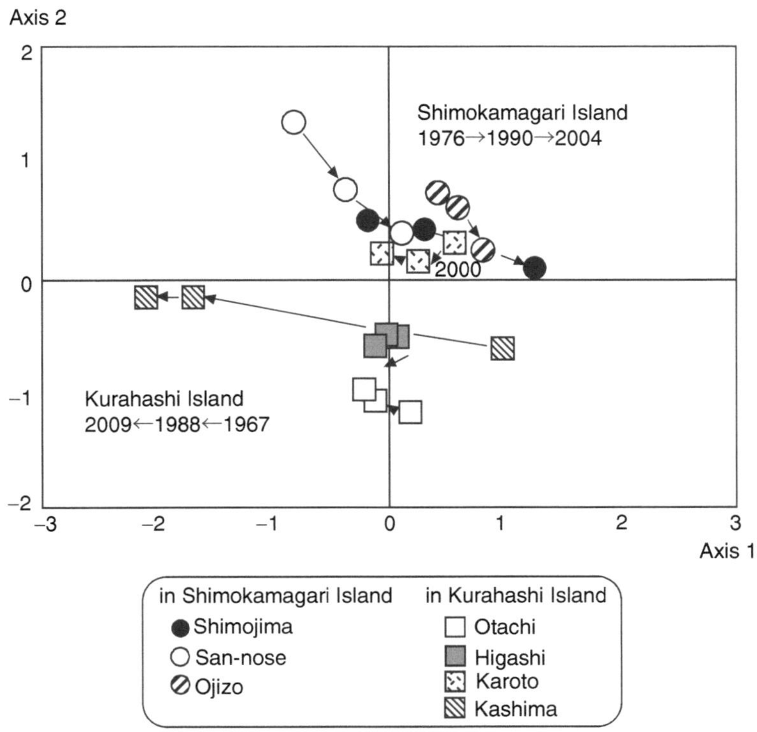

Fig. 12.6 Changes in vegetation composition in districts on Shimokamagari and Kurahashi Islands

axis shows the increase in full-time farmers. Most districts experienced a decline in agriculture, but the ways in which they had changed differed. Differences in vegetation composition among districts are shown in Fig. 12.6. The horizontal axis represents natural succession, and the vertical axis represents the decrease in farmland. The patterns of vegetation change were completely different between the two islands. However, site conditions were related to the abandonment of orchards in both islands.

Figure 12.7 shows the landscape elements and the factors affecting landscape change at the district scale. The society was characterized by its social and industrial structure and community history. The inorganic environment was characterized by the circumstances of the settlement. The vegetation was characterized by changes from orchards and red pine forests. Factors affecting landscape change at the district scale were similar to those at the municipal scale, but the building of infrastructure for sightseeing had an additional effect on the landform, The arrival of pine wilt disease also characterized human influence, and the site inclination influenced the orchards.

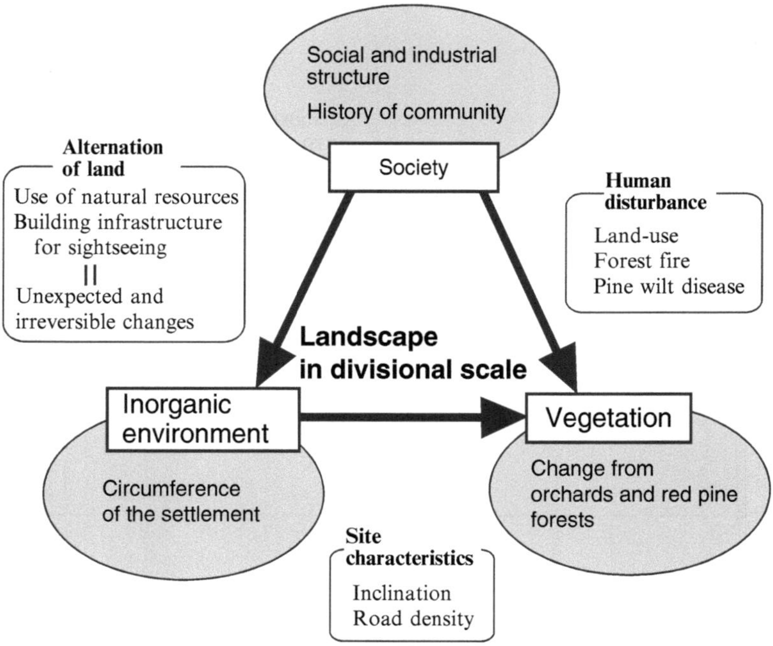

Fig. 12.7 Structure of the landscape and factors causing changes at the district scale

12.4 Influence of Landscape Changes on the Regional Ecosystem

In rural areas of Japan, communities of secondary vegetation have been maintained as a result of the disturbance created by the dominant industries, such as agriculture. Secondary forests have ecologically important functions, such as the conservation of biodiversity and landscapes, but are also an important forestry resource. People have used the land under specially defined management rules (Kamada and Nakagoshi 1990; Hong and Nakagoshi 1995), contributing to the maintenance of secondary vegetation. Farmers in particular have used various plant materials in forests for energy or feed (Kamada et al. 1991).

Human intervention maintained red pine forests in Japan for hundreds of years (Nakagoshi 1988). However, economic conditions changed greatly after the 1960s, when economical alternatives to wood as a fuel source and fertilizer became widely available. Many red pine forests were no longer actively managed and began to change into deciduous broad-leaved forests as a result of natural succession. On the other hand, terraced fields on steep slopes were abandoned (Nakagoshi and Ohta 2000). As a result, traditional landscapes, with pine trees and terraced orchards, disappeared rapidly.

Qualitative vegetation changes preceded quantitative changes as the vegetation sequence progressed (Ohta and Nakagoshi 2007). Figure 12.8 shows the similarities

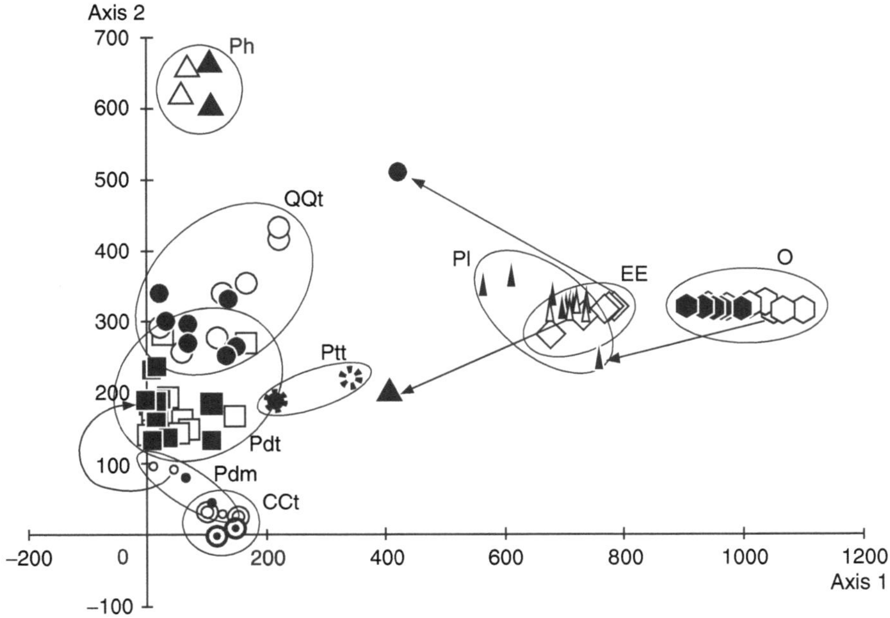

O: orchrds, EE: *Erigeron canadensis - E. smatrreis* community, Pl: Kudzu (Pueraria lobata) community, Ph: Bamboo (*Phyllostachys* sp.) stands, Ptt: Tall black (*Pinus thunbergii*) pine forest, Pdt: Tall red pine (*P. densiflora*) forest, Pdm: Medium red pine forest, QQt: Tall oak (*Quercus* sp.) forest, CCt: Tall Japanese chinquapin (*Castanopsis cuspidata*) forest.

Fig. 12.8 Qualitative changes in vegetation types during 1990–2002 on Shimokamagari Island

of composition between various vegetation types between 1990 and 2002. Axis 1 represents the stage of natural succession, from abandoned orchards to forest vegetation. Axis 2 represents species richness, ranging from bamboo plantation to tall *Castanopsis cuspidata* (Japanese chinquapin) forest communities.

The plots of red pine forests were more concentrated in 2002 than in 1990 (Fig. 12.8). The plots of deciduous broad-leaved forests showed the same trend. In addition, the distances between the plots of red pine forests and plots of deciduous broad-leaved forests became shorter in 2002. *Castanopsis cuspidata* forests, which are the climax forest vegetation in this region, form the central group.

On the other hand, the wide distances between plots of *C. cuspidata* forests and plots of deciduous broad-leaved forest suggest that deciduous broad-leaved forests cannot change into *C. cuspidata* forests by succession. The low supply of seeds from other sites contributed to this phenomenon. Landscape changes in the Seto Inland Sea region were determined by human activities and the natural environment, but conversely, did the changed landscape influence society and the environment?

Kudzu (*P. lobata*) communities appeared at many sites such as abandoned fields, construction fields, and road embankments. These communities and bamboo stands spread rapidly across the islands, and resisted change into other community

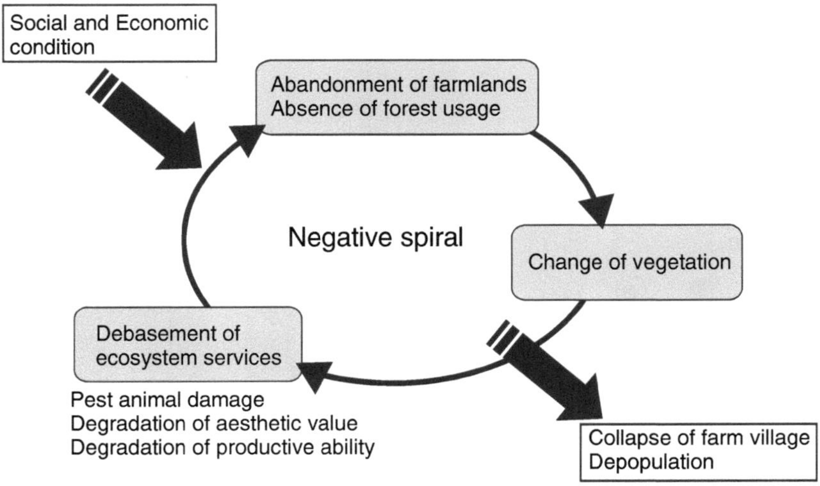

Fig. 12.9 Problems causing landscape change on Shimokamagari

vegetation types. They provide homes for pest animals and birds, such as Japanese raccoons, boars, and bulbuls. These animals attack agricultural fields, and push farmers to abandon their fields. This vicious cycle is shown in Fig. 12.9. Landscape changes include not only damage by agriculture, but also degradation of historic value and land conservation functions. Costanza et al. (1997) called this the degradation of ecosystem services. It affects the lives of residents and fosters emigration from the islands.

12.5 Conclusions

Landscape change is intimately linked with human activities. Because the available land is restricted on islands, residents use it intensively. Pine forests and deciduous broad-leaved forests are common secondary vegetation in the Seto Inland Sea region (Nakagoshi 1995). The predominance of red pine forests depends on natural site conditions and human impact.

Terraced orchards and red pine forests are widespread throughout Shimokamagari, and are important elements of the island's traditional and cultural landscapes (Nakagoshi and Ohta 2001). Qualitative changes in the vegetation have occurred. Through natural succession, some sites have progressed toward their original climax vegetation types. The remnant orchards and red pine forests are located on sites that were originally good places where they could maintain themselves. To protect landscapes that contain terraced orchards and red pine forests, we must carefully choose similar good sites and manage them intensively to prevent succession to undesirable plant communities.

At all three spatial scales we identified three broad categories of landscape elements: society, the inorganic environment, and vegetation. Their contributions differed slightly from scale to scale, and interrelationships were identifiable.

To determine whether the elements had the same relationships at all scales, we plotted their configurations (Fig. 12.10). At the regional scale, human activities included various industries and other characteristics. Agriculture strongly influenced landscape changes at the municipal scale. Community history and social structure, in addition to agriculture, were predominant at the district scale. Figure 12.10 shows that elements at the larger scale imposed limits on the elements at the smaller scale.

The inorganic environment had configurations in space. Because the site conditions were often integrated at the higher scales, the results at any scale had the same

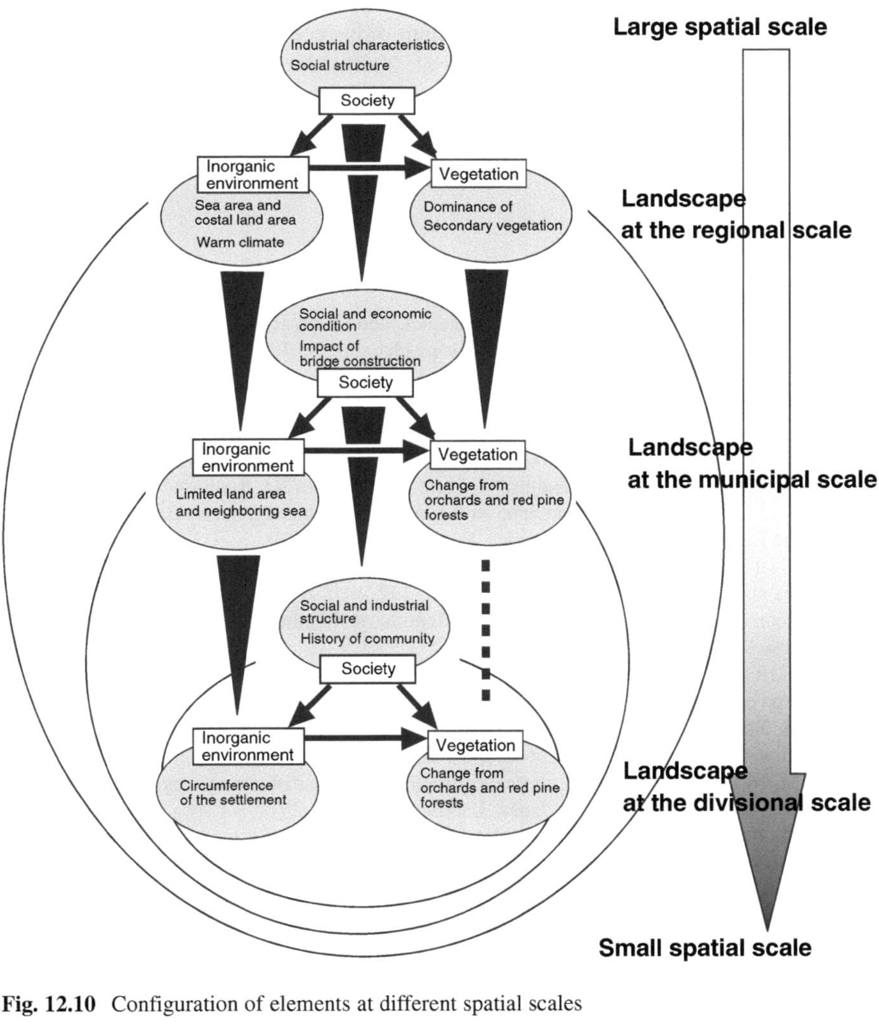

Fig. 12.10 Configuration of elements at different spatial scales

configuration. Secondary vegetation types were important at the regional scale, and dominated orchards and red pine forests (secondary vegetation types) at the district and municipal scales, which showed no configuration.

The results suggest that the elements at larger scales limited and defined the elements at smaller scales. Conversely, the superior elements were made up of the inferior elements. These results confirm the hierarchical concepts of O'Neill et al. (1986) and Turner et al. (2001). Moreover, conditions that were important at one scale were not necessarily important at another scale. This presents problems for ecological landscape studies, and may pose problems for regional landscape planning. In the Seto Inland Sea region, the best spatial scale for an analysis of the regional landscape was the municipal scale on account of the amount of information, the ease of defining spatial boundaries, and feasibility of the regional plans.

To understand landscape change, especially the loss of traditional and cultural landscapes, it is necessary to integrate social, economic, and ecological aspects. However, research into local planning began only recently, and there is as yet no agreement on the best approach (Dale et al. 2000). We suggest the importance of collecting information and reconstructing conservation systems along present lines in order to sustain rural management.

References

Ahern J (1999) Spatial concepts, planning strategies, and future scenarios: a framework method for integrating landscape ecology and landscape planning. In: Klopatek JM, Gardner RH (eds) Landscape ecological analysis, issues and applications. Springer, New York, pp 175–201

Costanza R, d'Arge R, de Groot R, Farber S, Grasso M, Hannon B, Limburg K, Naeem S, O'Neill RV, Paruelo J, Raskin RG, Sutton P, van den Belt M (1997) The value of the world's ecosystem services and natural capital. Nature 387:253–260

Dale VH, Brown S, Haeuber RA, Hobbs NT, Huntly N, Naiman RJ, Riebsame WE, Turner MG, Valone TJ (2000) Ecological principles and guidelines for managing the use of land: a report from the Ecological Society of America. Ecol Appl 10:639–670

Hong SK, Nakagoshi N (1995) Human impacts on pine-dominated vegetation in rural landscapes in Korea and western Japan. Vegetatio 116:161–172

Japan Meteorological Agency (2005) Annual Reports of Weather 2005. Japan Meteorological Bussiness Support Center, Tokyo, 271 pp

Jongman RHG, Ter Braak CJF, Tongeren FRV (eds) (1995) Data analysis in community and landscape ecology. Cambridge University Press, Cambridge, 299 pp

Kamada M, Nakagoshi N (1990) Patterns and processes of secondary vegetation at a farm village in Southwestern Japan after the 1960s. Jpn J Ecol 40:137–150

Kamada M, Nakagoshi N (1996) Landscape structure and the disturbance regime at three rural regions in Hiroshima Prefecture, Japan. Landsc Ecol 11:15–25

Kamada M, Nakagoshi N, Nehira K (1991) Pine forest ecology and landscape management: a comparative study in Japan and Korea. In: Nakagoshi N, Golly FB (eds) Coniferous forest ecology from an international perspective. SPB Academic Publishing, The Hague, pp 43–62

Nagashima K, Nakagoshi N (2001) Forestry expansion and land-use patterns in the Nelson Region, New Zealand. Landsc Ecol 16:719–729

Nagata O (1999) StatVIew – manuals for medical science and statistics. Shinko Trading Co. Ltd. Publication Department, Tokyo, 248 pp (in Japanese)

Nakagoshi N (1988) Pattern and process of secondary forests in Japan. Geogr Sci 43(3):9–14
Nakagoshi N (1995) Pine forests in East Asia. In: Box EO, Peet RK, Masuzawa T, Yamada I, Fujiwara K, Maycock PF (eds) Vegetation science in forestry. Kluwer, Dordrecht, pp 85–104
Nakagoshi N, Hong SK (2001) Vegetation and landscape ecology of East Asian 'Satoyama'. Glob Environ Res 5:171–181
Nakagoshi N, Ohta Y (1991) The eco-geography in Shimokamagari-cho. Furusato Shimokamagari, ser. 17. Shimokamagari-cho, Hiroshima, 55 pp (in Japanese)
Nakagoshi N, Ohta Y (1992) Factors affecting the dynamics of vegetation in the landscapes of Shimokamagari Island, southwestern Japan. Landsc Ecol 7(2):111–119
Nakagoshi N, Ohta Y (2000) Predicting future landscape of islands in the Seto Inland Sea, Japan. In: Mander Ü, Jongman RHG (eds) Landscape perspectives of land use changes. WIT Press, Southampton, UK, pp 83–106
Nakagoshi N, Ohta Y (2001) Landscape dynamics of the orange-producing islands in the Seto Inland Sea, Japan. In: van der Zee D, Zonneveld IS (eds) Landscape ecology applied in land evaluation, development and conservation. ITC, Enschede, The Netherlands, pp 95–113
Nan-nan-sya (ed) (2007) Dictionary of Seto Inland Sea. Nan-nan-sya, Hiroshima, 590 pp (in Japanese)
Naveh Z, Lieberman AS (1984) Landscape ecology: theory and application. Springer, New York, 356 pp
Nishida M (1995) Germination of new landscape viewpoints in the Seto Inland Sea of the late Edo Era. J Jpn Inst Landsc Archit 58(5):33–36 (in Japanese with English abstract)
Nishida M (1996) Modern Westerners' admiration of the Seto Inland Sea Landscape. J Jpn Inst Landsc Archit 59(4):298–309 (in Japanese with English abstract)
Nishida M (1999) Discovery and popularization of modern landscape in the Seto Inland Sea. J Jpn Inst Landsc Archit 63(1):17–24 (in Japanese)
Nishida M (2004) Changes of gaze at natural landscape and new viewpoints of landscape. J Jpn Inst Landsc Archit 68(2):130–133 (in Japanese)
O'Neill RV, DeAngelis DL, Waide JB, Allen TFH (1986) A hierarchical concept of ecosystems. Princeton University Press, Princeton, 262 pp
Ohta Y, Nakagoshi N (2003) Actual vegetation of Shimokamagari-cho, Hiroshima Prefecture. Mem Fac Integr Arts and Sci, Hiroshima Univ, Ser. IV 29:105–114
Ohta Y, Nakagoshi N (2006) Landscape changes in the Seto Inland Sea, Japan. Ekologia 25(Suppl 1):190–200
Ohta Y, Nakagoshi N (2007) Qualitative changes in the plant communities of Shimokamagari Island during a 12 – year period. Hikobia 15:37–52
Seto Inland Sea group of reporters of Chugoku Shimbun (ed) (1998) Walk in Seto Inland Sea region. Chugoku Shimbun, Hiroshima, 314 pp (in Japanese)
Sugata M (1995) Dictionary of islands in Japan. Sanko-sha, Tokyo, 495 pp (in Japanese)
Suyama S (2003) Quantitative regional division of islands. In: Hiraoka A (ed) Research into people, life and industry of Japanese islands. Kaiseisha Press, Ohtsu, 218 pp (in Japanese)
Turner MG, Dale VH (1991) Modeling landscape disturbance. In: Turner MG, Gardner RH (eds) Quantitative Methods in Landscape Ecology. Springer-Verlag, New York, pp 323–351
Turner MG, Gardner RH, O'Neill RV (2001) Landscape ecology in theory and practice: pattern and process. Springer, New York

Chapter 13
Creating Pondscapes for Avian Communities: An Artificial Neural Network Experience Beyond Urban Regions

Wei-Ta Fang

13.1 Introduction

Taoyuan Tableland farm ponds are ecologically significant because one-fifth of all the bird species find a home on these ponds (Fang 2004a). This tableland, at an area of 757 km^2, is an area of 2 898 ha of farm ponds on the northwestern portion of Taiwan. Located approximately 30 km from the capital city of Taipei, this rural area was easy to convert into urban land owing to the aggregated effects of urbanization and commercialization. Socioeconomic benefits are driving public opinion, which is urging the government to approve land-use conversion from farmlands into urban uses. Since Taiwan was accepted as a member of the World Trade Organization (WTO) in January 2002, agricultural products such as rice and flour may be imported at relatively lower prices compared with domestic products.

Meanwhile, the Council of Agriculture has assumed a policy of "precision agriculture" and is promoting the release of marginal rice fields for conversion into other urban uses. According to farmland release polices, in response the Water Resource Agency (WRA), the Ministry of Economic Affairs is considering reducing the water-supply ratio from a high of 69% for agricultural water to a moderate level of 59% of the total amount of water supplied, and then increasing the amount and ratio of industrial and residential water use in the future. Without farmland irrigation pressures demanding water supplies, the agency predicted that agricultural water use would drop from 12.4 to 11.8 billion tons/year from 2004 to 2020. This would cause farm ponds associated with farmlands to be rapidly converted into urban and industrial developments as a result of commercial driving forces.

Regarding farm-pond conversion, wintering bird refuges represent one of the multifunctional perspectives in the restoration of agro-ecosystems. Therefore, an organized survey was conducted for this study to resolve certain issues of the area per se hypothesis (MacArthur and Wilson 1967; Simberloff and Abele 1976; Connor and McCoy 1979). In their studies, this area per se hypothesis was also the

W.-T. Fang (✉)
Chung Hua University, Hsinchu 30012, Taiwan, Republic of China
e-mail: wawaf@hotmail.com

S.-K. Hong, et al. (eds.), *Landscape Ecology in Asian Cultures*, Ecological Research Monographs, DOI 10.1007/978-4-431-87799-8_13, © Springer 2011

major factor explaining species diversity, and these results were adopted to design water regimes for vernal pools and wetland bird refuges. Simberloff and Abele (1976) noted that as the size of a pond increases, the likelihood of discovering new creatures increases, thereby increasing its species diversity. In contrast, for species–area relationships, Forman (1995) argued that the ecological spatial form was not simply shaped by the effect of area per se. Some authors inferred that the causes of species diversity were affected by habitat heterogeneity. Apart from area, all the parameters of pond configuration can be condensed to fall into five categories that address eco-physical identities, these are shape, depth, edge, clustering, and connectivity in the landscape scale (Forman 1995; Linton and Boulder 2000; Leitão and Ahern 2002; Oertli et al. 2002; Fang et al. 2009).

This study attempts to investigate bird diversity and its relationship to the attributes of ponds. The results will be useful in ecological planning to solve the aforementioned problems created by anthropogenic driving forces. Hence, the objective of the study was to characterize the diversity of bird species associated with these ponds, and whose likelihood of survival was assessed along the gradient of land development intensities. Such characterization helps establish the decision criteria needed for designating certain ponds for habitat preservation and developing strategies for their protection.

13.2 Materials and Methods

13.2.1 Pondscape Analysis

Farm-pond areas are dominated by fields separated by hedgerows and windbreaks, and in Taoyuan there are scattered woods of various sizes. The distributions of avian species within such land mosaics are correspondingly discontinuous, depending on the locations of preferred habitats, density-dependent processes, and the quality of individual patches. The presence of trees and shrubs is a critical feature for migrants and residents, and densities of both hedgerows and woodlands are good predictors of avian communities. These configurations are surrounded by built-up areas, rivers, roads, and farmland. Given the remarkable diversity of farm-pond configurations and the complexities of their compositions, a *pondscape* is defined as "a series of water surfaces of ponds in association with various surrounding landforms, including farms, creeks, canals, roads, houses, woodlands, and other open spaces," according to the reports of some authors (Froneman et al. 2001; Francl and Schnell 2002).

Located between natural areas and urban areas, pondscapes in agricultural fields create ecotones from aquatic plants, windbreaks (i.e., the families Casuarinaceae and Meliaceae), anthropogenic crops, and fragmented woodland cover (i.e., the families Moraceae and Euphorbiaceae). These areas, which are associated with the rotation of crops in agricultural fields, include food and horticultural uses. In pondscape studies, selected parameters are used to measure the spatial arrangement of wooded and aquatic landscapes, and to compare any significant differences between them. These parameters have been used to measure temporal changes in actual landscapes, as well as changes in intensively used landscapes (Li and Reynolds 1994; Leitão and Ahern 2002). Many studies on pondscapes have focused on spatial configurations such as biodiversity, pond size, pond depth, pond shape, and pond sustainability. Pondscape parameters such as total area (TA), largest pond index (LPI), mean pond size (MPS), mean shape index (MSI), and mean pond fractal dimension (MPFD) should be considered. Descriptive statistics are used for statistical data processing to some extent in order to combine parameters in spatial analysis (Halsey et al. 1986; Evertsz and Mandelbrot 1992; Gustafson and Parker 1992; Cheng and Agterberg 1995; Kronert et al. 2001; McGarigal et al. 2002). During study periods, pond areas and the associated land were tested to determine whether there is a relationship between pond size and the numbers of species and individuals, and between the richness and diversity as determined by standardized sampling units (see Table 13.1).

Table 13.1 Definition and description of parameters used in the patch analysis of factors explaining remaining/disappeared farm ponds and the influences on birds

Types	Item acronym	Pondscape parameters (metrics/units)	Spatial pattern
Micro-scale	[a]PS	[b]Pond size	Pond size
	[a]LPI	[b,c]Largest pond index	Pond size
	[a]MPS	[b,c]Mean pond size (ha)	Pond size
	[a]NP	[b,c]Numbers of ponds	Pond size
	[a]MPFD	[b,c]Mean pond fractal dimension	Pond shape
	[a]MSI	[b,c]Mean shape index	Pond shape
	[a]ED	[b,c]Edge density (m/ha) of selected ponds (from TE)	Pond shape
	[a]TE	[b,c]Total parameters (in coverage units) of selected pond edges	Pond shape
	[a]FCA	[b]Foliage canopy area next to the waterfront edge of a pond (m^2)	Boundary delineation of disturbance
	[a,d]MUDA	[b]Mudflat area in a pond (m^2)	Boundary delineation of disturbance
	[a]%MUDA	[b]MUDA ÷ PS	
	[a,d]WASA	[b]Water surface area in a pond (m^2)	Boundary delineation of disturbance
Types	Item acronym	Pondscape parameters (metrics/units)	Spatial pattern
Meso-scale	[a]%FCA	[b]FCA ÷ PS	
	[a]%WASA	[b]WASA ÷ PS	
	[d]ELVA	[b]Elevation (m)	

(continued)

Table 13.1 (continued)

Types	Item acronym	Pondscape parameters (metrics/ units)	Spatial pattern
	[d]PERI	[b]Perimeter (m)	
	[d]TOPO	Waterfront topology (bold line represents pavement) Raised:  Level: 	
	[a]D-U	[b,e]Distance (m) to urban limit	Pond isolation from sources
	[a]D-C	[b,e]Distance (m) to coastline	Pondscape connectivity to sources
	[a]%FARM	[b,e]The ratio of farmland areas within a radius of 100 ha from the pond's geometric center (m^2)/ha	Pondscape isolation or connectivity from sources
	[a]%BUILD	[b,e]The ratio of permanent building areas within a radius of 100 ha from the pond's geometric center (m^2)/ha	Pondscape isolation from sources
	[a]%PONDS	[b,e]The ratio of multiple pond areas within a radius of 100 ha from the pond's geometric center (m^2)/ha	Pondscape connectivity to sources
	[a]%RIVER	[b,e]The ratio of all watercourse areas covered by rivers, creeks, channels, and ditches within a radius of 100 ha from the pond's geometric center (m^2)/ha	Pondscape connectivity to sources
	[a]%ROAD	[b,e]The ratio of all road and trail areas within a radius of 100 ha from the pond's geometric center (m^2)/ha	Pondscape isolation from sources

[a] Parameter measure obtained from a geographic information system (GIS) and other sources (Department of Land Administration, Ministry of the Interior 2002; Agricultural and Forestry Aerial Survey Institute 2003)

[b] The final results to explain pond and avian losses were due to being: (1) highly correlated ($r \geq 0.5$), (2) moderately correlated ($0.5 > r \geq 0.25$), or (3) slightly or not correlated ($r < 0.25$) due to a badly located or dysfunctional pondscape, as a result of which the pond had disappeared and/or the birds had gone

[c] Metrics and units: for calculation formulae see website at: http://www.umass.edu/landeco/research/fragstats/fragstats.html

[d] Parameter measure obtained from field measurements

[e] The mean values are expressed in percentages. The different land-use types were measured as the area of a circle as a percentage of an area of 100 ha (radius = 564.19 m) centered on each of the bird survey ponds ($n = 45$). The range of the percentage areas for each land-use type is also given

13.2.2 Study Areas

The Taoyuan Tableland lies between the northern border of the Linkou Tableland (23°05′N, 121°17′E) and the southern border of the Hukou Tableland (22°55′N, 121°05′E); it borders the town of Yinge in the east (22°56′N, 121°20′E) and the Taiwan Strait in the west (22°75′N, 120°99′E) (Department of Land Administration 2002) (Fig. 13.1). It has an elevation of 400 m (1 312 feet) above sea level, and is composed of tableland up to 303 m (994 feet) and hills with gentle slope from 303 to 400 m (994 to 1 312 feet). It runs in approximately a southeast-to-northwest direction, abutting mountains in the southeastern corner and the shore of the Taiwan Strait at the other end. With a high average humidity of 89%, the tableland is located in a subtropical monsoon region with humid winters and warm summers. January temperatures average 13°C, and July temperatures average 28°C. Annual average precipitation ranges from 1 500 to 2 000 mm (59.1 to 78.7 inches).

The tableland rose gradually approximately 1 80 000 years ago. At that time, the Tanshui River had not yet captured the flow from the ancient Shihmen Creek, which poured directly out of the northwestern coast forming alluvial fans. Eventually, foothill faults caused by earthquakes during the same era resulted in the northern region of Taiwan abruptly dropping by 200 m (656 feet), and thus the

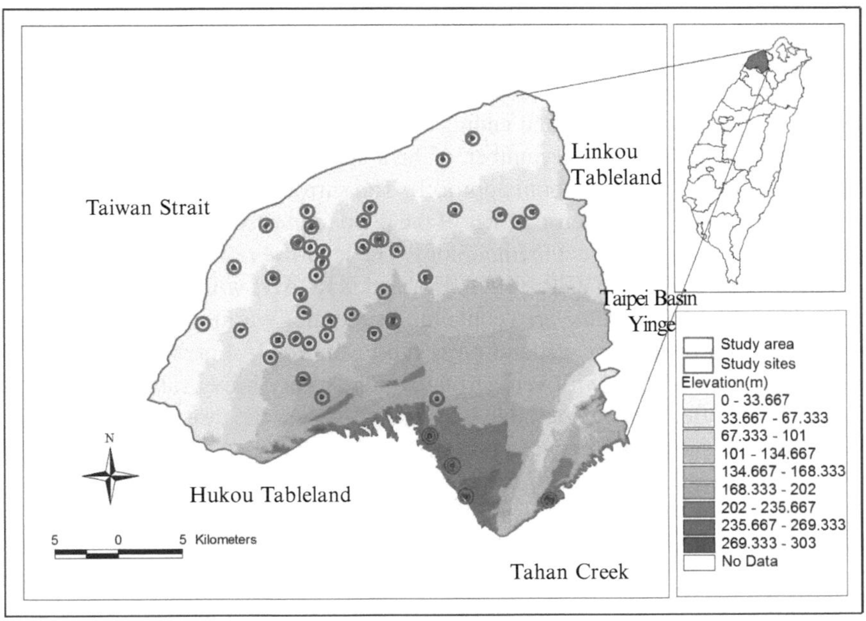

Fig. 13.1 Geographic contours of the Taoyuan Tableland at elevations from sea level to moderately sloping tableland up to 303 m (994 ft)

Taipei basin was born. Since the Taipei area had subsided, the ancient Shihmen Creek, which had meandered across the Taoyuan Tableland, was captured by northward-flowing rivers some 30 000 years ago. The middle streams changed their courses because of the subsidence in the Taipei basin. The resulting Tahan Creek became the upstream portion of the Tanshui River in the Taipei Basin. Due to the blockage of water sources, downstream areas on the Taoyuan Tableland were deficient in water. This caused high flushing and drops in water yields. Historically, it was difficult to withdraw and supply irrigated surface water from rivers owing to the tableland's unique topography, and this formed an obstacle to the development of agriculture (Huang 1999; Chen 2000).

This area has a population density of 2 331 people/km^2 (9 435 people/mile^2), and its population is increasing at a rate of approximately 2 000–3 000 per month. Population pressures have contributed to reductions in historical areas of farmlands and farm ponds (Fang 2001). Losses of farm ponds and farmland habitats have had serious effects on a range of avian communities as well as other fauna and flora (Fang and Chang 2004). On the Taoyuan Tableland, agricultural practices are intensifying, which is reducing the heterogeneity of the original landform, and adding pollutants as well as industrial practices.

13.2.3 Survey Method

In this survey project, I organized an intensive 4-month bird survey in which simultaneous censuses were carried out four times at 45 ponds from November 2003 to February 2004 (for pond plots see Fig. 13.1).

All surveys were conducted by 45 experienced bird observers starting at the same moment before sunrise and ending at 10:00 a.m. on the same day. Each pond was surveyed and coded for number of bird species and number of individuals within 30 min with a point-count approach. The surrounding area at a basal radius of 564.19 m from the geometric center of the pond (comprising a 100-ha circle) was also surveyed by line-transect methodology. The land-use ratios of basal areas (e.g., %FARM, %BUILD, %RIVER, %PONDS, and %ROAD) within the 100-ha circle were also calculated, and the area containing structures was used to infer attributes which were unsuitable for wetland birds (i.e., waterfowl, shorebirds, birds of the water's edge, etc.) Days with very strong winds or very heavy rain were avoided. To reduce the effects of any possible bias by bird-observers, three to four observers were grouped and rotated between ponds. The observers counted birds that were in any habitat. Birds belonging to the families Apodidae (swifts) and Hirundinidae (swallows) were also included using counts in flight. The Shannon–Wiener diversity index (H') was calculated to determine bird diversity, and the results are discussed in the following sections. The proportion of avian species (i) relative to the total number of species (P_i) was calculated, and was then multiplied by the logarithm of this proportion ($\log_2 P_i$). The resulting product was summed across species and multiplied by –1

$$H' = -\sum_{i=1}^{S} P_i \log_2 P_i, \tag{13.1}$$

where S is avian species richness and Pi is the percentage of species i in the avian community.

13.2.4 Data Mining

From the center of each selected pond, we drew a circle with a radius of 564.19 m to delineate an area of 100 ha, and measured the cover ratio of five major land-use types (i.e., ponds, watercourses, farmlands, roads, and structures) (Fig. 13.2), and three microhabitat types (i.e., water surfaces, mudflats, and foliage canopy areas). The land-use plots were identified based on field surveys and Taiwan's Geographic Aerial Map at a 1:5000 scale (Department of Land Administration, Ministry of the Interior 2002) and aerial photographs at a 1:5000 scale from 2003 (Agricultural and Forestry Aerial Survey Institute 2003). We also measured four parameters related to pond elevations, perimeters, and built-up topologies of waterfronts by GPS and field surveys. In addition, information on consolidated areas, as well as distances measured from sources that contained the study sites, was derived from the Taiwan Geographic Aerial Map at a 1:5000 scale (Department of Land Administration, Ministry of the Interior 2002).

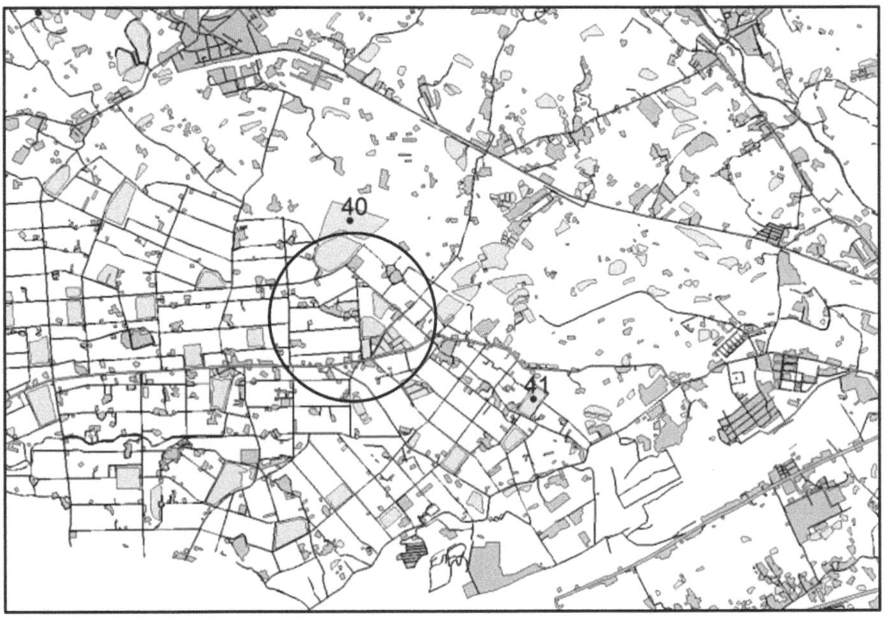

Fig. 13.2 A circle with a radius of 564.19 m delineates an area of 100 ha

13.3 Results

13.3.1 Avian Grouping

According to the recordings, 94 species and 15 053 individuals were detected in 4 months, and their microhabitats were categorized into seven guilds: air feeders (10 species), waterfowl (9 species), shorebirds (14 species), waterside birds (22 species), woodland birds (20 species), scrubland birds (13 species), and grassland birds (6 species). The value of dissimilarities was divided in accordance with distance (marked at a distance of 0.25) into seven groups. If this classification was adopted, the low similarities (marked at a distance of 0.75) could be divided into four guilds: waterfowl (9 species), shorebirds (14 species), waterside birds (22 species), and land birds (49 species). The likelihood of species occurrence was surveyed and categorized into a concentric pattern, such that the gradients ran from the pond's core to its edge: (1) interior pond species consisting of waterfowl (families Anatidae and Podicipedidae) and shorebirds (families Charadriidae and Scolopacidae); (2) waterside species (families Ardeidae et al.); and (3) external pond species of land birds (i.e., species detected in such microhabitats as grasslands, scrublands, and woodlands, of families Corvidae, Frigillidae, Laniidae, Passeridae, Pycnonodidae, Sylviidae, and Zosteropidae, et al.). These birds were dominant in their respective microhabitats.

I then proposed that waterfowl, shorebirds, and waterside birds (families Anatidae, Charadriidae, Scolopacidae, and Ardeidae) in microhabitats were associated with the distribution patterns of "interior species–edge species." Therefore, pond areas mainly provide shelter and foraging for wintering migrants and residents. The structural influences of microhabitat on bird distribution patterns can be classified into pond core, edge, and surrounding land uses (Fang 2004b; Fang et al. 2004). On farm ponds, I detected more than half of the species richness of land birds, and these species did not change much over the 4 months (Table 13.2). In order to relate the richness,

Table 13.2 Number individuals detected and species richness of avian guilds by month

Guild/dates	11.15.2003	12.20.2003	01.31.2004	02.28.2004
Air feeders	96(5)	248(7)	90(6)	79(4)
Waterfowl	85(6)	209(6)	157(7)	132(5)
Shorebirds	240(6)	261(10)	212(10)	94(6)
Waterside birds	2 192(10)	1 776(14)	1 775(11)	1 465(15)
Grassland birds	31(4)	127(3)	9(2)	12(4)
Scrubland birds	233(11)	213(9)	354(9)	296(8)
Woodland birds	844(18)	1 438(18)	1 303(17)	1 082(17)
No. of individuals (Species richness)	3 721(60)	4 272(67)	3 900(62)	3 160(59)

Notes: Number of individuals detected in each group; the value in parentheses is species richness abundance, and diversity of guilds to environmental variables, correlation analyses were carried out. The species–pondscape relationships were investigated between ecological groups and areas surrounding the ponds. Such analyses were restricted to pooled guild data in order to reduce confounding area effects. Parameters which were found to be important as local determinants of community structures were the amount of farmland and the amount of urban environment. Specific land uses which were found to be important were low-rise residential houses and high-density apartments.

13.3.2 Modeling Formation

Modeling was applied to the avian population of the Taoyuan Tableland, Taiwan. One parameter was selected to describe its structure: Shannon–Wiener's diversity index (*H*′) of the same waterside bird guild. Four environmental parameters were selected as explanatory parameters: pond size (PS), pond shape (MPFD), proportion of farmland area in peripheral areas (%FARM), and proportion of areas with structures in peripheral areas (%BUILD). Correlations between observed values and values estimated by ANN models of the four dependent parameters were moderately significant. The ANN models were developed from 35 of the 45 farm ponds sampled. These 35 were chosen at random and were validated using the 10 remaining farm ponds sampled. The role of each parameter was evaluated by inputting fictitious configurations of independent parameters and checking the response of the model. The resulting habitat profiles depict the complex influences of each environmental parameter on the biological parameters of the population, and the nonlinear relationships between the dependent and independent parameters.

I compared the following community characteristics with the corresponding ratio of constructed area values associated with pond configurations at each site for waterside species. Pondscape was a strong and/or moderate correlate for waterside birds, but not for land birds or air feeders. The presence of adjoining natural and/or farmland habitats was probably the most important determinant of wetland avifauna in farm-pond areas. Regarding this detailed study, there may be a number of reasons why some farm ponds do not become refuges for the more sensitive species. First, there is too little ornamental vegetation cover on the surrounding areas, and therefore there may be a small insect population. Second, anthropogenic structures are usually made of concrete with no native trees, and this may make such areas unattractive to waterside species that require an intact shrub layer, dead wood, or generally undisturbed microhabitats. Third, the small pond size associated with a curvilinear shape is not optimal to support, preserve, and attract waterside birds and other avifauna.

13.3.3 Modeling Application

In contrast to a lower information status from a linear regression model ($r<0.28$), I determined the uses of nonlinear models by ANN, with the number in the hidden

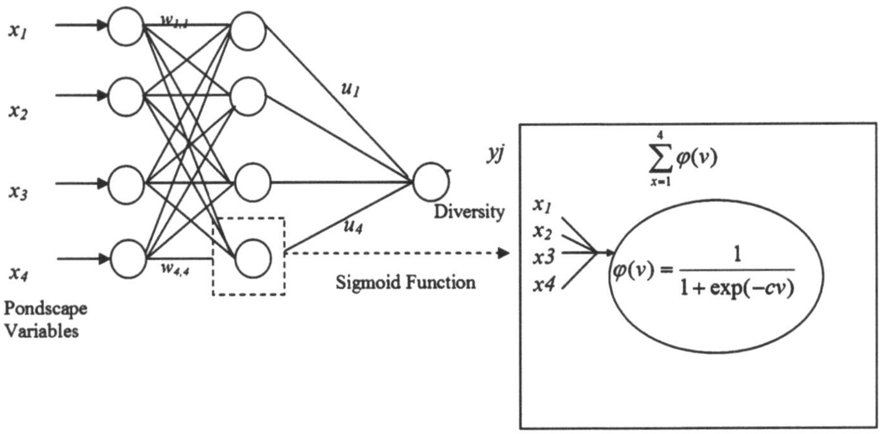

Fig. 13.3 Structure of the neural network used in this study. The input layer was composed of four cells representing each of the four pondscape parameters *Xi* ($i=1,4$). The hidden layer was composed of four neurons which calculated the dot products between its vector with weightings $w_j=[w_{ji,}\ i=1,4]$ and $x=[x_{i,}\ i=1,4]$

layer for four neurons (2^2), without choosing two or eight neurons. The correlation coefficient (r) for four neurons was found to differ between the training set ($r=0.725537$, $n=35$) and the validation set ($r=0.722752$, $n=10$).

The H' predicted by the ANN was able to meet the underlying rules embedded in the true H'. The diversity of waterside birds was predicted throughout the exercise using the back-propagation (BP) algorithm with a three-layered neural network. The first layer, called the input layer, was composed of four cells representing each of the environmental parameters. The second layer, or hidden layer, was composed of a further set of neurons whose number depends on the best-calculated unbiased results. Since the BP algorithm was trained by the least mean square method, the least mean square training could reduce the error, or the distance between the actual output and the desired output, by adjusting the weightings. Training cases were presented sequentially, and the weightings were adjusted. I determined the number of second-layer neurons through a series of interactions which varied from two and four to eight neurons. I calculated the correlation coefficients between true values of H' and the predicted value of H' from the ANN. In my study, a network with one hidden layer of four neurons was selected. It emphasized a stable fit which avoided over-training. The BP neural network architecture is shown in Fig. 13.3, and consists of four layers of neurons connected by weightings.

I calculated a refining simulation model for ten extra values of H' (for a total (n) of 55). The information was captured by the network when input data passed through the hidden layer of neurons to the output layer. The weightings connecting from neuron one to neuron four were denoted as w_{ji}. The output of each neuron was calculated based on the amount of stimulation it received from the given input vector,

x_i, where x_i was the input of neuron i. The net input of a neuron was calculated as the weights of its inputs, and the output of the neuron was based on some sigmoid function which indicated the magnitude of this net input. So the net output, u_j, from a neuron can be indicated as

$$u_j = \sum_{i=1}^{p} w_{ji} x_i \tag{13.2}$$

and

$$y_j = \varphi(u_j - \theta_j), \tag{13.3}$$

where

w_{ji} is the incremental change in the weighting from x_i to u_j,

θ_j is a threshold to be passed through by the nonlinear activation function, $\varphi(\cdot)$,

x_i is the ith pondscape parameter,

u_j is the jth neuron from an outgoing signal to the magnitude of all observations,

$\varphi(\cdot)$ is the activation function, and

y_j is the output of the jth neuron in any layer.

For this research, I chose the continuous sigmoid as the basic function:

$$\varphi(v) = \frac{1}{1+\exp(-cv)}, \tag{13.4}$$

where v is the net effect, and c is a constant.

13.3.4 Modeling Scenarios

I used several scenarios to illustrate wintering bird refuges. For example, I selected different values for the likelihood of a pond loss rate (0.25–0.75) from different land-use intensities. Plots were calculated using expectations of the historical model: the smaller the pond size, the greater likelihood of pond loss; and the shorter a parameter of a pond, the greater likelihood of pond loss. If the lower value of the likelihood of pond loss is equal to 0.25, overall ponds noted as threatened red spots (pond size > 0.996 ha, TE_{km} > 0.997 km) need to be conservatively protected owing to the slight likelihood of their loss. If the high value of the likelihood of pond loss was equal to 0.75, all ponds noted as threatened by the red spots (pond size > 0.2666 ha, TE_{km} > 0.371 km) need to be intensively protected owing to the likelihood of their loss. The extrapolated map (r=0.72, n=55) of waterside bird diversity H' suggests raising the priority of designating waterside bird refuges in patches against the potential pond-loss rate overlaid by threatened red spots (Hpool: pond size > 0.2666 ha, TE_{km} > 0.371 km) (diversity H': 0.4–0.6; 0.6–0.8; 0.8–1.0; 1.0–1.5; 1.5–1.741; distance (km), 12) (r=0.72) (Fig. 13.4).

13.4 Conclusion

ANN is one of the tools that can resolve prediction problems, and this ANN property is now well understood. The validation results were satisfactory with a four-neuron model, and confirmed the nonlinearity of the relationship among the parameters. The training set (r=0.725537, n=35) and validation set (r=0.722752, n=10) were surprisingly close in meeting the underlying rules embedded in the real values of the true H'. To analyze the results of this study, GIS pondscape data were coded, calculated, and compared with data on avian communities. A simulation model was developed to describe the cumulative influences of species–area and species–habitat relationships, and to produce a detailed simulation map for the

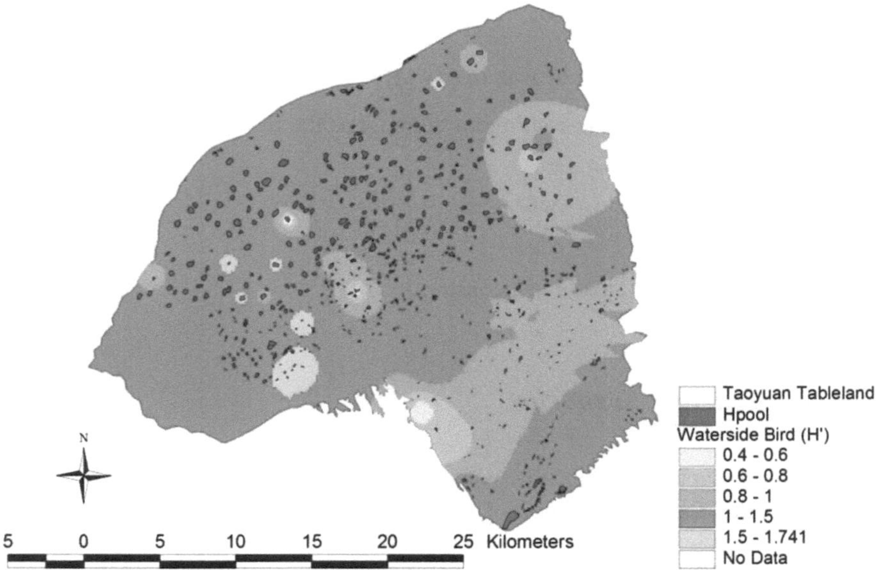

Fig. 13.4 Scenario refined by the ANN model for an intensive land-use pattern. The simulation did not cover the pond areas in the southwestern portion adjacent to the Houkuo Tableland located in Hsinchu County. Potential waterside bird refuges were determined from priorities of gradients with the references of pond-loss likelihood overlaid by threatened red spots (Hpool, pond size > 0.2666 ha, TE_{km} > 0.371 km) (diversity H': 0.4–0.6; 0.6–0.8; 0.8–1.0; 1.0–1.5; 1.5–1.741; distance (km) 12)

distribution of waterside bird diversity (*H′*). This study offers recommendations for better management practices for wintering bird refuges in Taiwan.

References

Agricultural and Forestry Aerial Survey Institute (2003) Aerial Photographs, 1:5, 000 of scale in digital debase forms. AFASI, Taipei

Chen C-J (2000) The change about culture landscape of irrigation reservoir and pond in the Taoyuan terrace, and the establishment of sustainable environment waterways. National Science Council, Taipei (in Chinese)

Cheng Q, Agterberg FP (1995) Multifractal modeling and spatial point processes. Math Geol 27:831–845

Connor EF, McCoy ED (1979) The statistics and biology of the species-area relationship. Am Nat 113:791–832

Department of Land Administration, Ministry of the Interior (2002) Taiwan's Geographic Aerial Map, 1:5,000 of scale in digital database forms. Taipei, Taiwan, ROC. (in Chinese)

Evertsz CJG, Mandelbrot BB (1992) Multifractal measures. In: Peitgen H-O, Jurgens H, Saupe D (eds) Chaos and fractals: new frontiers of science. Springer, New York, pp 921–953

Fang T-Y (2001) The study of the spatial structure change of water land in Taoyuan terrace. Master Thesis, Department of Bioenvironmental Systems Engineering, National Taiwan University, Taiwan, ROC (in Chinese)

Fang W-H (2004a) Threaten birds of Taiwan, 2nd edn. Wild Bird Federation Taiwan, Taipei (in Chinese)

Fang W-T (2004b) The ecological drawdown assessment for avian communities in Taoyuan's farm ponds. In: Yeh S-C (ed) The proceeding of 2004 annual symposium of environmental education. Chinese Society for Environmental Education, Taipei, pp 861–869 (in Chinese, abstract in English)

Fang W-T, Chang T-K (2004) The scientific exploring of Ecoscape-design on farm ponds in Taoyuan. In: Lee C-J (ed) The proceeding of water source management of Taoyuan Main Canal. Taoyuan Irrigation Association, Taoyuan, pp 345–369 (in Chinese, abstract in English)

Fang W-T, Loh D, Chang T-K (2004) Spatial analysis of wintering bird habitats on farm ponds, Taoyuan Tableland. In: Tsai M-H (ed) The proceeding of 2004 annual symposium of agricultural engineering. Chinese Society of Agricultural Engineering, Taipei, pp 1179–1190 (in Chinese, abstract in English)

Fang W-T, Chu H, Cheng B-Y (2009) Modelling waterbird diversity in irrigation ponds of Taoyuan, Taiwan using an artificial neural network approach. Paddy and Water Environment 7:209–216

Forman RTT (1995) Landscape mosaic: the ecology of landscape and regions. The University of Cambridge, Cambridge

Francl KE, Schnell GD (2002) Relationships of human disturbance, bird communities, and plant communities along the land-water interface of a large reservoir. Environ Monit Assess 73:67–93

Froneman A, Mangnall MJ, Little RM, Crowe TM et al (2001) Waterbird assemblages and associated habitat characteristics of farm ponds in the Western Cape, South Africa. Biodivers Conserv 10:251–270

Gustafson EJ, Parker GR (1992) Relationships between landcover proportion and indices of landscape spatial pattern. Landsc Ecol 7:101–110

Halsey TC, Jensen MH, Kadanoff LP, Procaccia I, Shraiman BI et al (1986) Fractal measures and their singularities: the characterization of strange sets. Phys Rev A 33:1141–1151

Huang B-J (1999) Assessment and application of groundwater resource in Taoyuan Area. Master Thesis, Department of Bioenvironmental Systems Engineering, National Taiwan University, Taiwan, ROC, Taipei (in Chinese)

Kronert R, Volk M, Steinhardt U et al (2001) Landscape balance and landscape assessment. Springer, Berlin

Leitão AB, Ahern J (2002) Applying landscape ecological concepts and metrics in sustainable landscape planning. Landsc Urban Plan 59:65–93

Li H, Reynolds JF (1994) A simulation experiment to quantify spatial heterogeneity in categorical maps. Ecology 75:2446–2455

Linton S, Boulder R (2000) Botanical conservation value related to origin and management of ponds. Aquat Conserv Mar Freshw Ecosyst 10:77–91

MacArthur RH, Wilson EO (1967) The theory of island biogeography. Princeton University Press, Princeton

McGarigal K, Cushman SA, Neel MC, Ene E et al (2002) FRAGSTATS: spatial pattern analysis program for categorical maps, computer software program produced by the authors at the University of Massachusetts, Amherst, Massachusetts, USA. http://www.umass.edu/landeco/research/fragstats/fragstats.html

Oertli B, Joye DA, Castella E, Juge R, Cambin D, Lachavanne J et al (2002) Does size matter? The relationship between pond area and biodiversity. Biol Conserv 104:59–70

Simberloff DS, Abele LG (1976) Island biogeographic theory and conservation practice. Science 191:285–286

Chapter 14
Integrating Geographic Information into Scenic Assessments of Middle Landscapes

Sampei Yamashita

14.1 Introduction

Suburban areas, or middle landscapes, can serve as ideal human habitats and demonstrate how human–environment synergy can be sustained. Middle landscapes tend to vary in type (farmland, suburbia, garden city and garden, model town, etc.) and are susceptible to either urban or natural encroachment, a fact which makes the assessment and management of their aesthetic quality quite difficult (Marx 1964; Tuan 1998).

There are two different approaches to landscape quality assessment: the ecologist's approach and the engineer-architect's approach (Turner et al 2001; Shinohara 1982). The former, landscape ecology, emphasizes and explores the causes and effects of the spatial heterogeneity of a region, whereas the latter, landscape architecture/engineering, focuses on investigating perceptual features of the region for its design and management. Landscape ecology often uses and analyzes aerial photographs and satellite images. By contrast, landscape architecture/engineering establishes a designated viewpoint on the ground which represents ordinary, "down-to-earth" experiences of landscapes. Landscape ecology addresses a wide-ranging area of the region, whereas landscape architecture/engineering emphasizes a relatively small locale encompassing the viewpoints.

Combining these contrasting approaches enables environmental planners and managers to address the geographic information relevant to ordinary perceptions and aesthetic assessments of landscapes. It also allows us to appreciate the aesthetic attributes of the region and examine places beyond the vicinity of the designated viewpoint (Daniel 2001). The combination of these two approaches may thus contribute to identifying elusive middle landscapes and assessing them aesthetically in the geographic context of the region.

This study will identify the middle landscapes in a catchment basin of a Japanese river and obtain information relevant to the assessment and management of the basin by integrating GIS information on the basin into an aesthetic assessment of

S. Yamashita (✉)
Kyushu Sangyo University, 2-3-1 Matsukadai, Higashi-ku, Fukuoka 813-8503, Japan
e-mail: samp@ip.kyusan-u.ac.jp

S.-K. Hong, et al. (eds.), *Landscape Ecology in Asian Cultures*, Ecological Research Monographs, DOI 10.1007/978-4-431-87799-8_14, © Springer 2011

the river landscape. The study will thus explore the roles the middle landscape plays in the perception and evaluation of landscapes.

14.2 Method

14.2.1 Site Description

The selected catchment basin is of the Mikasa River, which runs from upstream, natural areas to downstream, intensely urbanized/densely populated areas. The urbanized and/or residential areas account for over 60% of the catchment, whereas the farmland/rice paddy fields cover less than 10% of the basin. Mountains and forests that constitute the upstream areas of the catchment are registered as a prefecture natural park, and excellent conservation is practiced toward the natural scenery. The catchment area is 94 km^2 and the length of the main stream is 24 km.

14.2.2 Land Use and View-Shed Area

To combine geographic information with aesthetic data, a 3-D land-use map of the study site was created, and view-shed areas classified by land use were generated as described below.

A digital surface model of the catchment area (cell size 4 m) was created using aerial photographs and integrated with a digital land-use map of the area, which had 14 land-use categories (scale 1:10000): road site, public open space, residential area, public facilities area, mountains and forests, industrial district, farmland, other open space, rice paddy field, other natural area, water zone, commercial zone, transportation facilities area, and other public facilities area (Fig. 14.1).

The top of a bridge can serve as an observatory for a fine view (Shinohara 1982). Thus panorama video images were taken both upstream and downstream from all the 66 bridges along the main stretch of the Mikasa River for landscape aesthetic quality assessment (Fig. 14.2). The viewpoint was set at 1.5 m above the top of each bridge. In taking an image, the video camera was turned horizontally to cover as much of the scenery as possible. First, the direction of view was set parallel to the flow of the river and fixed for 3 s. Next, the direction was changed by turning the camera slowly to the left for 4 s, and then it was fixed again for 2 s. It was then turned to the right for 4 s at the same rate as for the left turn, and then fixed again for 2 s. The direction of view was then changed again by turning the camera to the left at the same rate and then fixing it in the direction of flow for 2 s.

The horizontal angle that this process covered was 191°, and the vertical angle for the picture field was 50.8°. The vertical angle was slightly less than 60°, which is known to give an effective field of view (Gibson 1986). However, the horizontal angle was much larger than 60°, thus complementing the vertical angle to represent the scenic experience on site.

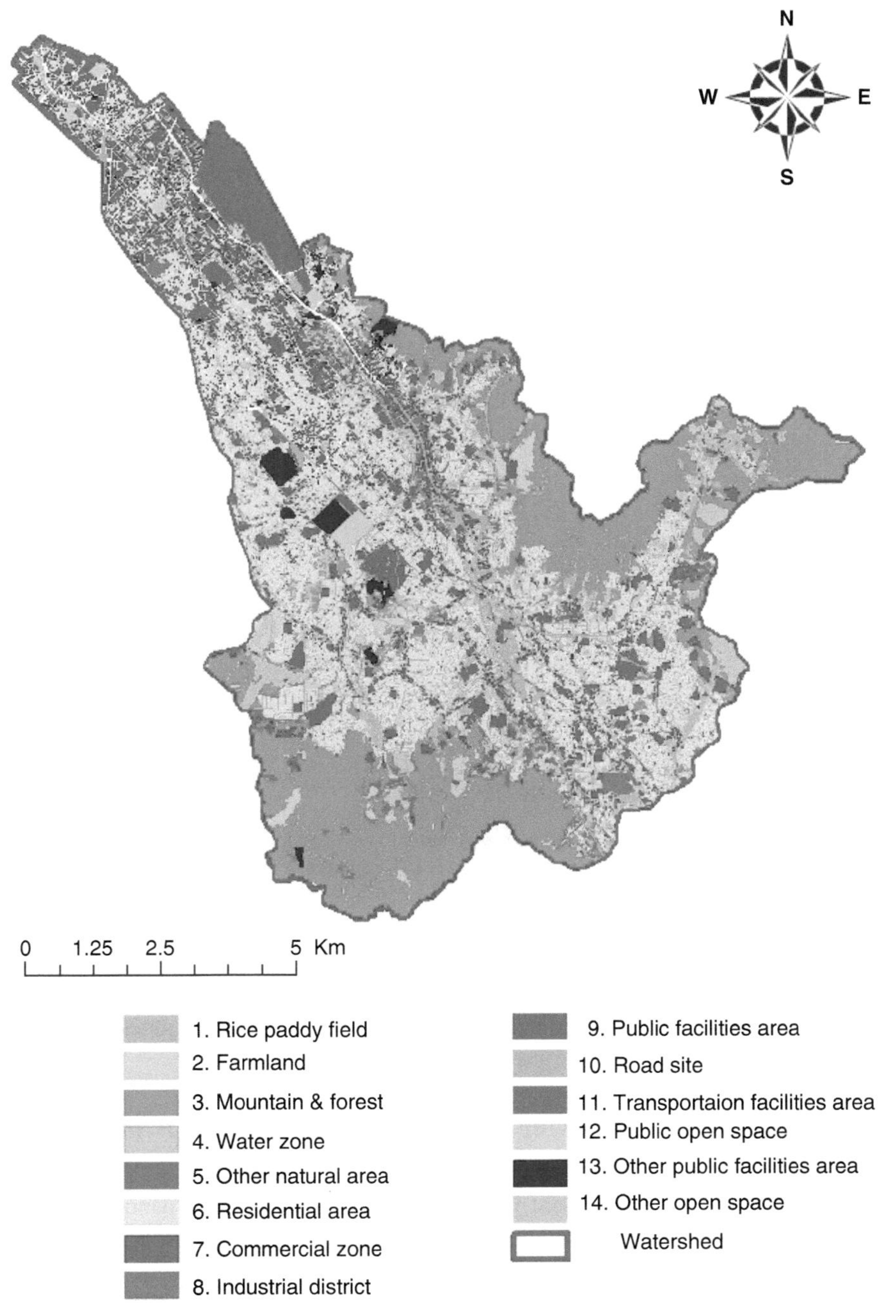

Fig. 14.1 The watershed of the Mikasa River and its 14 categories of land use

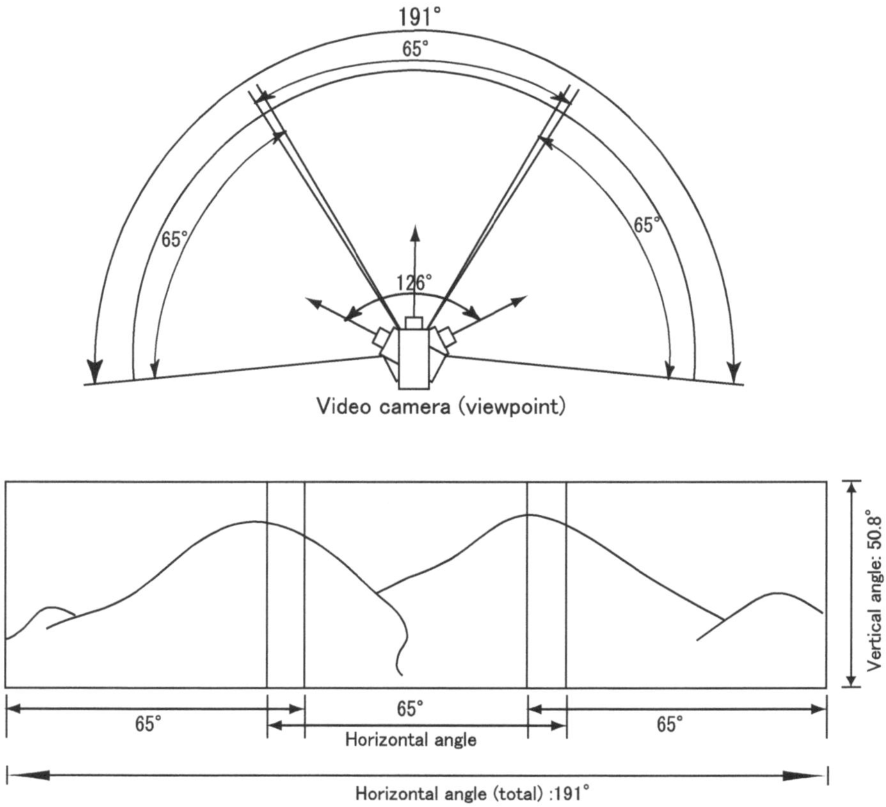

Fig. 14.2 The video camera was turned horizontally to cover as much of the scenery as possible. The horizontal angle that this process covered was 191°, and the vertical angle for the picture field was 50.8°

By feeding the 3-D land-use map data and the 191°×50.8° frame to the GIS system, separate view-shed areas were generated both upstream and downstream from the 66 bridges, and these were analyzed individually for the 14 land-use categories (Fig. 14.3).

14.2.3 Aesthetic Assessment of the Scenes from the Bridges

An aesthetic quality assessment was carried out using the 132 panorama video images taken both upstream and downstream from the 66 bridges along the Mikasa River. Twenty-two male and 19 female college students took part in the assessment. They were classified into 3 groups: (1) 10-member male group, (2) 12-member female group, and (3) group of 9 females and 10 males. Nine images were randomly chosen from the 132 images to be shown to all the groups as a "baseline." From the remaining 123 images, 41 images were randomly selected, and the 9 baseline images were added to make a set of 50 images. Three sets of 50 images were created following the same procedure, and each set was assessed by one of the

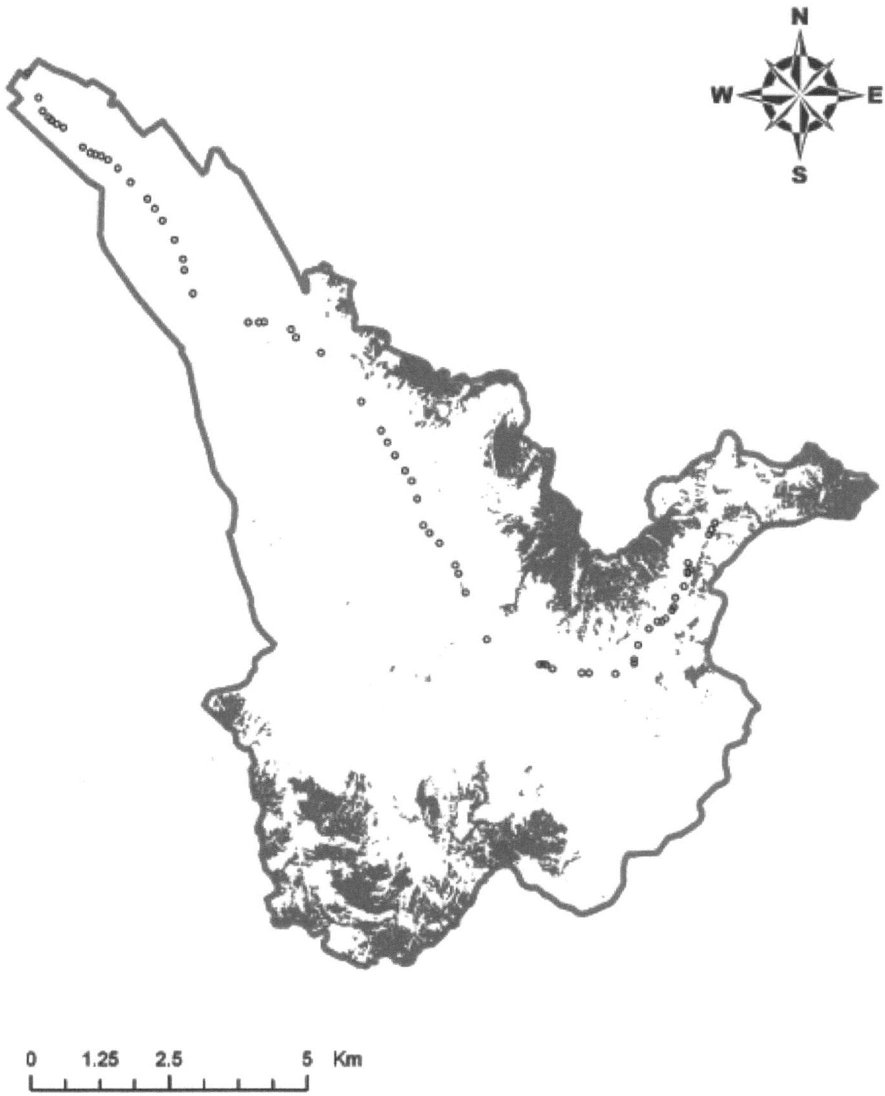

Fig. 14.3 Collective view-shed of mountains and forests in the watershed. The *small circles* in the watershed represent the 66 viewpoints on the bridges

three groups. This procedure helped reduce the assessment time (about 20 min in total) and lower the fatigue effect on the result.

The participants rated each image using a ten-point preference scale, with 0 as the lowest in preference and 9 as the highest. Using the baseline images, the ratings obtained were standardized between groups and between individuals. This method is called the scenic beauty estimation (SBE) method and is known to be an effective and reliable procedure for aesthetic quality assessments (Brown and Daniel 1990).

Some of the video images involved construction work on river development. Although such work was temporary, it could significantly affect the landscape assessment. Therefore, those images were excluded from the analysis. The number of video images and view-sheds that were used in the analysis thus came to 108.

14.2.4 Analysis Framework

Middle landscapes have aspects of both artificial cities and wild nature and vary in type. This study first attempts to identify typical middle landscapes that are perceived from the bridges and that involve multiple land uses. The study then explores the relationship between the attributes of typical middle landscapes and aesthetic assessments of the landscapes.

14.3 Results

14.3.1 Typical Middle Landscapes

14.3.1.1 Type

It is expected that the view-sheds categorized by land use inter-relate as they were generated within the views from the bridges. To find potential variables explaining the relationships between the view-sheds, a factor analysis was carried out, and as a result, three factors with eigenvalues of over 1.0 were obtained (Table 14.1). The three factors all involved both urban and natural aspects and helped identify three different types of middle landscape.

Table 14.1 Inter-relations among view-shed areas categorized by land use (factor analysis). Factor loadings with eigenvalues of over 1.0

Land-use category	Mountain/forest	Country/nature	Developed waterfront
Road sitePublic open space	0.905[a]	0.170	0.139
Residential area	0.873[a]	−0.121	0.119
Public facilities area	0.802[a]	0.379	0.120
Mountains and forests	0.767[a]	0.323	−0.139
Industrial district	0.750[a]	0.439	0.024
Farmland	0.730[a]	0.047	0.516
Other open space	0.345	0.849[a]	0.015
Rice paddy field	0.237	0.823[a]	0.102
Other natural area	0.013	0.765[a]	−0.108
Water zone	0.548	0.720[a]	0.017
Commercial zone	0.218	0.064	0.775[a]
Transportation facilities area	0.384	0.139	0.761[a]
Other public facilities area	−0.218	−0.012	0.593
	−0.017	0.483	0.198

[a] >0.7

The first factor had relatively large correlations (loadings over 0.7) with "mountains and forests," "residential area," "road site," "industrial district," "public facility area," and "public open space." This factor is interpreted as an indicator of "mountain/forest type," i.e., a perspective of distant mountains and forests with anthropogenic structures in the foreground. The second factor was highly correlated (loadings over 0.7) with "rice paddy field," "farmland," "other natural zone," and "other open space," which represents "country/nature type." The third factor had high correlations (loadings over 0.75) with "water zone" and "commercial zone" and represents the "developed waterfront type."

14.3.1.2 Location and Direction

To make out the locations and directions of the three types of middle landscape, "mountain/forest type," "country/nature type," and "developed waterfront type," a cluster analysis (Ward method) was conducted using the respective factor scores assigned to the images. As a result, the 108 images or scenes were classified into four clusters. (1) Cluster 1: scenes located in midstream and upstream catchments and directed both upstream and downstream from the bridges. (2) Cluster 2: scenes located in upstream catchments and directed downstream. (3) Cluster 3: scenes located from upstream through downstream catchments. (4) Cluster 4: scenes located in midstream catchments and directed upstream.

It is also clear that the mean values of the factor scores relevant to "developed waterfront type," "country/nature type," and "mountain/forest type" are remarkably high among the scenes consisting of Clusters 1, 2, and 4, respectively, and thus the relationships between the landscape types and the clusters are evident (Table 14.2, Fig. 14.4). However, Cluster 3 had negative values for all mean scores and showed no specific characteristics in this analysis.

14.3.2 Landscape Type and Assessment

The SBE obtained in the landscape aesthetic assessment is an index representing preference for the scene. The mean values of the SBE for Cluster 2, which is associated with "country/nature type," and Cluster 4, "mountain/forest type," are relatively high and are thus preferred, whereas those for Cluster 1, the "developed waterfront type" and Cluster 3 are relatively low and are less preferred (Table 14.3).

Table 14.2 Average factor scores of the middle landscapes in each cluster

Cluster	Mountain/forest	Country/nature	Developed waterfront
1	−0.3	0.1	1.1[a]
2	0.3	6.1[a]	−0.9
3	−0.1	−0.2	−0.6
4	3.7[a]	−0.8	0.2

[a]Maximum

Fig. 14.4 An image of the mountain/forest type (**a**), country/nature type (**b**) and developed waterfront type (**c**) perceptual middle landscape

Table 14.3 Relations between scenic beauty estimation (SBE) and the clusters/middle landscapes

Cluster	Average SBE	Number	Standard deviation
1	−68.3	35	96.2
2	4.1	2	28.9
3	−101.1	66	103.6
4	61.3	5	61.1

Table 14.4 Correlations between scenic beauty estimation (SBE) and the three types of factor score

Mountain/forest	Country/nature	Developed waterfront
0.412**	0.097	0.131

** $P<1\%$

Moreover, the correlations between SBE values and the factor scores representing the three types of middle landscapes reveal that SBE is significantly related to "mountain/forest type" ($r=0.412$, $p<0.001$; Table 14.4). This result indicates that if urbanized view-shed areas in the foreground increase without dwarfing the perspective of mountains and forests in the background, the aesthetic assessment of the landscape can still improve.

14.4 Concluding Remarks

This study integrated geographic information into the scenic assessment of landscape and investigated middle landscapes in a watershed involving both urban and natural aspects. As a result, three types of middle landscape were identified. (1) Mountain/forest type: a perspective of distant mountains and forests with anthropogenic structures in the foreground, located in midstream catchments and directed upstream from the bridges. (2) Country/nature type: views consisting of rice paddy fields, farmlands, natural areas, and/or vacant land, located in upstream catchments and directed downstream. (3) Developed waterfront type: views of a water surface with commercial facilities, located in midstream and upstream catchments and directed both upstream and downstream. The aesthetic assessment test revealed that the visibility of middle landscapes, particularly the perspective of distant mountains and forests, has a positive impact on the view from the river, even if conspicuous anthropogenic structures are present in the foreground. Conversely, the combination of commercial facilities with water tends to be rated relatively low.

It has often been pointed out that areas of nature, man-made structures, and water surfaces in a scene tend to have a significant impact on the aesthetic quality assessment of the landscape (Shafer and Mietz 1970; Steinitz 1990; Yu 1995; Purcell and Lamb 1998; Daniel 2001; Yamashita 2002). As these indices are always examined using landscape photographs alone and are closely related to the framing of the photographs, it is hard to incorporate them into the planning and management of a regional landscape (Palmer 2004). However, these indices, when combined with geographic information about the catchment basin, may illuminate the three types of middle landscapes much more and account for the scenes that are not categorized into middle landscapes in this article. Examining this point will be the next step in the study of middle landscapes.

References

Brown TC, Daniel TC (1990) Scaling of ratings: concepts and methods. USDA Forest Service Research Paper RM-293, Rocky Mountain Forest and Range Experiment Station, Fort Collins, CO

Daniel TC (2001) Wither scenic beauty? Visual landscape quality assessment in the 21st century. Landsc Urban Plan 66:267–281

Gibson JJ (1986) The ecological approach to visual perception. Lawrence Erlbaum Associates, London

Marx (1964) The machine in the garden: technology and the pastoral ideal in America. Oxford University Press, New York

Palmer JF (2004) Using metrics to predict scenic perception in a changing landscape: Dennis, Massachusetts. Landsc Urban Plan 69:201–218

Purcell AT, Lamb RJ (1998) Preference and naturalness: an ecological approach. Landsc Urban Plan 42:57–66

Shafer EL, Mietz J (1970) It seems possible to quantify scenic beauty in photographs. USDA Forest Service Research Paper NE-162, Northeastern Forest Experiment Station, Upper Darby, PA

Shinohara O (1982) Doboku Keikan Keikaku (Landscape planning in civil engineering). Gihodo Shuppan, Tokyo

Steinitz C (1990) Toward a sustainable landscape with high visual preference and high ecological integrity: the loop road in Acadia National Park, USA. Landsc Urban Plan 19:213–250

Tuan Y-F (1998) Escapism. Johns Hopkins University Press, Baltimore

Turner MG, Gardner RH, O'Neill RV (2001) Landscape ecology in theory and practice: pattern and process. Springer, New York

Yamashita S (2002) Perception and evaluation of water in landscape: use of photo-projective method to compare child and adult residents' perceptions of a Japanese river environment. Landsc Urban Plan 62:3–17

Yu K (1995) Cultural variations in landscape preference: comparisons among Chinese sub-groups and Western design experts. Landsc Urban Plan 32:107–126

Chapter 15
Sustainable Management of Satoyama Bamboo Landscapes in Japan

Shigeo Suzuki and Nobukazu Nakagoshi

15.1 Introduction

Rural Japan has seen changes in vegetation composition in recent years due to a reduced need for woodlands and grasslands for resource production (Nakagoshi 1995). For example, Hong et al. (1995) reported that secondary forests composed mainly of *Pinus densiflora* Sieb. et Zucc., which historically were harvested for firewood and fertilizer, have been succeeded by deciduous oaks; furthermore, this shift in species composition corresponded to changes in human resource use. In addition, Torii and Isagi (1997) reported that deciduous broad-leaf forests have been replaced by bamboo groves, also in accordance with social changes. Finally, Fukamachi et al. (2001) discovered that vegetation changes in rural Japan were associated with differences in accessibility from nearby villages, and varied with topography and land ownership.

The development of management plans for these seminatural landscapes is a current research theme in landscape ecology (Turner et al. 2001). In this chapter, we discuss sustainable management methods for bamboo groves that incorporate considerations of the Japanese cultural landscape (*satoyama*, as discussed by Nakagoshi and Hong (2001)).

Bamboo is one of the most commonly used plants in Japan. With the exception of Northern Japan, where the climate is too cold, bamboo groves are a fundamental component of the *satoyama* landscape (Shibata 2001). Historically, people used bamboo culms for building materials and other commodities, and the edible shoots were used as seasonal ingredients in Japanese cuisine. However, increasingly, wood materials are being replaced by plastic or metal, and less expensive bamboo shoots

S. Suzuki (✉)
Faculty of Geo-Environmental Science, Rissho University, 1700, Magechi, Kumagaya, Saitama 360-0194, Japan
e-mail: suzusige_1980@yahoo.co.jp

N. Nakagoshi
Graduate School for International Development and Cooperation, Hiroshima University, 1-5-1, Kagamiyama, Higashi-Hiroshima 739-8529, Japan

S.-K. Hong, et al. (eds.), *Landscape Ecology in Asian Cultures*, Ecological Research Monographs, DOI 10.1007/978-4-431-87799-8_15, © Springer 2011

are being imported from China. Thus, many Japanese bamboo groves have been abandoned. As a result, bamboo groves have expanded rapidly as new culm recruitment occurs via the clonal propagation of existing plants (Okutomi et al. 1996; Isagi and Torii 1998).

The expansion of abandoned bamboo groves has become a significant problem for rural landscape planning and management. Within these groves, the forest canopy is composed exclusively of bamboo species, and the understory and forest floor are very dark. Thus, the expansion of bamboo groves results in reduced biodiversity (Searashi et al. 1989; Okutomi 2005). Moreover, bamboo groves retain less water (Ishiga et al. 2001) and preserve less soil (Hiura et al. 2004) than broad-leaved forests. Studies examining bamboo-grove expansion have focused on the relationship between the expansion of bamboo groves and topography (Torii 1998) as well as adjacent vegetation (Nishikawa et al. 2005). However, these studies have not addressed management planning for bamboo landscapes.

In this chapter, we review the importance of bamboo landscapes in Japan from both cultural and ecological viewpoints. In addition, using the results of our analysis, we discuss the expansion characteristics of bamboo groves with respect to vegetation characteristics and human-use trends. Finally, based on our results, we propose a plan for the sustainable management of bamboo landscapes.

15.2 Methods

We started by researching the history of bamboo species and their cultural uses in Japan, to clarify the position of bamboo in Japanese culture. Next, we examined the ecological characteristics of bamboo species. First, we surveyed the expansion rate of bamboo groves in Hirasawa (near the town of Otaki, in Chiba Prefecture, Eastern Japan) and Kofuki (near the city of Takehara, in Hiroshima Prefecture, Western Japan). We compared these rates of expansion with those reported in previous studies. We used a stepwise procedure based on a multivariate analysis, as described by Hirasawa and Kofuki, to define the rate of expansion of a bamboo grove (Suzuki and Nakagoshi 2008). Specifically, we calculated the original bamboo grove area (AREA), the slope aspect (the cosine of slope aspect, SN; and the sine of slope aspect, SE), and the slope inclination (SI), and calculated the percentage of adjacent vegetation according to height and land use (high vegetation, PH; short vegetation, PS; human use, PU). We also considered the distance from the nearest road (DR), and the shipment of bamboo shoots (SBS) in each bamboo grove. The annual expansion rate (AER) of *Phyllostachys pubescens* Mazel ex Houzeau de Lehaie groves was defined as the dependent variable.

Second, we evaluated whether the expansion of bamboo groves was correlated with the height of neighboring vegetation. To do this, we compared the models for *P. pubescens* invading adjacent vegetation of various heights in the city of Oda, in Shimane Prefecture, Western Japan (Suzuki et al. 2006). We compared the culm density of the bamboo grove with the age structure of plants taller than 18 m in an abandoned graveyard and secondary woodland.

To measure the effects of bamboo-grove expansion on plant-species diversity, we surveyed the culm density of *P. pubescens* and vegetation structure in 23 plots within the Kamo and Uchihama river basins, Takehara City, where different geologic surface conditions exist. We then assessed the relationship between culm density and plant species diversity. Finally, we developed a potential bamboo landscape planning method based on both cultural and ecological data, and conducted a trial of the method using experimental cutting (Suzuki et al. 2008).

15.3 History of Bamboo and Its Use in Japan

Three tall bamboo species grow throughout Japan, with the exception of the northern regions (Table 15.1). *Phyllostachys bambusoides* Sieb. et Zucc. and *Phyllostachys nigra* Munro var. *henonis* Stapf ex Pendle are known to have existed in Japan at least 800 years ago. *P. pubescens* was introduced during the eighteenth century from China (Ogura 1988), and thus has only grown in Japan for a few hundred years.

All of these bamboo species, including *P. pubescens*, have played substantial roles in Japanese culture. Bamboo culms (mainly *P. bambusoides*) were traditionally used as building materials and other commodities. For example, many bamboo-made items, such as teaspoons (*chasaji*) and bamboo whisks (*chasen*; Fig. 15.1), are used during the traditional Japanese tea ceremony (*sado*). In addition, edible bamboo shoots (mainly from *P. pubescens*) are fundamental to Japanese cooking, especially during the spring. For this reason, *P. pubescens* was planted throughout most of Japan for food production after the early nineteenth century (Ogura 1988). Some owners of bamboo groves practiced management schemes for edible bamboo shoots or bamboo material production. The most important aspect of bamboo grove management is adjusting bamboo culm density (Fig. 15.2). The owner would therefore cut down old and poor quality culms every autumn (Ueda 1968).

Since the 1960s, many bamboo products have been replaced by products made from plastic or light metals, resulting in the abandonment of many *P. bambusoides* groves. Beginning in the late 1980s, the consumption of inexpensive edible bamboo shoots imported from China has increased rapidly, resulting in a decrease in domestic production and the abandonment of *P. pubescens* groves. These abandoned bamboo groves then expanded quickly via new culm recruitment by clonal propagation (Shibata 2001).

Table 15.1 Three tall bamboo species that grow in Japan and their origins

Scientific name	*Phyllostachys pubescens*	*Phyllostachys bambusoides*	*Phyllostachys nigra*
Native or introduced	Introduced in 1736 from China	Native, but probably introduced from China before 1 000 years ago	Introduced from China about 800 years ago
Use	Edible shoots, fishing materials	Building materials, industrial arts	Edible shoots, industrial arts

Fig. 15.1 A bamboo whisk (*chasen*)

Fig. 15.2 A *Phyllostachys pubescens* grove used for the production of edible bamboo shoots in Kofuki. Thin *P. pubescens* groves in this region are under continuous management. The dense *P. pubescens* groves towards the back of this region represent abandoned groves

15.4 Trends in Bamboo Grove Expansion

We surveyed bamboo-grove expansion rates in Hirasawa and Kofuki. Our results indicate that the area of bamboo groves expanded roughly 1.5-fold in Hirasawa, from 105.5 ha in 1984 to 154.4 ha in 2001 (Suzuki and Nakagoshi 2008, Fig. 15.3). In Kofuki, bamboo groves increased 1.2-fold, from 8.0 ha in 1986 to 10.6 ha in 2006. Previous evaluations of other sites, mainly in Western and Southern Japan, show that bamboo groves, particularly those of *P. pubescens*, have invaded quickly (Fig. 15.4).

To identify factors contributing to the expansion of *P. pubescens* groves, we used multiple linear regression analysis based on a stepwise procedure, and we determined whether site differences existed for data obtained from Hirasawa and Kofuki (Suzuki and Nakagoshi 2008).

We found that SBS was a positive factor for *P. pubescens* grove expansion, whereas PS, SN, and AREA negatively affected AER in Hirasawa ($R^2=0.683$):

$$\mathrm{AER} = -7.133\,\mathrm{PS} - 3.606\,\mathrm{SN} - 2.980 + 2.175\,\mathrm{SBS} + 3.567 \tag{15.1}$$

In Kofuki, DR and SE positively affected expansion, whereas SI was a negative factor ($R^2=0.942$):

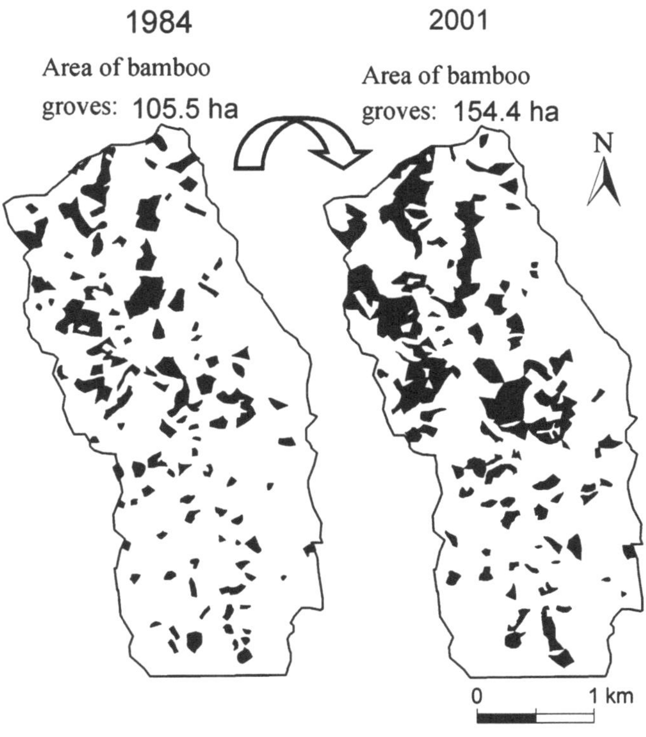

Fig. 15.3 The distribution of bamboo groves (*black*) in 1984 and 2004 in Hirasawa, Otaki, Chiba, Japan

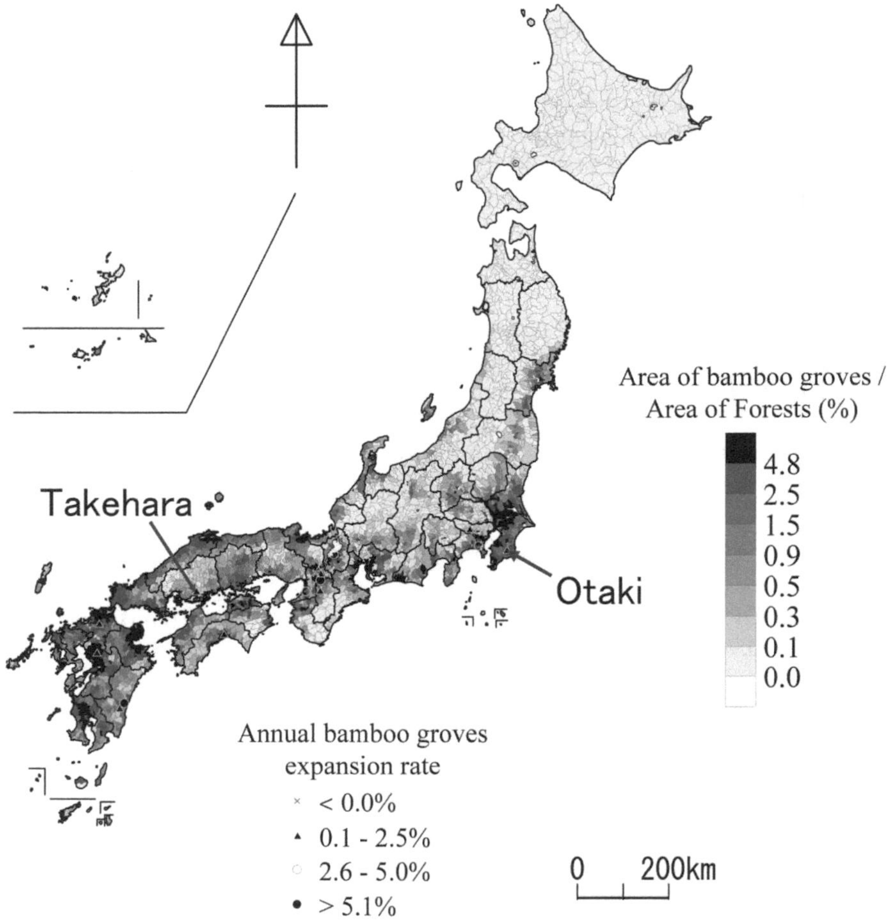

Fig. 15.4 Distribution of bamboo groves in Japan and their annual expansion rate

$$\mathrm{AER} = 0.190\,\mathrm{DR} + 8.991\,\mathrm{SE} - 0.395\,\mathrm{SI} + 4.162 \tag{15.2}$$

Thus, we showed that both natural conditions and management factors affect *P. pubescens* grove expansion. In addition, we found regional differences in natural and anthropogenic effects on bamboo grove expansion.

We also surveyed the effects of adjacent vegetation on *P. pubescens* invasion. At the abandoned graveyard, the density of new culms of *P. pubescens* was higher where *P. pubescens* shoots were invading (the expansion front). However, in a secondary woodland (plants exceeding 18 m), the density of *P. pubescens* culms was lower in areas near to the expansion front. *P. pubescens* culms were as high as the stems of *Magnolia obovata* Thunb. and *Quercus serrata* Thunb. in secondary forests. Thus, adjacent broad-leaf trees in secondary forest hindered the invasion of *P. pubescens* culms by limiting the amount of light available for photosynthesis.

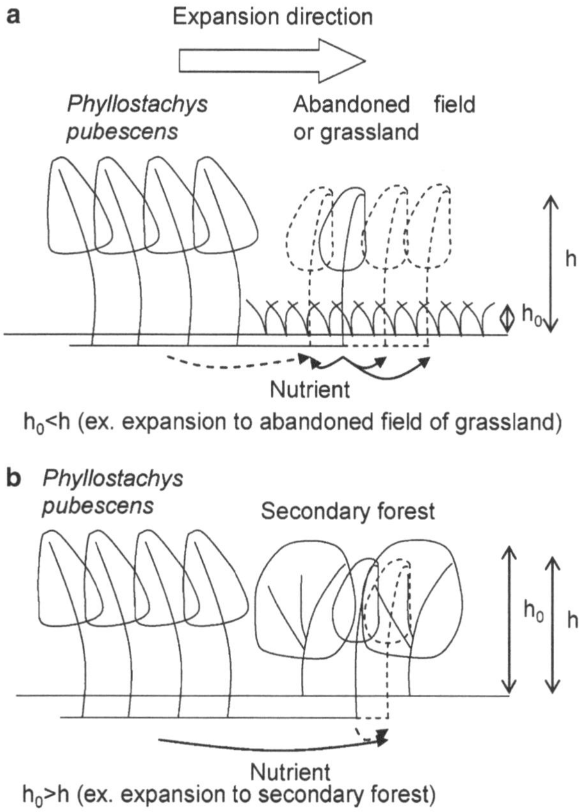

Fig. 15.5 An expansion model of *Phyllostachys pubescens* (Suzuki et al. 2006). Assuming that the nutrient supply (via rhizomes) of *P. pubescens* is limited within a narrow range (Ito and Yamada 2005), expansion would be faster when the height (h) of the invading culms is greater than that of the invaded vegetation (h_0). When h is lower than h_0, the expansion speed tends to decrease because of the limited light available to invading culms for photosynthesis

As a result, a smaller number of new culms was able to grow around the existing culms each year.

Assuming that the nutrient supply (via rhizomes) of *P. pubescens* is limited within a narrow range (Ito and Yamada 2005), expansion will be faster when invading culms are taller than the original vegetation. When invading culms are shorter than the original vegetation, the expansion rate will tend to decrease due to limited primary production in the invading culms (Fig. 15.5).

15.5 Ecological Impacts of Bamboo Grove Expansion

Searashi et al. (1989) reported that species diversity decreases due to bamboo invasion. We surveyed the culm density of *P. pubescens* and plant species diversity in the Kamo and Uchihama river basins. Plant species diversity did indeed decrease in

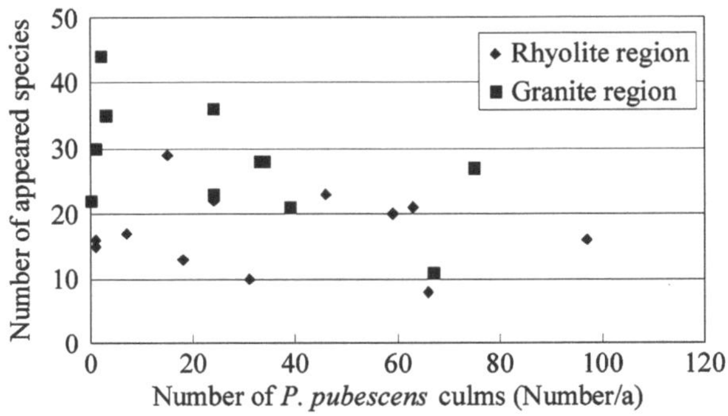

Fig. 15.6 Relationship between the number of *Phyllostachys pubescens* culms and species diversity

association with the increased density of *P. pubescens* culms, especially in the granite region (Fig. 15.6). We confirmed a negative correlation between the number of species and the number of *P. pubescens* culms, which is consistent with the findings of Searashi et al. (1989).

In the rhyolite region, however, species diversity did not decrease in association with *P. pubescens* culm density (Fig. 15.6). Plant species diversity was lower than that in the granite region before the onset of bamboo grove expansion, most likely because the tall broad-leaved species had already produced a dark forest floor in the rhyolite region. Thus, succession was faster because the existing species were already resistant to shaded conditions. Species composition was not altered in the rhyolite region, and plant species diversity did not decline when bamboo was introduced to the forest. Thus, our findings suggest that *P. pubescens* invasion has more significant species diversity effects if the forest is less mature.

15.6 Discussion

Japanese bamboo groves, which are mainly composed of *P. pubescens*, an introduced species, are expanding quickly. Natural conditions and anthropogenic effects have facilitated this expansion. However, bamboo, including *P. pubescens*, is an important plant in Japanese culture; therefore, the complete removal of this invasive foreign species is not an ideal option. Instead, the development of suitable management methods for bamboo groves is required.

We suggest a management plan based on the expansion characteristics of bamboo groves. Unmanageable groves can be found on steep slopes or far from roads. Such groves are expected to undergo very rapid expansion, which is facilitated by the fact that bamboo invades vegetation more readily when the vegetation is lower than the height of the bamboo itself. Thus, bamboo groves situated in unmanageable locations or adjacent to low or transitional vegetation require more immediate attention.

In contrast, mature secondary forests are more resistant to decreases in plant species diversity as a result of bamboo invasion. Therefore, bamboo groves that are

Table 15.2 A suitable bamboo-grove management plan based on the characteristics of bamboo invasion

	Expansion speeds of bamboo groves and effects	
Management labors	Large: adjacent to low or transitional vegetation	Small: adjacent to high and mature forest
Easy, gentle slope (<25°) Easy to access (less than 150 m from roads)	Control bamboo groves	Bamboo groves for sustai nable use with moderate control
Hard, steep slope (>25°) Difficult to access (more than 150 m from roads)	High-priority bamboo groves	Low-priority bamboo groves

adjacent to vegetation that is tall, or in mature stages of growth, require less immediate management attention.

We developed a bamboo-grove management plan that incorporates ease of use while reducing the risk of declining biodiversity in neighboring vegetation (Table 15.2). In Table 15.2, we distinguish between bamboo groves that are in need of moderate control and high-priority bamboo groves, based on the characteristics of the adjacent vegetation. We recommend that high-risk groves, such as those adjacent to low vegetation or transitional vegetation, be cut down.

In addition, we suggest a technique to eradicate such undesired bamboo groves. After a grove is logged, new shoots grow very quickly. Indeed, Suzuki et al. (2008) reported that 28 bamboo shoots (maximum height 7.2 m) were present 2 years after bamboo culms were logged. Therefore, to completely remove the bamboo, the bamboo shoots that appear in spring for several years after cutting must be removed. Moreover, few trees grow naturally in cleared bamboo sites (Suzuki et al. 2008). Therefore, planting desired tree species is required.

Conversely, if bamboo groves are to be used sustainably, clear-cutting bamboo culms is not an ideal management approach; the thin shoots that appear during the year following clear-cutting (Suzuki et al. 2008) are not useful as building materials. Therefore, thinning culms is superior to clear-cutting for the sustainable use of bamboo for materials.

Acknowledgments We extend our thanks to a number of people for their help in preparing this Ph.D. thesis. We are particularly indebted to Dr. Sun-Kee Hong, of Mokpo National University, who provided the opportunity to conduct this study.

References

Fukamachi K, Oku H, Nakashizuka T (2001) The change of a satoyama landscape and its causality in Kamiseya, Kyoto Prefecture, Japan between 1970 and 1995. Landsc Ecol 16:703–717

Hiura H, Arikawa T, Bahadur DD (2004) Risk of sediment related disasters due to the abandoned expanding bamboo stands at the foot of slopes surrounding city area. J Jpn Landslide Soc 41:323–334 (in Japanese)

Hong S-K, Nakagoshi N, Kamada M (1995) Human impacts on pine-dominated vegetation in rural landscapes in Korea and western Japan. Vegetatio 116:161–172

Isagi Y, Torii A (1998) Range expansion and its mechanisms in a naturalized bamboo species, *Phyllostachys pubescens*, in Japan. J Sustain For 6:127–141

Ishiga H, Dozen K, Kodera Y, Haito K (2001) Effects of bamboo invasion on the soils of broadleaf forests and their potential environmental impact. Geosci Rep Shimane Univ 20:83–86 (in Japanese)

Ito T, Yamada A (2005) Actual condition of intrusion of bamboo forests and charactaerstics of the growth of moso bamboo (*Phyllostachys heterocycla* (Carr.) Mitf.). Bull Agric Food Environ Sci Res Cent Osaka Prefect 41:11–18 (in Japanese)

Nakagoshi N (1995) Changing cultural landscape in Japan. In: Von Droste B, Plachter H, Rössler M (eds) Cultural landscapes of universal value. Gustar Fisher, Jena

Nakagoshi N, Hong S-K (2001) Vegetation and landscape ecology of East Asian 'Satoyama'. Glob Environ Res 5(2):171–181

Nishikawa R, Murakami T, Yoshida S, Mitsuda Y, Nagashima K, Mizoue N (2005) Characteristic of temporal range shifts of bamboo stands according to adjacent landcover type. J Jpn For Soc 87:402–409 (in Japanese)

Ogura J (1988) The transition of bamboo forest in the suburbs of Kyoto in and after modern times. Kino-Hyoron Kyoto Seika Univ 19:25–41 (in Japanese)

Okutomi K (2005) Bamboo groves. In: Fukushima T (ed) Vegetation Management. Tokyo, Asakura Publishing, pp 79–86 (in Japanese)

Okutomi K, Shinoda S, Fukuda H (1996) Causal analysis of the invasion of broad-leaved forest by bamboo in Japan. J Veg Sci 7:723–728

Searashi T, Maru M, Omori M, Nishii T (1989) Special characters of the bamboo forest on the structure and the succession – a succession from the summer-green forest to the bamboo forest. Bull Fac Educ Kanazawa Univ Nat Sci 38:25–40 (in Japanese)

Shibata S (2001) Bamboo. In: Ogata M, Shibata S (eds) Cat and Bamboo. Iwanami Publishing, Tokyo, pp 67–135 (in Japanese)

Suzuki S, Nakagoshi N (2008) Expansion of bamboo forests caused by reduced bamboo-shoot harvest under different natural and artificial conditions. Ecol Res 23:641–648

Suzuki S, Takahachi Y, Kikuchi A, Nakagoshi N (2006) Process of *Phyllostachys pubescens* culm invasion at expansion front. Hikobia 14:477–482 (in Japanese)

Suzuki S, Kikuchi A, Nakagoshi N (2008) Structure and species composition of regenerated vegetation after clear cutting of *Phyllostachys pubescens* culms. Landsc Ecol Manag 12(2):43–51 (in Japanese)

Torii A (1998) Estimation of range expansion rate of bamboo stands using aerial photographs – case study on Mt. Hachiman, Shiga Prefecture, and Mt. Otoko, Kyoto Prefecture, Japan. Jpn J Ecol 48:37–47 (in Japanese)

Torii A, Isagi Y (1997) Range expansion of bamboo species in southern areas of Kyoto Prefecture, Japan. Jpn J Ecol 47:31–41 (in Japanese)

Turner MG, Gardner RH, O'Neill RV (2001) Landscape ecology in theory and practice: pattern and process. Springer, New York

Ueda K (1968) The bamboos. Mainichi-shinbunsha, Tokyo (in Japanese)

Chapter 16
Characteristics of the Spatial Distribution, Vegetation Structure, and Management Systems of Shrine/Temple Forests as Urban Green Space: The Case of Kitakyushu City

Tohru Manabe, Keitaro Ito, Daisuke Hashimoto, Dai Isono, Takashi Umeno, and Shuji Iijima

16.1 Introduction

Green spaces within urban areas (urban green space; UGS) could have ecological functions for wildlife, such as habitat, conduit, sink, and source (Manabe et al. 2007a), as well as providing some ecosystem services for residents, such as the regulation of environmental conditions, prevention of disasters, amenities, and so on (Pickett et al. 2001). Techniques for the re-vegetation of the rooftops and walls of buildings, and the creation of artificial biotopes, have been studied in order to increase such ecological functions and their ecosystem services in urban areas with little UGS.

Shrine/temple forests (S/T forests), which are forests owned and managed by shrines and temples, still exist in urban areas of Japan. Further, some urban S/T forests have been free from human disturbance for a long time because they themselves have been regarded as sacred, and therefore they have some characteristics of the primeval vegetation of the region. From technological and economic viewpoints, it is clearly more effective to use these primeval S/T forests as the centers for networks of UGS than to create new green spaces, although many S/T forests are isolated in the urban landscape (Fujita and Kumagai 2004; Hashimoto et al. 2006).

T. Manabe (✉)
Kitakyushu Museum of Natural History and Human History, Kitakyushu 805-0071, Japan
e-mail: manabe@kmnh.jp

K. Ito and D. Isono
Department Civil Engineering, Faculty of Engineering, Kyushu Institute of Technology, Kitakyushu 804-8820, Japan

D. Hashimoto and T. Umeno
Graduate School of Civil Engineering, Kyushu Institute of Technology, Kitakyushu 804-8550, Japan

S. Iijima
Graduate School of Human-Environment Studies, Kyushu University, Fukuoka 812-8581, Japan

S.-K. Hong, et al. (eds.), *Landscape Ecology in Asian Cultures*, Ecological Research Monographs, DOI 10.1007/978-4-431-87799-8_16, © Springer 2011

Vegetation structures, as well as some micro-environmental conditions such as irradiance, temperature, moisture, and wind velocity, of many S/T forests in an urban landscape may be altered, as seen in many isolated forests in various types of landscape (Saunders et al. 1991; Chen et al. 1992; Murcia 1995; Williams-Linera et al. 1998; Zipperer 2002; Tojima et al. 2004). Indeed, some studies have reported the effects of isolation on urban S/T forests. For example, the internal micro-environmental conditions of the S/T forests are less stable in urban than in rural areas (Takeda et al. 1979; Sakamoto et al. 1989; Ishii et al. 2004, 2005). The species richness of the urban S/T forests decreases as the area of the forests decreases (Sakamoto et al. 1990; Ishida et al. 1998; Murakami and Morimoto 2000; Murakami et al. 2005; Imanishi et al. 2007). Changes in the matrix surrounding the S/T forests also bring about changes in the micro-environments (Takeno and Ishii 2007) and community structures of the forests (Manabe et al. 2003). Further, community structures of the urban S/T forests are affected by forest management systems and social and political factors (Sakamoto et al. 1985; Shimizu et al. 2004; Fujita et al. 2005; Iwasaki and Ishii 2005).

In this chapter, we report a case study to evaluate the factors affecting the ecological functions for wildlife and the ecosystem services of the urban S/T forests in Kitakyushu City.

16.2 Kitakyushu City

Kitakyushu City, which had an area of 485.55 km^2 and a population of about one million in 2005, was created in 1963 by merging five cities; the cities were Moji, Kokura, Tobata, Wakamatsu, and Yahata (Fig. 16.1).

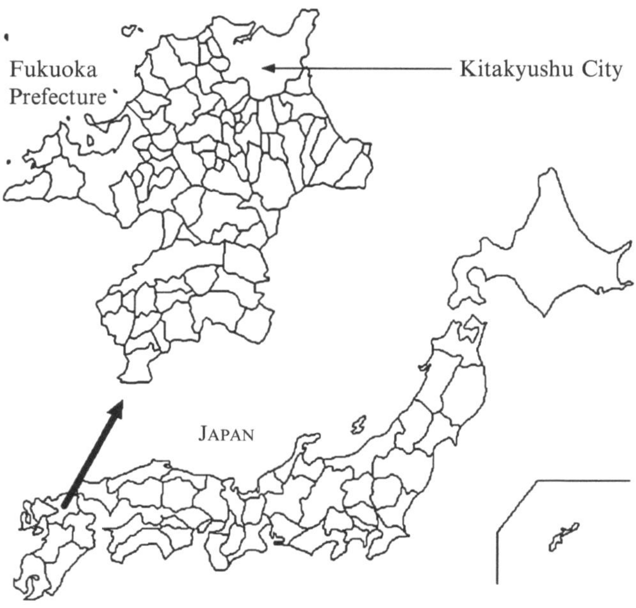

Fig. 16.1 Study site (Kitakyushu City)

The first public iron-manufacturing foundry was constructed in the city (ex-Yahata City) in 1901; since then, the city has developed as "the town of iron foundries." However, many secondary broad-leaved coppices of various areas, shapes, and connectivity remained in the city (Suzuki et al. 2004; Manabe et al. 2007b). The potential natural vegetation in the city area is evergreen broad-leaved forest, dominated by evergreen canopy species such as *Castanopsis cuspidata*, *Quercus salicina*, *Distylium racemosum*, and *Persea thunbergii*, and evergreen sub-canopy species such as *Cinnamomum japonicum*, *Camellia japonica*, and *Cleyera japonica*. Some regulation systems play an important role in conserving areas of UGS in the city (Manabe et al. 2007c).

16.3 Spatial Distribution

How large are the urban S/T forests remaining in the city? How isolated are they? In this section, we describe the spatial distribution patterns of the S/T forests in the city.

There are datasets giving the areas of the precincts of some shrines and temples in Kitakyushu City. About 64% of 222 shrines and about 83% of 415 temples had no forests within their precincts. However, the forests covered more than half of the precincts at most of the shrines and temples which did have forests within their precincts.

We could not find complete datasets on the areas of the precincts of all the shrines and temples in the city. Therefore, spatial distribution patterns were evaluated, not only of the shrines, temples, and urban parks, but also of the forests surrounding them. The digital point data of the shrines, temples, and urban parks were developed using information such as available on the Zenrin Z-Map (http://www.zmap.net/) and the natural environmental information GIS (http://www.biodic.go.jp/kiso/gisddl/gisddl_f.html) using a GIS system. Five circular buffer zones were generated at each point (shrines, temples, and urban parks). The width of each buffer zone was 50 m, and the distances from the point to the outside edge of the buffer zone (the distance classes) were 50, 100, 150, 200, and 250 m. The forest in a buffer zone (FAB) index was calculated as the ratio of the area of forest contained in each buffer zone to the area of the buffer zone for each distance class for each point. This index represents the isolation patterns of the points. For example, large forests are indicated as remaining around the point in the cases when the values of the index at each of the five distance classes are high. However, index values which decrease rapidly as the distance classes increase show that those forests tend to be isolated. The numbers of shrines, temples, and urban parks analyzed were 412, 608, and 1 572, respectively. The FAB index was also used to cluster the shrines, temples, and urban parks with similar spatial distribution patterns using the Ward clustering method. See Hashimoto et al. (2006) for more details.

The values of the FAB index for the shrines gradually decreased as the distance classes increased, and the values were higher for shrines than for either temples or urban parks in all distance classes (Fig. 16.2). These results indicate that forests around the shrines remained relatively larger than those around the temples in this city.

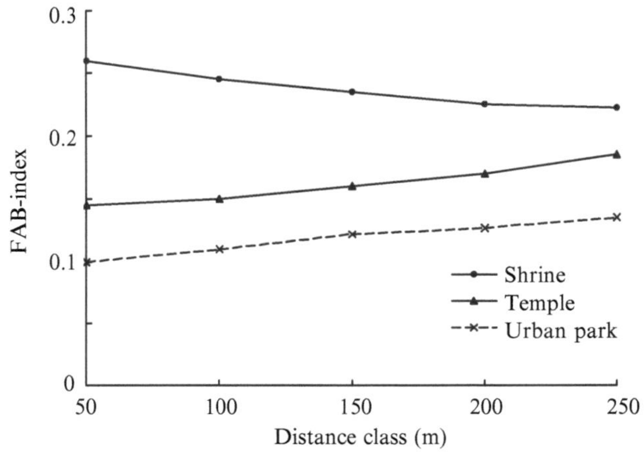

Fig. 16.2 FAB index of the shrines, temples, and urban parks in Kitakyushu City

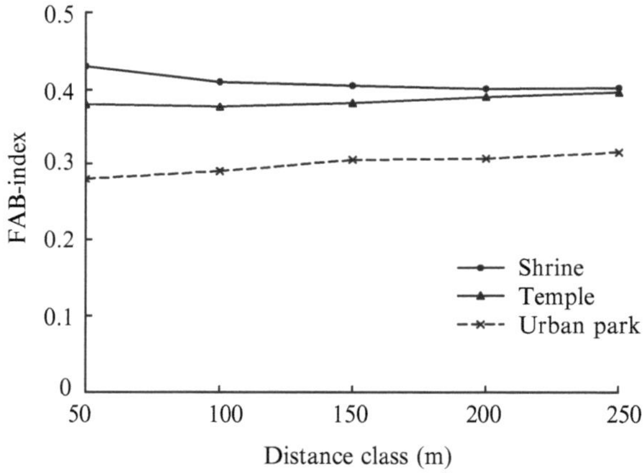

Fig. 16.3 FAB index of the shrines, temples, and urban parks in the urbanization control areas in Kitakyushu City

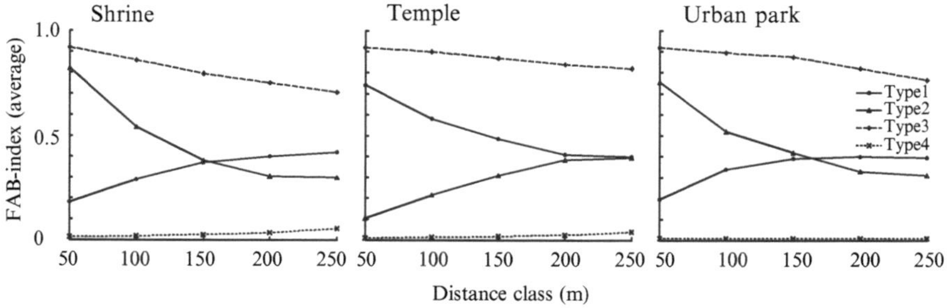

Fig. 16.4 Classification of the shrines, temples, and urban parks based on the FAB index by the Ward clustering method

Table 16.1 The number of shrines, temples, and urban parks classified into four types by the Ward method

	Type 1	Type 2	Type 3	Type 4
Shrine	71	51	64	226
Temple	113	35	54	406
Urban park	258	170	200	2083

The shape of the FAB index for the shrines differed from that of the temples and the urban parks in the whole city. The shapes of the index, however, were similar for the shrines and the temples in the urbanization control areas, which should have restrained urbanization (Fig. 16.3). These results show that the larger forests remained around the temples located at the foot of mountains, since the urbanization control areas occur mainly at the foot of the mountains in the city (Umeno et al. 2006; Manabe et al. 2007c).

The shrines, temples, and urban parks were classified into four groups by the Ward clustering method (Fig. 16.4). In Type 1, the average values of the index were relatively small in the smaller distance classes, and tended to converge toward ca. 0.4. For Type 2, the values were relatively large in the smaller distance classes, and tended to converge to ca. 0.3–0.4, In Type 3, most of the values were large (0.8–1.0), whereas in Type 4, most of the values were small (almost 0.0). More than half of the shrines, temples, and urban parks have no forests (Type 4) or small (Type 1) forests, although about 12% of the shrines have large forests (Type 2, Table 16.1). Therefore, the S/T forests classified as Type 2 might be expected to have the highest ecological functions for wildlife among the four types. These roles are more to be expected for the shrine forests than for the temple forests.

16.4 Managements of the S/T Forests

To evaluate the relationships between the actual management of the S/T forests and the consciousness of their priests, qualitative data were collected by direct interviews from 12 managers of shrines in Kitakyushu City. Those data were analyzed using the cumulative KJ Method (Kawakita 1986). See Hashimoto et al. (2007) for more details.

Most of the priests wished to preserve the S/T forests. However, what was considered ideal management differed slightly among the priests, and was generated from ideas concerning the "function" of the forests and the "future" of the forests, which were based on the religious background of the priests (Fig. 16.5). The function of the S/T forest derives from ideas that the forest itself is sacred; the forest brings ease and peace of mind, and so on. The future of the forests is based on the idea that the forest must be protected for ever. Ideal management concepts affected the actual management patterns. For example, the construction of a fence was to protect the forest from damage by people who entered without permission (=function of the forests). Some priests cut the understory vegetation and branches of trees because such management was necessary to keep visitor's safe, and some planted trees such as *Chamaecyparis obtusa*, *Cryptomeria japonica*, and *Cleyera japonica* (one of the important trees for the Shinto religion) to make the forest more sacred (=future of the forests).

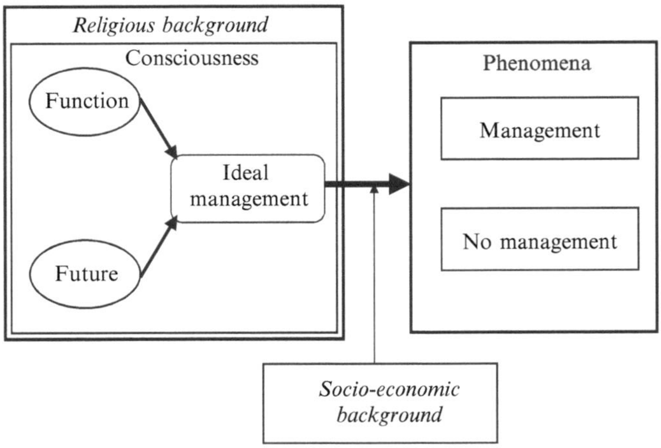

Fig. 16.5 Relationships between the actual management of the S/T forests and the consciousness of their priests

Some priests have done nothing because they thought that the S/T forests were sacred and had to be free from human impact. On the other hand, other priests could not carry out their ideal management plans because of lack of funds. Therefore, the ideas of the priests influence the vegetation structures of the S/T forests.

16.5 Vegetation Structure

Some well-preserved S/T forests were thought to have the characteristics of the probable primeval vegetation in the region. What, then, is the vegetation structure of the isolated S/T forests in the city?

A phytosociological survey was conducted on 32 plots at 24 shrines and on six plots at six temples in the study site located in the middle of Kitakyushu City (Fig. 16.6). In addition, the phytosociological data at five typical secondary coppices in the study site, which are dominated by deciduous oak (*Quercus serrata*) (Mitsuda et al. 2003), were also used in the following analysis. The area of each plot was between 100 and 400 m^2.

The similarity index (R) was used to analyze the similarity among the 43 plots. That is,

$$R = 2 * \min (x_i, y_i)/2 * (x_i, y_i)$$

where x_i and y_i are the coverage of species i at plot x and plot y for each layer. Then the similarity index was used to obtain clusters of similar plots by the Ward clustering method.

The plots contained a total of 179 species, of which 152 were in the S/T forests and 106 were in the secondary coppices. The number of species was largest in the secondary coppices (average 50.6, range 41–57), followed by the shrine forests (27.6, 14–44) and the temple forests (23.0, 18–34).

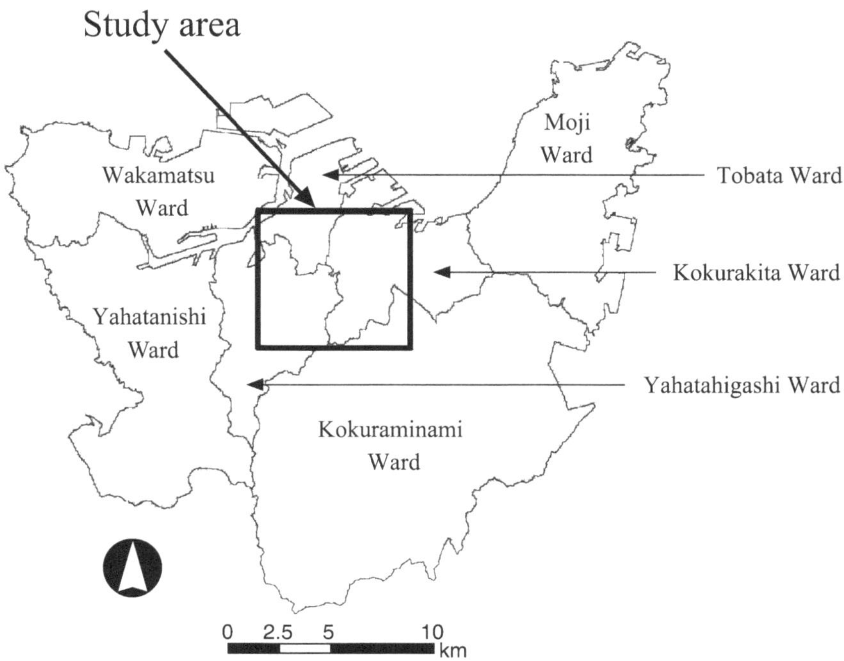

Fig. 16.6 Study area (in the middle of the northern region of Kitakyushu City)

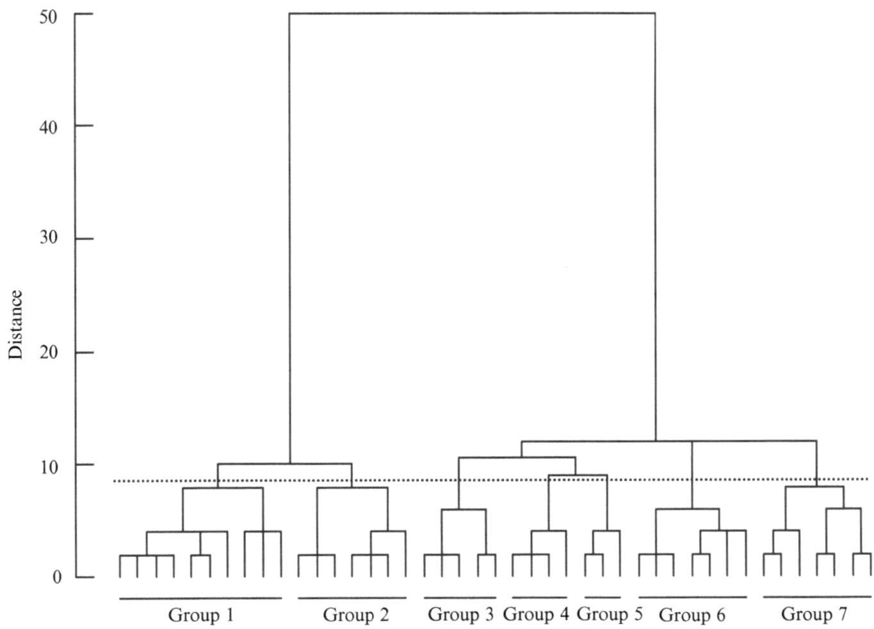

Fig. 16.7 Dendrogram for 43 plots at the shrines, temples, and secondary deciduous coppices based on cluster analysis using Ward methods

Table 16.2 Dominant species in the canopy layer at each group divided by the Ward method

Species	Group 1	Group 2	Group 3	Group 4	Group 5	Group 6	Group 7
Persea thunbergii	●	●	●		●		
Castanopsis cuspidata		●	●				
Ilex rotunda			●				
Cinnamomum camphora			●				
Cinnamomum japonicum					●		
Myrica rubra						●	
Rhus succedanea	●						
Quercus serrata						●	
Chamaecyparis obtusa							●
Cryptomeria japonica							●
Phylostachys pubescens				●			

Dominant species in the canopy layer varied considerably among the plots (see Appendix). *P. thunbergii*, *C. cuspidata*, and *Rhus succedanea*, which are potential components in the natural vegetation of the region, occurred frequently in the canopy. On the other hand, *Phyllostachys pubescens*, which is an invasive alien plant in Japan, and plantings of *C. obtusa*, *Cryptomeria japonica*, *Ginkgo biloba*, and *Cinnamomum camphora* were also observed in this layer. Evergreen species such as *Camellia japonica*, *Quercus glauca*, *P. thunbergii*, *Symplocos lucida*, and *Neolitsea sericea* often occurred in the sub-canopy layer. In the understory, *Fatsia japonica* and *Trachycarpus fortunei*, which are typical invasive species, and typical pioneers such as *Zanthoxylum ailanthoides* and *Mallotus japonicus*, grew in some plots. Further, saplings and seedlings of *N. sericea* and *Ficus erecta* were very common in the understory.

A dendrogram based on cluster analysis shows that the plots were divided into two types (Fig. 16.7). Type 1 was formed by the plots with components of the potential natural vegetation of the region in their canopy layer, such as *P. thunbergii* and *C. cuspidata*. Type 2 contained plots with no dominants. Type 1 is thought to have a community structure relatively similar to the old-growth forests in the region.

Types 1 and 2 were subdivided into two and five groups, respectively. Groups 1 and 2 were constructed with ten and seven plots, respectively, and both were dominated by *P. thunbergii* in their canopy layers (Table 16.2). In Group 2, *C. cuspidata* was dominant in the canopy layer, indicating that the group is more similar to the old-growth forests than Group 1. This has *R. succedanea* in the canopy layer, which is one of the pioneer trees in the region. Group 3 contained eight plots, and the canopy layer of the group was dominated by one of the components in the range of potential natural vegetation of the region, such as *P. thunbergii* or *C. cuspidata*. However, the group also had *C. obtusa* and *C. camphora* in the canopy layer; these were planted by people. Group 4 (four plots) was characterized by a high coverage of *P. pubescens*. Group 5 (three plots) included *Cinnamomum japonica* and *N. sericea*, which produce bird-dispersed seeds, in canopy and subcanopy layers.

Plots of this group might be considerably affected by their isolation, since a dominance of *N. sericea* was often observed in the secondary evergreen broad-leaved forests (Manabe et al. 2007b), especially the isolated ones in the region (Manabe et al. 2003). The forests classified in Group 6 contained seven plots. The group consists of relatively young secondary forests, because it contained all five plots of the secondary coppices dominated by deciduous oak. Group 7, which also contained seven plots, was largely dominated by *C. obtusa* and *Cryptomeria japonica*, indicating that the S/T forests were artificial ones.

Thus, the vegetation structure of some S/T forests was relatively highly natural, while some forests were affected by such factors as the degree of isolation and various kinds of human impact such as planting.

16.6 Concluding Remarks

The S/T forests vary in their area and their degree of isolation in Kitakyushu City. Many shrines had relatively large forests. On the other hand, some temples at the foot of mountains also had large forests, suggesting that the spatial distribution of temples was affected by the geographic features of the city. Different patterns were reported in Tokyo. The shrine forests were distributed randomly, but were strongly associated with geographic features, whereas temple forests displayed clustered distributions and a weak association with geographic features. Parks also displayed random distributions, but had no association with geographic features (Fujita and Kumagai 2007).

The vegetation structures of the S/T forests, as well as the spatial distribution patterns, were affected by their management systems. Indeed, some S/T forests had relatively highly natural vegetation structures, whereas others had habitat and/or sink functions for edge species, pioneer species, and invasive aliens, as has also been seen in the Kansai region in Japan (Murakami and Morimoto 2000; Iwasaki and Ishii 2005).

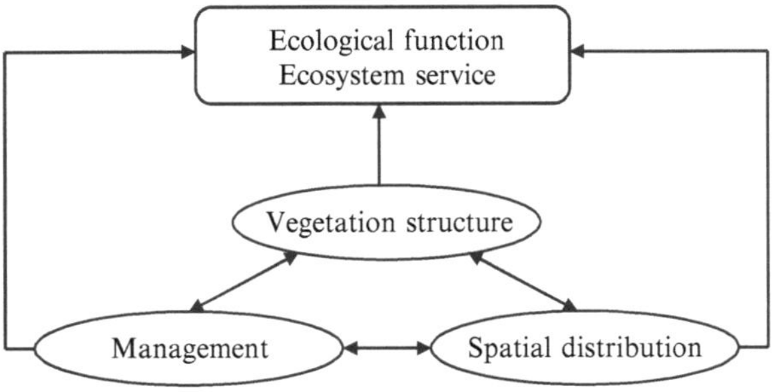

Fig. 16.8 Factors affecting the ecological functions and ecosystem services in an urban landscape

These findings suggest that the spatial distributions, management systems, and vegetation structures of the urban S/T forests are related to each other, and their ecological functions for wildlife and their ecosystem services were affected by these conditions (Fig. 16.8). It is therefore important to set up goals based on scientific evaluations of spatial distribution and management patterns, as well as the vegetation structures, of the S/T forests, and to build a consensus among planners, the priests of the forests, and the residents around the forests in order to make effective network plans of the UGS in urban landscapes, with the S/T forests as the center of them.

Acknowledgments We thank Dr. Ishii, Dr. Imanishi, and Dr. Fujita for their helpful comments on the manuscript. This study was supported by a Grant in Aid of Science Research (155700266 and 19300264) from the Japanese Society for the Promotion of Science and contributions to projects for the region from the Ministry of Education, Culture, Sports, Science and Technology and the Kyushu Institute of Technology.

16.7 Appendix

Tree species in the canopy layer on 43 plots

		Number of plots	
Species	Life form[a]	S/T forest	Secondary coppice
Persea thunbergii	Eb	16	1
Castanopsis cuspidata	Eb	14	1
Rhus succedanea	Db	12	1
Ilex rotunda	Eb	7	1
Cinnamomum camphora	Eb	7	
Neolitsea sericea	Eb	7	
Cinnamomum japonicum	Eb	7	
Quercus glauca	Eb	5	
Chamaecyparis obtusa	G	4	
Phyllostachys pubescens	B	4	
Myrica rubra	Eb	3	3
Quercus serrata	Db	2	4
Symplocos lucida	Eb	2	1
Ginkgo biloba	G	2	
Cryptomeria japonica	G	2	
Cleyera japonica	Eb	1	1
Prunus jamasakura	Db	1	1
Ternstroemia gymnanthera	Eb	1	1
Daphniphyllum macropodum	Eb	1	1
Firmiana simplex	Db	1	1
Carpinus tschonoskii	Db	1	
Acer palmatum	Db	1	
Celtis sinensis var. *japonica*	Db	1	
Myrsine seguinii	Eb	1	

(continued)

16.7 Appendix (continued)

Species	Life form[a]	Number of plots	
		S/T forest	Secondary coppice
Pasania glabra	Eb	1	1
Ligustrum japonicum	Eb	1	
Symplocos myrtacea	Eb	1	
Evodia meliaefolia	Db	1	
Robinia pseudo-acacia	Db	1	
Aphananthe aspera	Db	1	
Ilex macropoda	Db		1
Quercus phillyraoides	Eb		1
Zanthoxylum ailanthoides	Db		1
Picram quassioides	Db		1
Rhamnus crenata	Db		1

[a]*Eb* evergreen broad-leaved species, *Db* deciduous broad-leaved species, *G* evergreen gymnosperms, *B* evergreen bamboo

References

Chen J, Franklin JF, Spies TA (1992) Vegetation responses to edge environments in old-growth Douglas-fir forests. Ecol Appl 2:387–396

Fujita N, Kumagai Y (2004) Landscape change and uneven distribution of urban forest in center of Tokyo. Landsc Res J 67(5):577–580 (in Japanese with English abstract)

Fujita N, Kumagai Y (2007) Comparative study between distribution of shrine, temple and park in urban areas analyzed by GIS. Landsc Ecol Manage 12:9–21 (in Japanese)

Fujita N, Ono R, Kumagai Y (2005) Meanings and their position of "Shasoh" on the preservation system for national monuments. Landsc Res J 68(5):417–420 (in Japanese with English abstract)

Hashimoto D, Ito K, Manabe T, Isono D, Umeno T (2006) Basic study on distributional patterns and ecological characteristics of shrine/temple forests in Kitakyushu City. Kyushu J For Res 59:56–59 (in Japanese)

Hashimoto D, Ito K, Iijima S (2007) The consciousness structure of the Shinto priests for the management the shrine forests in urban area. Landsc Ecol Manage 12(1):45–52 (in Japanese)

Imanishi A, Murakami K, Imanishi J, Hashimoto H, Morimoto Y, Satomura A (2007) Conservation of isolated urban green spaces for plant species: characteristics of shrine and temple forests and precincts. Landsc Ecol Manage 12(1):23–34 (in Japanese with English abstract)

Ishida H, Hattori T, Takeda Y, Kodate S (1998) Relationship between species richness or species composition and area of fragmented lucidophyllous forests in southeastern Hyogo Prefecture. Jpn J Ecol 48:1–16 (in Japanese with English abstract)

Ishii H, Iwasaki A, Sato S (2004) Seasonal variation of edge effects on the vegetation, light environment and microclimate of primary, secondary and artificial forest fragments in southeastern Hyogo Prefecture. In: Proceedings of the IUFRO international workshop on landscape ecology 2004. Forestry and Forest Products Research Institute, Tsukuba

Ishii H, Sato S, Iwasaki A (2005) Microclimate mitigation in shrine/temple forests of southeastern Hyogo Prefecture. Nat Hum Activities 9:47–56

Iwasaki A, Ishii H (2005) Vegetation structure of fragmented shrine/temple forests in southeastern Hyogo Prefecture – estimation of edge-effect distance and minimum conservation area. Hum Nat 15:29–42 (in Japanese with English abstract)

Kawakita J (1986) KJ method – seeking order out of chaos. Chuokouron-sha, Tokyo (in Japanese)

Manabe T, Kashima H, Ito K (2003) Stand structure of a fragmented evergreen broad-leaved forest at a shrine and changes of landscape structures surrounding a suburban forest, in northern Kyushu. J Jpn Soc Reveg Tech 28(3):438–447

Manabe T, Ishii H, Ito K (2007a) Evaluation of shrine forests as urban green space. Landsc Ecol Manage 12(1):1–7 (in Japanese with English abstract)

Manabe T, Ito K, Hashimoto D, Naito K, Watanabe S (2007b) Community structure of secondary evergreen broad-leaved forests in the urban area of Kitakyushu City, western Japan. Bull Kitakyushu Museum Nat Hist Hum Hist Ser A 5:39–48

Manabe T, Ito K, Isono D, Umeno T (2007c) The effects of the regulation system on the structure and dynamics of green space in an urban landscape. In: Hong SK et al (eds) Landscape ecological applications in man-influenced areas. Springer, Dordrecht

Mitsuda Y, Manabe T, Ito K, Kashima H, Suzuki T (2003) The methods of making the digital vegetation maps by using digital orthophotograhs. Bull Kitakyushu Museum Nat Hist Hum Hist Ser A 1:57–65 (in Japanese with English abstract)

Murakami K, Morimoto Y (2000) Landscape ecological study on the woody plant species richness and its conservation in fragmented forest patches in Kyoto City area. J Jpn Soc Reveg Tech 25(4):345–350 (in Japanese with English summary)

Murakami K, Maenaka H, Morimoto Y (2005) Factors influencing species diversity of ferns and fern allies in fragmented forest patches in the Kyoto city area. Landsc Urban Plan 70:221–229

Murcia C (1995) Edge effects in fragmented forests: implications for conservation. Trend Ecol Evol 10:58–62

Pickett STA, Cadenasso ML, Grove JM, Nilon CH, Pouyat RV, Zipperer WC, Costanza R (2001) Urban ecological systems: linking terrestrial ecological, physical, and socioeconomic components of metropolitan areas. Ann Rev Ecol Syst 32:127–157

Sakamoto K, Kobayashi T, Ikeuchi Z (1985) A case study on the structure of the stand in the forest surrounding Shimogamo Shrine, Kyoto. J Jpn Inst Landsc Arch 48(5):175–180 (in Japanese with English summary)

Sakamoto K, Ishihara S, Chiba K (1989) Study on the woods in shrine or temple precincts of Okayama, Japan (I) – total structure of the woods in shrine or temple precincts, in urban and suburban areas. J Jpn Soc Reveg Tech 15(2):28–35 (in Japanese with English summary)

Sakamoto K, Ishihara S, Chiba K, 3 (1990) Species diversity of the woods in shrine or temple precincts – a case study on the woods in shrine or temple precincts of Okayama City. J Jpn Soc Reveg Tech 15(3):39–47 (in Japanese with English summary)

Saunders DA, Hobbs RJ, Margules CR (1991) Biological consequences of ecosystem fragmentation: a review. Conserv Biol 5:18–32

Shimizu M, Tahara N, Tanikatu N, Kamihogi A (2004) A study on the change process of historic greenery at shrines and temples precincts in Osaka Drawn in meizho-zue. Pap Environ Inf Sci 18:171–176 (in Japanese with English abstract)

Suzuki T, Manabe T, Ito K, Umeno T (2004) Analysis of landscape changes by using digital vegetation map in mid-northern region in Kitakyushu City. Bull Kitakyushu Museum Nat Hist Hum Hist Ser A 2:79–85 (in Japanese with English abstract)

Takeda A, Umebayashi M, Yamashita Z, Matsuura M (1979) Studies on environmental interaction between urban area and shrine forest 4: Microclimatic conditions on the margin of Tsu-Hachimangu shrine forest and diurnal changes of carbon dioxide concentration in the air and in the soil atmosphere of that forest. Rep Environ Sci, Mie University 4:27–45 (in Japanese with English abstract)

Takeno S, Ishii H (2007) Forest community dynamics over a 27-year-period in Hino Shrine. Hyogo Pref Hum Nat 18:21–28 (in Japanese with English abstract)

Tojima H, Koike F, Sakai A, Fujiwara K (2004) Plagiosere succession in urban fragmented forests. Jpn J Ecol 54:133–141 (in Japanese with English abstract)

Umeno T, Ito K, Manabe T (2006) Research on the influence of city planning system for change in land covering with satellite images –a case study in Kitakyushu City. Kyushu J For Res 59:56–59 (in Japanese with English abstract)

Williams-Linera G, Domînguez-Gastelú V, García-Zurita M (1998) Microenvironments and floristics of different edges in a fragmented tropical rainforest. Conserv Biol 12:1091–1102
Zipperer WC (2002) Species composition and structure of regenerated and remnant forest patches within an urban landscape. Urban Ecosyst 6:271–290

Chapter 17
Conservation and Management of the Coastal Pine Forest as a Cultural Landscape

Michiro Fujihara, Mariko Ohnishi, Hiroyuki Miura, and Yoshihiro Sawada

17.1 Introduction

Coastal forests, which are located between the sea and human settlement, are one of the representative cultural landscapes in East Asia in general, and Japan in particular. Plenty of coastal forests are distributed along the coastlines of the Japanese archipelago (Kawai 1992). The main component of the coastal forests in Japan is Japanese black pine (*Pinus thunbergii* Parl.). The coastal pine forests were established and have been maintained by the local residents for a long time (Oda 1992). These forests perform many functions, including protection against wind, sand, and natural disasters, providing food and materials for fuel and compost, habitat for animals and plants, etc. In addition, some of these forests have been conserved for their scenic beauty. Coastal pine forests are closely related to the daily lives of residents, and are maintained as a cultural landscape (Tanaka 1992). However, the functions of these forests have been changing since the 1960s. In recent years, the fallen leaves of and deadwood in coastal forests have ceased to be used. This is because the main energy resource has changed from wood fuel to fossil fuel, and the favored agricultural resource has changed from compost to chemical fertilizer (Taoda 1988). The number and variety of edible plants on the forest floor has decreased in these four decades. In addition, the coastal pine forests have suffered severely from pine wilt disease. Knowing this, good forest management is absolutely essential, because coastal forests are necessary and indispensable to the local people (Fujihara and

M. Fujihara (✉), M. Ohnishi, H. Miura, and Y. Sawada
Awaji Landscape Planning and Horticulture Academy, Awaji, Hyogo, Japan
e-mail: fujihara@awaji.ac.jp

M. Fujihara and Y. Sawada
Graduate School of Landscape Design and Management, University of Hyogo, 954-2 Nojimatokiwa, Awaji, Hyogo 656-1726, Japan

M. Ohnishi
Department of City Planning, Itami, Japan

H. Miura
Okinawa Environment Club, Naha, Okinawa, Japan

S.-K. Hong, et al. (eds.), *Landscape Ecology in Asian Cultures*, Ecological Research Monographs, DOI 10.1007/978-4-431-87799-8_17, © Springer 2011

Iwasaski 2006). Some management of coastal forests already exists, but it is focused on even-aged artificial forests and conducted by the national and local forestry sector (Sakamoto et al. 2007). Because the coastal forest is considered to be a regional resource and a cultural landscape, it is best maintained by local residents and local government. Moreover, researchers propose that management techniques should incorporate long-term forecasts, because a cultural landscape does not appear in a short period of time. In this chapter, we describe the historical changes in the relationship between the coastal forest and local residents, and we propose that the conservation and management of coastal forests as a cultural landscape should be improved.

This research was partially supported by the Ministry of Education, Science, Sports and Culture, Grant-in-Aid for Scientific Research (A), 2008–2010, 20241009.

17.2 Method

17.2.1 Site Description

The study sites (Fukiagehama and Keinomatsubara) are located in the southern part of Awaji Island, Hyogo Prefecture, Japan (Fig. 17.1). Fukiagehama is located on the southern part of Awaji Island. Pine forests at Fukiagehama have been removed as a result of changes in land use. Keinomatsubara is located on the southwestern part of the island. It has been maintained as a beautiful scenic area with large and old pine trees. This forest belongs to Seto Inland Sea National Park.

Fig. 17.1 Study sites Keinomatsubara and Fukiagehama are located on southern part of Awaji Island, Hyogo Prefecture, Japan

17.2.2 Land Use and Landscape Structure

The relationships between land use and landscape structure were analyzed in the area of Fukiagehama. In order to clarify the relationship between land use and landscape structure, aerial photographs and topographical maps were used. Interviews with elderly people who lived near the coast helped to clarify the changes in the land used by the local residents.

17.2.3 Management of Coastal Pine Forest After Pine Wilt Disease

Ancient literature and drawings helped to clarify the changes in the landscape structure at Keinomatsubara. In order to maintain pine forest suitably on an individualized basis, information about growth conditions, size distribution, and land-use management is necessary. The location, diameter at breast height, height of the lowest branch, height of the lowest leaves, and diameter of the crown were measured, as shown in Fig. 17.2 (Fujihara and Iwasaski 2006). The stem height to diameter ratio (tree height (cm)/diameter at breast height (cm)) was calculated. The widths of annual rings were measured for dead pine trees.

17.2.4 Maintenance Management Including Coastal Sand Vegetation

The number of people using the coastline was counted, and the species composition of sand dune vegetation distributed from the coastline to the pine forest was investigated. Vegetation and land-use maps were drawn up and compared in order to describe the most suitable land-use system at Keinomatsubara.

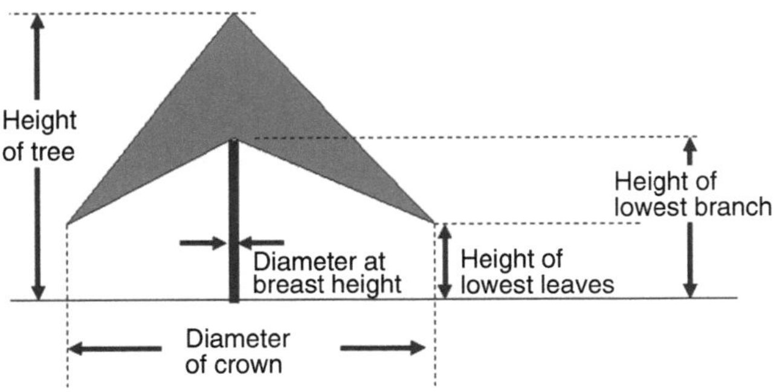

Fig. 17.2 Some measurements of pine trees

17.3 Results

17.3.1 Relationship Between Land Use and Landscape Structure

Over the past seven decades, landscape structure and the use of coastal forest by local residents have changed in Fukiagehama, Japan. In the 1930s, a wide variety of landscape element types was available, and was located from the coastline to inland areas (Ohnishi, unpublished data). Before the 1930s, the coastline, sand dunes, pine forests on sand dunes, embankments, housing (residential areas), paddy and arable fields, private woodland, and common woodland were distributed according to natural disturbances (Figs. 17.3 and 17.4). Pine forests were distributed between the sand dunes, and tall pine trees were planted along the main street. However, artificial levees have been built between the coastline and the sand dunes, and cultivated land and residential areas had expanded before 1975 (Figs. 17.3 and 17.4).

Interviews with the elderly people who lived near the coast helped clarify the changes in the land used by the local residents (Ohnishi, unpublished data). In the 1930s, residents used coastal land for various purposes such as collecting food, fuel, and materials, conducting recreation, hosting outings for pupils, hosting entertainment, conducting religious rites, working, walking, etc. (Table 17.1). Some mushrooms, such as false-truffle (*Rhizopogon rubescens* (Tul. & C. Tul.) Tul. & C. Tul.) and hatsutake (*Lactarius hatsudake* Tanaka), grow on the floor of coastal pine forests and are collected for cooking by the local residents. People used to play in

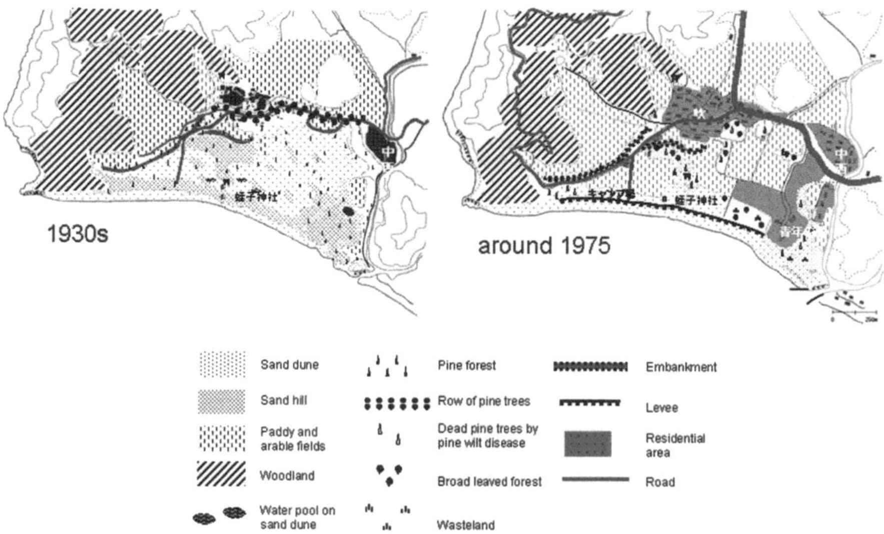

Fig. 17.3 Change in the horizontal structure of the coastal landscape at Fukiagehama, Awaji Island, Hyogo Prefecture, Japan (Ohnishi, unpublished data)

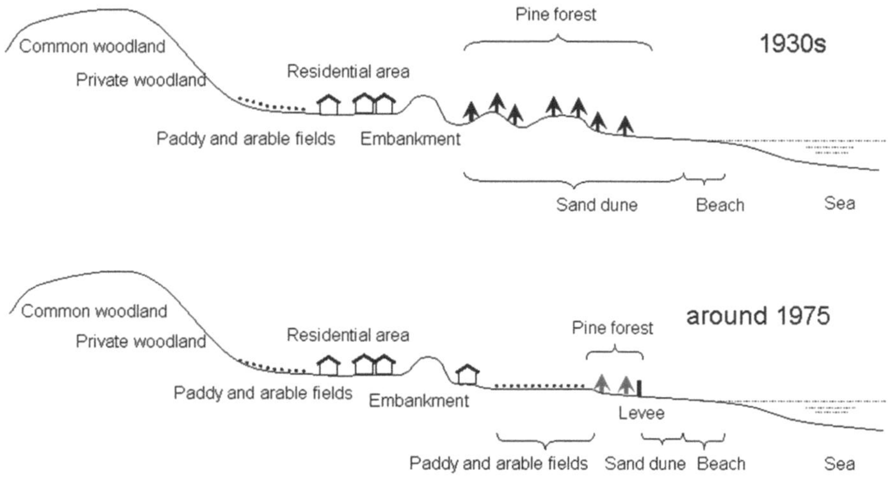

Fig. 17.4 Change in the vertical structure of the coastal landscape at Fukiagehama, Awaji Island, Hyogo Prefecture, Japan (Ohnishi, unpublished data)

the pine forest, using a pine-tree branch as athletic equipment. In the 2000s, the frequency of land use for most of these purposes had decreased or disappeared altogether, but the pine forest's function of protection against wind, sand, and natural disasters remained. According to changes in land use, landscape structure changed between the 1930s and 1975. Landscape structure is closely related to land use by the local residents. The coastal pine forest is a major element of both the coastal landscape and the cultural landscape. However, there was nothing on record about land use, because these uses were very common for the local residents. These land uses were preserved only in the memory of each individual person. It is important to derive these memories from local people in order to clarify the relationship between land use and landscape structure.

17.3.2 Management of Coastal Pine Forest After Pine Wilt Disease

17.3.2.1 Historical Change in the Relationship Between Coastal Forests and Residents

We should understand the historical changes in the relationship between coastal pine forest maintenance and local residents. Keinomatsubara has been maintained as a beautiful scenic area with large and old pine trees (Fig. 17.5). It is presumed that coastal pine trees with low branches were distributed by the 1600s, because descriptions of coastal pine forest appear in the literature from that time (Kikukawa 1981). "Tens of thousands of trees are distributed on Keinomatsubara, and its

Table 17.1 Change of land use by residential people at Fukiagehama, Awaji Island, Hyogo Prefecture (Ohnishi, unpublished data)

			Frequency	
Type	Activity	Place	Past	Present
Collecting food	Collecting edible plants	Sand dune, pine forest	High	None to low
	Collecting fish and shell	Beach	High	Low
	Collecting sea grass	Tide pool	High	Low
Collecting fuel	Collecting leaves and fallen branches of pines	Pine forest, row of pine trees	High	None
	Collecting driftwood	Sand hill	High	
Collecting materials	Collecting plants	Sand dune	Occasional	Rare
	Collecting stones	Sand dune	High	High
	Collecting coconut	Beach, Tide pool	Rare	None
Play	Sea bathing	Sea	High	Infrequent than past
	Swimming	Water pool on sand dune		None
	Baseball	Sand dune		Infrequent than past
	Catching birds	Shrubs on sand dune		None
	Playing the horizontal bar by using a branch of pine tree	Pine forest		None
		Sand hill		None
School	School trip	Sand dune pine forest	High	Infrequent than past
Entertainment	Party on the beach	Sand dune pine forest	Periodic	None
Religious rites	Festival	Sand dune Sand hill	Periodic	Periodic
Works	Construction work	Levee	Occasional	None
	Shipping onion	Sand dune and pier		
	Maintaining ship	Beach		
	Catching fish by dragnet	Beach		
Others	Walking	Sand dune	Unknown	Occasional
	Fishing	Beach		
	Tanking photographs	Sand dune		

scenery is quite unique. It would be a rest place for noble people, if it was located near the Capital Kyoto." (Saikaku Ihara: friend of the remaining influences). One hundred and fifty years ago, the area of pine forest was larger than at present, i.e., 6 km from south to north and 4 km from east to west. All this area was owned by the national government in ancient days, but some of the areas were sold to the private sector. Pine forest which was sold was clear-cut and changed to farmland

Fig. 17.5 A characteristic large old pine tree as a component of the scenic beauty at Keinomatsubara, Awaji Island, Hyogo Prefecture, Japan

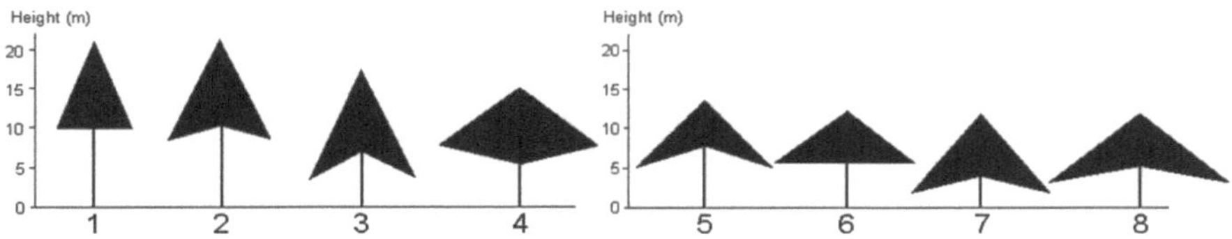

Fig. 17.6 Shape of pine trees in each coastal pine forest (after Asami et al. 2003). 1–4, tall tree type; 5–8, low tree type. 1. Irinomatsubara (Irino cho, Kochi Pref.). 2. Takadanomatsubara (Echizenta-kada city, Iwate Pref.). 3. Amanohashidate (Miyazu city, Kyoto Pref.). 4. Nijinomatsubara (Karatsu city, Saga Pref.). 5. Kehinomatsubara (Tsuruga city, Fukui Pref.). 6. Mihonomastubara (Shizuoka city, Shizuoka Pref.). 7. Keinomatsubara (Minamiawaji city, Hyogo Pref.). 8. Tsu-danomastubara (Tsuda cho, Kagawa Pref.)

(Committee of Education of Seidan town 2001). The rest of the pine forest remains in its original location, where it is preserved as national governmental land.

17.3.2.2 Form of Pine Trees

The main component of coastal forests in various regions in Japan is Japanese black pine. The tree shape varies in each region, as shown in Fig. 17.6 (Asami et al. 2003). Forest maintenance should vary according to each region. The heights of the trees and of their lowest branch were low for pine trees at Keinomatsubara. In general, trees with a high height/diameter (H/D) ratio (over 60) are apt to be damaged by wind storms and snow storms. The H/D ratio of old and large pine trees established in Keinomatsubara worked out to about 20, and these trees have low branches and low leaves. This bold shape is one of the important characteristics of the cultural landscape (Fujihara et al. 2007a).

17.3.2.3 Density and Distressed Condition of Pine Trees

In order to obtain information about forest maintenance, we investigated the distribution, size structure, and distressed condition of pine trees constituting a coastal forest. The number and density of large pine trees were 120 and 4.0/ha, respectively. The number and density of medium-sized pine trees were 189 in 9.3 ha and 20.3/ha, respectively. The H/D ratio of pine trees decreases as the height of the trees increases. The size structure of the trees in the present study was quite different from that in other coastal pine forests composed of even-aged trees. Here, 68.3% of large and 69.3% of medium-sized pine trees were stunted by small pine trees, which prevent sunlight from penetrating to the leaves of large and medium-sized trees (Table 17.2). It is necessary to remove these trees in order to restore good light conditions.

After the 1970s, many large pine trees died because of pine wilt disease, and bare ground appeared after the removal of these dead trees. Many pine saplings have been planted by local groups and individuals, not only on bare ground but also under other pine trees. These planting events are frequent, but selective cutting for density control has not been conducted. Also, there was no control of tree density following planting in many cases. Most of these planted trees grow without self-density selection and selective cutting, which results in a low height-to-diameter ratio, a high "lowest" branch, and higher "lowest" leaves. Many large pine trees in the coastal pine forests, which should be preserved as characteristic trees, are stunted by the planted trees (Fig. 17.7). It is necessary to free these trees and allow them more space in order to keep the shape of each tree. Local people should be given information about which planted trees might be cut down. In addition, it is important to conserve the growth conditions of relatively small trees, because large trees could die from several causes, including, for example, pine wilt disease and aging.

17.3.2.4 Growth Rates of Coastal Pine Trees

In the past, the growth condition and age of living large-sized pine trees were estimated by the radial growth of trunks of dead, fallen Japanese black pine trees in a coastal pine forest preserved for its scenic beauty (Fujihara et al. 2007a).

In general, there is a strong correlation between a tree's age and diameter. The age of a tree can be presumed according to the diameter of remaining trees using the relationship between tree age and diameter. It is considered that the correlation between a tree's age and its diameter at ground level is high, as shown in Fig. 17.8. The ages of living, old pine trees were estimated from the number of annual rings

Table 17.2 Ratio of suppression of large and medium-sized pine trees (after Fujihara and Iwasaski 2006)

	Nonsuppressed tree (well grown tree)	Suppressed tree	Ratio of suppression
Large-sized pine tree	41	82	66.7
Middle-sized pine tree	60	131	68.6

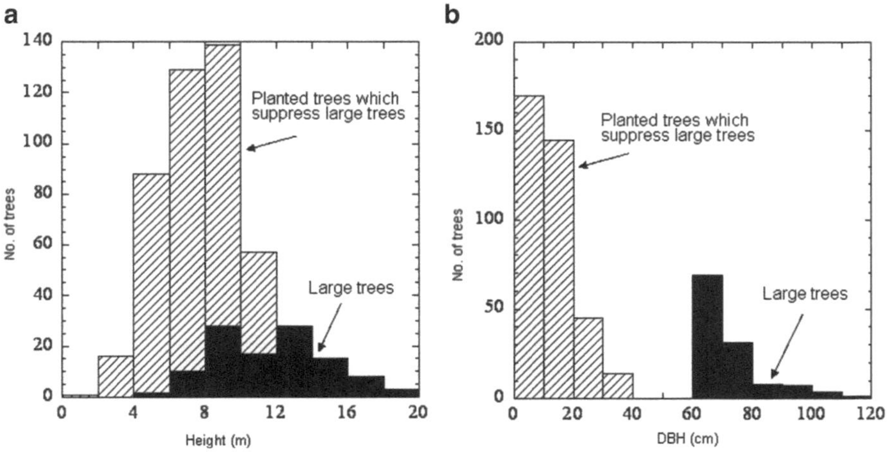

Fig. 17.7 Frequencies of height (**a**) and diameter (**b**) of large pine trees and overlapping trees, which stunt large trees (after Fujihara and Iwasaski 2006)

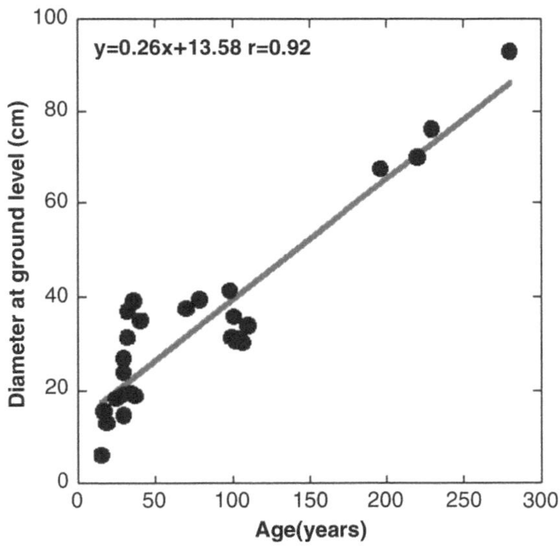

Fig. 17.8 Relationship between age and diameter at ground level of pine trees (Fujihara et al. 2007a)

and the diameters of disks of fallen Japanese black pine trees. Figure 17.9 shows the distribution of the frequencies of the presumed ages of pine trees of 60 cm or more at breast height. The number of trees aged between 170 and 180 years was 45, and the number of trees decreased according to age. The ages of the oldest trees were estimated to be about 350 years, which means that the tree was planted around 1650. An ancient document describes a pine forest as existing in the 1600s, and it is presumed that the pine forests have continued at least since the 1650s.

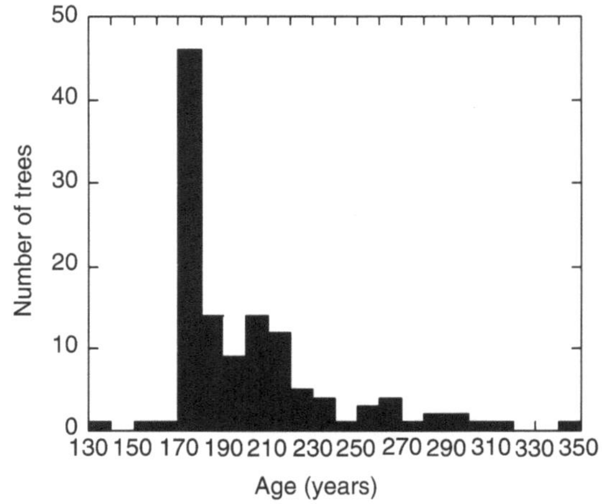

Fig. 17.9 Distribution frequency of age estimated from diameter at breast height (after Fujihara et al. 2007a)

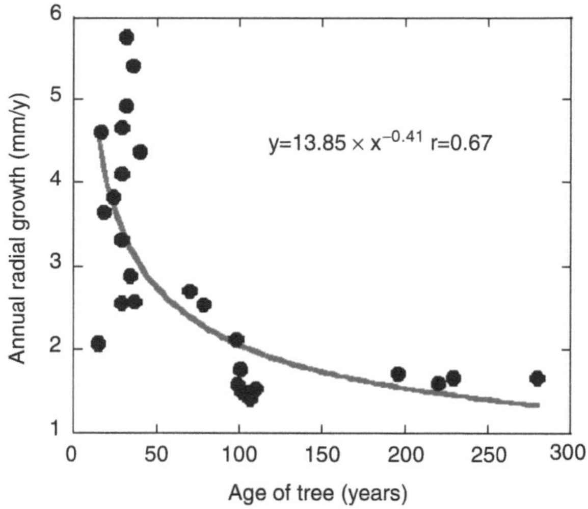

Fig. 17.10 Relationship between age and annual radial growth of pine trees (Fujihara et al. 2007a)

The average annual ring width of old trees and young trees differs greatly. Figure 17.10 shows the relationship between age and annual ring width, e.g., $y = 13.85 \times x - 0.41$ ($r = 0.67$). The annual radial growth of pine trees younger than 80 years old ranged from 2.0 to 6.0 mm. However, the annual radial growth of pine trees older than 95 years ranged from 1.4 to 2.2 mm. The correlation between age and diameter was extremely high for dead pine trees older than 95 years. Growth conditions seem to have been severe from the 1650s to 1900s, and young pine trees were introduced from outside the coastal pine forests. This is another reason why the growth rate is different, and this also led to differently shaped trees. The regional strain of trees also seems to be important. It is necessary to plant endogenous trees in order to maintain the characteristics of a coastal pine forest as a cultural landscape.

17.3.2.5 Habitat for Pine Seedlings

The most suitable habitat for seedlings derived from old and large-sized Japanese black pine trees in the coastal pine forests has been estimated (Fujihara et al. 2007b). The forest floor was divided into three types: sandy soil, mossy floor, and herb-covered floor. Many seedlings of pine trees were found on a mossy floor, but all of them were spindly. On the other hand, there were fewer seedlings on sandy soil than on a mossy floor, but they were generally sturdy (Fig. 17.11). The number of individuals was extremely low, although the seedlings of pine trees on sandy soil were stronger than those in a mossy area.

Seeds from old and large-sized pine trees have a high germination rate. It was presumed that using these seeds and planting out the seedlings would be an effective way of maintaining pine forests that consisted of old and large-sized pine trees. An open habitat covered with sand was most suitable for establishing small, sturdy pine seedlings, although the actual number of seedlings was low (Fujihara et al. 2007b).

The germination rate of the seeds was assumed to be about 95%. Moisture is lost rapidly in the closing phase of seed maturity, and loss of water decreased the germination rate by up to 5%. The best way to maintain beach pine forests is to gather the seeds from old, large-diameter trees, to store them safely, and to cultivate the saplings which are thought to be the strongest.

Broad-leaved trees invade coastal pine forest from surrounding forests and woods in residential areas (Takahashi and Kamitani 2004). It is important to prevent

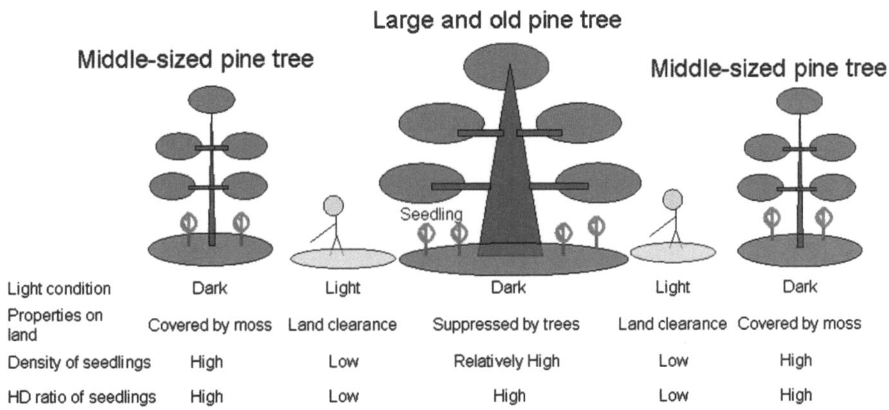

Fig. 17.11 Habitat for seedlings of pine trees (after Fujihara et al. 2007b)

this invasion in order to maintain traditional pine forests, since pine trees used to dominate all areas of Keinomatsubara.

17.3.3 Maintenance Management Including Coastal Sand Vegetation

Vehicles have been be used to collect fallen leaves and branches in order to maintain the coastal forests. In addition, machines such as beach cleaners have been used on the coast for garbage removal (Fig. 17.12). This artificial disturbance with such heavy equipment exerts severe stress on the coastal forest floor and vegetation. The coastal sand vegetation, which stretches from the coastline to the pine forest, is very valuable, because the natural environment and biodiversity could be damaged all along the coastline on a nationwide basis. Coastal pine forests and sand dunes are closely related to each other in terms of function and structure. It is important to control the maintenance of the coastal forest in order to conserve coastal sand vegetation. On the other hand, the beach is important as a place of natural environmental recreation, such as sea bathing. It is important to preserve the coexistence of natural vegetation and artificial land use. Land clearance is inevitable in order to maintain the coastal pine forest. Appropriate zoning by knowledgeable authorities is necessary (Fig. 17.13).

Before the 1960s, management practice for coastal pine forests took the form of collecting daily food, fuel, compost, and materials, but now forest management is carried out for its own sake. A knowledge of ecology and relevant techniques are necessary to maintain the pine forest for the long term. The training of administrators, local residents, and various concerned parties is important, as is site-specific and adaptive management. There are many studies concerning the management of coastal forests, but most of them focus on artificial even-aged forests. There are few reports

Fig. 17.12 Coastal vegetation disturbed by wheel tracks

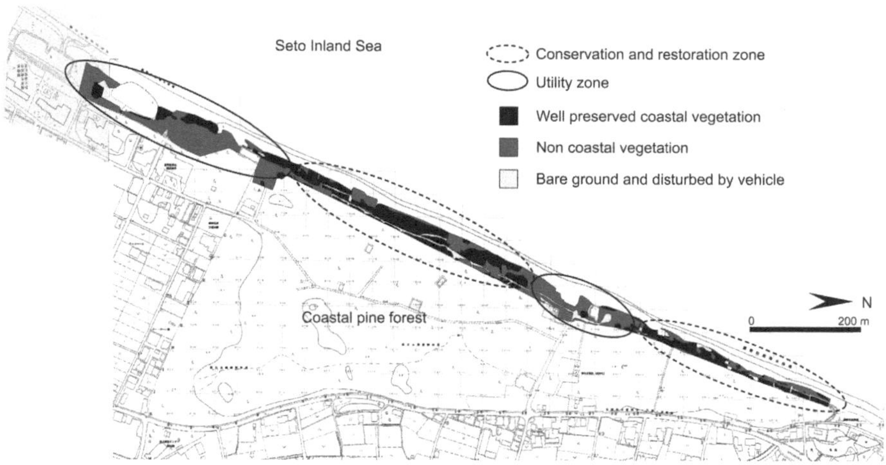

Fig. 17.13 Zoning map for the management and conservation of coastal pine forests and coastal vegetation by local residents (Miura, unpublished data)

concerning natural regeneration. Therefore, it is necessary to accumulate various case studies of maintenance methods for uneven forests, where the only goal was maintaining the region as a cultural landscape.

17.4 Concluding Remarks

Coastal forests are mainly populated by Japanese black pine trees. Forest maintenance should vary according to region. Beginning in the 1970s, many large pine trees died of pine wilt disease, and bare ground appeared after the removal of these dead trees. Planting events were frequent, but selective cutting for density control was not carried out. Coastal sand vegetation is an important remnant of natural vegetation; but on the other hand, beaches are places of recreation that utilize the natural environment. It is important for artificial land use and the maintenance of natural vegetation to coexist.

References

Asami K, Akamatsu K, Matsumura T, Tsuji H, Tamura K, Hattori T (2003) The effect of management on the conservation of vegetation scenery of pine woods. J Landsc Res Inst 66:555–558 (in Japanese with English abstract)

Committee of Education of Seidan town (ed) (2001) Plan for conservation and management of Keinomatsubara as a scenic beauty. Committee of Education of Seidan town (in Japanese)

Fujihara M, Iwasaski Y (2006) Distribution, size-structure and suppressed condition of Japanese black pine (*Pinus thunbergii* Palm.) trees constituting in a coastal forest preserved as a place of scenic beauty. Landsc Ecol Manage 10:81–88 (in Japanese with English abstract)

Fujihara M, Iwasaki Y, Oyabu T, Sawada Y (2007a) The relationship between tree ring widths and age of the large-sized Japanese black pine trees established on the coastal sand pine forest preserved as a scenic beauty. J Jpn Soc Coast For 6:19–22 (in Japanese with English abstract)

Fujihara M, Oyabu T, Sawada Y, Iwasaki Y, Yamamoto S (2007b) The suitable habitat for establishing of seedlings derived from old sand large-sized Japanese black pine trees on the coastal sand pine forest. J Jpn Soc Coast For 7:25–30 (in Japanese with English abstract)

Kawai E (1992) Distribution and change. In: Murai H, Ishikawa M, Endo J, Tadaki Y (eds) Coastal forests in Japan. Soft Science Co. Ltd., Tokyo (in Japanese)

Kikukawa K (1981) Nature and culture of Seidan town. Committee of Education of Seidan town (in Japanese)

Oda T (1992) Conservation, density control and regrowth treatment. In: Murai H, Ishikawa M, Endo J, Tadaki Y (eds) Coastal forests in Japan. Soft Science Co. Ltd., Tokyo (in Japanese)

Sakamoto T, Ogino H, Noguchi H, Shimada K (2007) Regulation of number of trees for coastal pine forests. J Jpn Soc Coast For 6:1–6 (in Japanese with English abstract)

Takahashi K, Kamitani T (2004) Effect of dispersal capacity on forest plant migration at a landscape scale. J Ecol 92:778–785

Tanaka K (1992) Change of coastal forests. In: Murai H, Ishikawa M, Endo J, Tadaki Y (eds) Coastal forests in Japan. Soft Science Co. Ltd., Tokyo (in Japanese)

Taoda H (1988) Succession of Japanese black pine forest on coastal dunes, Hitotsuba Coast, Kyushu, Japan. Hikobia 10:119–128

Chapter 18
How to Conserve Japanese Cultural Landscapes: The Registration System for Cultural Landscapes

Nobukazu Nakagoshi

18.1 Introduction

The conservation of cultural landscapes has recently become a growing environmental issue in Japan (Nakagoshi et al. 1998). Landscapes have developed through interactions between people and nature, thus forming terraced rice paddies and the coppiced woodlands "Satoyama" (Nakagoshi and Hong 2001). These agricultural landscapes are the closest that most people come to nature, and they are acknowledged as being indispensable in environmental conservation and disaster prevention. While being recognized as homeland landscapes, they are now serving a new role, i.e., providing venues to encourage communication between urban and rural lives through various forms of tourism.

The Ministry of Agriculture, Forestry, and Fisheries (MAFF) has introduced new measures, in particular to conserve terraced paddy fields, and gradually the land-owning governments have followed these rules and taken action. The terraced paddies have become widely known today, and many people are interested in visiting them to enjoy the beautiful environment and atmosphere. In addition, some local governments have enthusiastically implemented the "ownership system," in which urban residents rent rural rice paddies and experience rice farming with assistance from local farmers. As a result, terraced paddies are facilitating the desired interaction between urban and rural people.

Since the "Rice Terraces of the Philippine Cordilleras" were added to the World Heritage List of the UNESCO World Heritage Convention, the category of cultural landscape has gained acknowledgment globally. Aided by recent trends in World Heritage registrations, and coupled with contemporary changes in social structures and public awareness in Japan, in 1994, I was able to request that officials of the Agency for Cultural Affairs (ACA) should take all necessary measures for the conservation of cultural landscapes. This was soon after the UNESCO meeting of 1993, which was held in Germany (Nakagoshi 1995).

N. Nakagoshi (✉)
Graduate School for International Development and Cooperation, Hiroshima University, Higashi-Hiroshima 739-8529, Japan
e-mail: nobu@hiroshima-u.ac.jp

S.-K. Hong, et al. (eds.), *Landscape Ecology in Asian Cultures*, Ecological Research Monographs, DOI 10.1007/978-4-431-87799-8_18, © Springer 2011

In response to international and domestic movements, in October 2000 the ACA installed the provisional "Committee on the Preservation, Development, and Utilization of Cultural Landscape Associated with Agriculture, Forestry, and Fisheries." The purpose was to investigate issues regarding the preservation, development, and utilization of cultural landscapes in Japan, particularly in relation to the status of cultural properties represented by "Places of Scenic Beauty" designated under the Law for the Protection of Cultural Properties (the LPCP). The provisional committee consisted of 14 members (as an ecologist, I was one of the members), including the chairperson, and compiled two reports, one in Japanese and the other, its translation in English. This report carries the results from the five meetings held during the period of office (2001–2003) of the Committee. This chapter is based on the above-mentioned English report. First, the question of how cultural landscapes have become the focus of attention both globally and domestically is outlined, and then the purpose of the study delivered by the committee is described. Next, the history and procedure of the research is briefly explained, and the definitions and categories of cultural landscapes are presented. This short-lived committee adopted the concept of cultural landscapes, accompanied by the criteria and results of the selection of important areas. Finally, a new protection system is proposed while taking the characteristics of cultural landscapes into consideration, with the basic principles of development and utilization.

When the Landscape Act was established in 2004, the ACA became the corresponding body on cultural landscapes. Then the sites targeted for conservation were named as "important cultural landscapes." Immediately in the following year, the ad hoc Committee of Cultural Landscapes (of which I am a member) was re-organized within the Investigation Committee of the Council for Cultural Affairs, with the aim of examining all important cultural landscapes. This Investigation Committee of the Council appraised the specific definition of cultural landscapes and classified them into nine categories. Meetings of the Investigation Committee of the Cultural Council were held twice a year in the period between April 2005 and April 2009, a proposal to the Commissioner for Cultural Affairs was endorsed, and 11 sites were designated as "important cultural landscapes."

This chapter is a summary of the two reports (the ACA 2005) on the cultural landscapes of Japan which were compiled (Committee on the Preservation, Development and Utilization of Cultural Landscapes Associated with Agriculture, Forestry and Fisheries 2003; Takashima city 2009).

18.2 Background and Requirements of the Research

The details of the outcome of the research delivered by the provisional committee, and the history and current situation of cultural landscapes at a national level are described below.

18.2.1 Background

How scientific research into cultural landscapes has been conducted is described, together with coverage of international trends in attitudes toward cultural landscapes which were highlighted at UNESCO's World Heritage Convention (Nakagoshi 1995).

In addition, the trends in cultural landscape conservation are briefly explained, paying particular attention to the protection of "cultural properties" and the "cultural heritage" in Japan, as well as the promotion of agriculture, forestry, and fisheries.

18.2.1.1 Academic Research into Cultural Landscapes

Both the destruction of the natural environment and the active development of suburban areas have caused a continuous decrease of agricultural land. Natural scientists have recognized that land associated with agriculture, forestry, and fisheries plays an important role in maintaining ecosystems, because such land provides habitats for a diversity of species. Encouraged by the suggestions made by landscape ecologists, cultural landscapes started to receive more attention than ever before (Nakagoshi and Kim 2007). Research findings included the positive aspects of human interventions (Fukamachi et al. 2001) that have been practiced repeatedly through agricultural, forestry, and fisheries activities. A study has revealed that a certain degree of human disturbance to ecosystems contributes to maintaining habitats and their biodiversity in a suitable condition (Kamada and Nakagoshi 1996). For example, the role of water surfaces in rice paddies and agricultural water channels which can provide passages to animals are extremely important. Animals and plants with high scientific value have been designated as natural monuments, and thus their habitats are officially protected.

The baseline of recent studies is to maintain such ecosystems (e.g., Nakagoshi 1995; Ohta and Nakagoshi 2006). Proposals to realize desirable images of cultural landscapes in agricultural, forestry, and fisheries societies were developed to maintain such ecosystems. Research in both civil and ecological engineering has been implemented to uncover the most appropriate technologies and techniques for creating living environments suitable for diverse species.

18.2.1.2 Public Demand for Conservation of Cultural Landscapes

Public demand for the conservation of cultural landscapes in Japan has increased (e.g., Nakagoshi and Ohta 1992; Kamada and Nakagoshi 1996; Fukamachi et al. 2001). Enthusiastic citizens have undertaken activities throughout Japan in order to conserve terraced paddies and Satoyama. Abandoned farmland was increasing, particularly in semi-mountainous regions. Therefore, it was pointed out that due attention should be paid to the landscape's multi-faceted functions, which consist of land conservation, water retention, natural environmental conservation, good landscape formation, etc., in order to prevent the further abandonment of agricultural land and to maintain and improve agricultural land in the future.

One of the multi-functions of cultural landscapes is the cultural value inherited in them. The perspectives of cultural properties/heritage have been recognized by the recently established ad hoc committee on cultural landscapes. The need to take active measures to conserve such landscapes has been officially acknowledged and is stated in their recommendations.

18.2.1.3 Action Taken by the Agency for Cultural Affairs

The ACA drew up a Cultural Promotion Master Plan on March 25, 1998, based on the recommendations made in July of the previous year by the Cultural Policy Promotion Committee as well as the advisory body for the Commissioner for Cultural Affairs. The plan states that it is absolutely essential to succeed, develop, and expand the scope for the conservation of cultural properties, and to protect historic cultural settings. It also states that there is a need to review the existing protection system, and consider the establishment of a new protection system to ensure the preservation and utilization of historic cultural landscapes. In addition, it calls for extensive discussion on the comprehensive conservation of designated cultural properties and their surrounding environments, as prescribed in the concept of buffer zones in the World Heritage scheme. The possibility of an amendment should be considered for the Law for the Protection of Cultural Properties (the LPCP).

With regard to the conservation of cultural landscapes, the above report points out that some areas are not covered by the existing cultural property protection system, such as the designation of "places of scenic beauty," and it therefore proposes that a new framework should be set up at the national level. It emphasizes the need to consider establishing legal instruments enabling local governments to take the major role in the protection of such places, whilst cooperating fully with the national government. The report notes some cases of similar individual cultural properties, which have been valued below set thresholds, being distributed in a certain area in relation to each other, as well as cases of different types of individual property gathering together in a certain relation to each other, thus calling attention to the possibility that such properties could exhibit new values when evaluated as one collective. The report points out the need to consider assessing such properties as an ensemble which is eligible for protection.

The report concludes that the preservation and utilization of such cultural heritage needs a broader range of effort, and should consider its future scope, possible conservation methods, and targeted objects for selection. While recognizing that some such properties are already covered within the existing LPCP framework, or protected through the flexible operation of the law, nevertheless, there could still be cases in which sufficient care cannot be provided. It is therefore essential to study to what extent the current structure of the law can accommodate such heritage, and to consider amendments to the law to meet such needs. Furthermore, it suggests the need to establish a new system, rather than a new law, with a view to extending the current protective measures by comprehensive coverage of a wide range of cultural heritages that can not be fully covered by the law.

The key recommendations introduced above were reflected and incorporated into a "Basic Policy Concerning the Promotion of Culture and Art," which was compiled on December 5, 2002, by the Council for Cultural Affairs, and which was later adopted as a Diet Decision on December 10, 2002.

When a property is nominated for inclusion on the World Heritage List, it is necessary that buffer zones of an appropriate size should be provided around the site. At the World Heritage Sites of Japan that have been nominated or are already included, buffer zones of an appropriate size are provided by the Natural Parks Law, the Forestry Law, and supplementary national laws, as well as the landscape ordinances and other by-laws of local governments. These surround the buildings, other tangible cultural properties, historic sites, and other cultural properties designated by the state government under the LPCP. In these buffer zones, cultural landscapes such as rice paddies, farmland, forests where forestry activities are undertaken, ponds, lakes, and marine areas where fishing is practiced, are contained and protected through appropriate control measures as components of the surrounding environments of cultural heritage sites of outstanding universal value, and which are worthy of being on the World Heritage List.

In this way, through the process of establishing buffer zones for sites which are nominated for, or included in, the list of World Heritage Sites, local governments have become more actively and frequently involved in the conservation of cultural landscapes and the surrounding environment of cultural properties. This fact needs to be fully noted when a comprehensive framework for the protection of cultural properties is discussed in the future.

18.3 Research Processes for Cultural Landscapes

The research for the report was carried out from October 2000 to March 2003, a period of approximately 2.5 years. It progressed from two initial investigations through to the selection of "important" areas, as well as a basis for more detailed investigations. First, the concept of cultural landscapes was defined, and according to that definition, the second phase involved 502 areas being selected from 2,311 which were considered. Then after further screening, 180 areas were selected as the most important (the ACA 2005, Table 18.1).

Table 18.1 Selection process of cultural landscapes in a number of areas investigated

Stages	First stage	Second stage	Important areas
Number of sites	2,311	502	180

18.3.1 *Definition of a Cultural Landscape*

Cultural landscapes were defined in the first phase of the study as:

> A landscape of high value which exists in harmony, utilizing the local characteristics of the surrounding nature, history, and culture of agricultural, forestry, and fisheries communities, as well as having a close relation with traditional industries and modes of life, and which embraces a unique land use or natural features representative of the area.

18.3.2 *Early Stages of the Research*

According to the above definition, the first phase of the study examined the nation-wide distribution of such places in Japan, and identified 2,311 cultural landscapes. Of these, 502 were selected as areas satisfying two or more of the four criteria shown below. The second phase of the study identified the existing condition of these sites (Table 18.1).

1. Landscapes should be closely associated with agriculture, forestry, and fisheries, and possess unique characteristics or components. Consideration should be given to those that are not yet widely known but are deemed to be important.
2. Having been selected for a national list of "100 selected landscapes," or included in a suitable publication, so that it can be seen that its scenic value is widely recognized.
3. Activities such as agriculture, forestry, and fishing are still practiced, and thus the landscape is well maintained.
4. Not being significantly affected by recent development pressures, so that it can be seen that its fundamental value has been retained.

18.3.2.1 Categorization and Selection of Important Areas

The second phase of the study resulted in dividing the cultural landscapes into four categories as follows:
Landscapes which are:

I. Associated with land use
II. Associated with natural features
III. Unified with and encompassing cultural properties representing traditional industries or modes of life
IV. A combination of I, II, and/or III

Of the 502 cultural landscapes selected in the second phase, 180 important areas were further screened to reflect the conditions below (Table 18.1).
Landscapes which:

I. Have a close relation with traditional industries and modes of human life, both of which should be related to agriculture, forestry, or fisheries activities. Furthermore, they must represent unique natural features or land use, which in addition should have been preserved in either printed or illustrated documents.

II. Have a close relation with traditional industries and modes of human life, both of which should be related to agriculture, forestry, or fisheries activities. Furthermore, their natural features should be unique and exclusive to the region.

III. Include the surroundings of a "cultural property" or a group of "cultural properties," as well as the surroundings of buildings belonging to representative traditional industries or lifestyles. Further, the cultural properties and their surrounding areas must have a value in respect of their inseparability.

IV. Represent the local characteristics, and which are in a condition which combines I, II, and/or III.

18.3.2.2 Viewpoints for Selection

Important areas were further selected to reflect the following four viewpoints:

1. Cultural landscapes should be positively influenced by the characteristics of the region. The prototypical landscape particular to the region should favorably influence human emotions by providing aesthetic values as well as feelings of relief.
2. Cultural landscapes should contribute to the natural ecosystems by providing habitats for a diversity of precious fauna and flora, including endangered species.
3. In light of the fact that cultural landscapes that represent specific local characteristics can be distributed widely in various places, and that social changes are driving many cultural landscapes to the verge of extinction, representation and rarity should be taken into consideration.

Table 18.2 Proposed categories of sites which are important as cultural landscapes

Category	Sub-category	Items
I	1	Rice paddy landscape
	2	Farmland landscape
	3	Grassland landscape
	4	Managed forest landscape
	5	Fishing ground, fishing port, and sea coast landscapes
	6	River, pond, lake, and waterway landscapes
	7	Landscape associated with settlements
II	1	Landscape which has traditionally been the object of worship or a destination of tourism
	2	Landscape which has traditionally been the motif of artistic works or the source of creative inspiration
	3	Landscape which occurs as a result of unique climatic conditions
III		Landscape surrounding cultural properties which represent traditional industries and modes of life
IV		A combination of I, II, and/or III

4. In the light of the close relation with traditional industries and modes of life associated with agricultural, forestry, and fisheries communities, conservation and utilization measures, such as concerted efforts by the residents and the local governments, must be assessed to ensure that the landscape can be expected to be maintained in the future.

Some of the landscapes which provide a venue for early morning markets or local festivals were included during the first and second phases, but they were excluded from the selection of important areas because their major components were human activities which last for only a limited period. Therefore, it would not be appropriate to evaluate them from the perspective of monuments. After selecting important cultural landscapes based on these criteria, they were categorized as shown in Table 18.2.

18.4 Proposed Cultural Landscapes

18.4.1 Characteristics and Problems

The most appropriate example of a cultural landscape is the terraced field landscape, which accounts for about 70% of the rice paddy landscape. The rest includes landscapes associated with archeological remains of historic value, such as a grid system accompanied by the unique agricultural facilities known Hasaki (trees planted on which to dry harvested rice ears), or remains showing unique structures reflecting geographic features or agricultural activities such as Yatsuda (valley-bed rice fields) and Kurumada (round-shaped rice fields), as well as the large-scale rice fields existing in the plain transformed by the official agricultural land consolidation projects started in the modern era (Education Board of Akita Prefecture 2009).

Although local people's understanding and awareness of the need for protection is increasing due to successful publicity efforts such as the "100 selected terraced paddies of Japan," there are also serious challenges to the protection of the terraced fields, e.g., aging farmers and a lack of heirs. More than ever, modern materials and techniques are introduced to maintain the rice field levees, the most detrimental being that traditional techniques like stonework have been disappearing. The traditional values of the cultural landscape have been significantly disturbed.

When the rice paddy landscapes have some subterranean archeological grid structure, people fail to comprehend the relationship between the landscape and the archeological remains beneath the ground. The lack of understanding of their value makes it difficult to implement active protection measures, and even endangers the preservation of such values. Residents are not aware of the values threatened by the destruction caused by the official agricultural land consolidation project, which merges many precious traditional paddies into a unified mass of land.

The next example is farmlands. Farmland landscapes include the terraced farmland developed on steep slopes (Nakagoshi 1995; Nakagoshi and Ohta 1992), while on moderate hills and in the plains, ordinary types of farmlands are developed.

From ca. 1912 to 1945, large-scale mono-cultivation of taro was practiced nationwide in a form of mass production. Later, in the post-war period, fruit orchards became prevalent, taking advantage of the abundant sunlight. In many farmlands on the plains and on gentle slopes, traditionally grown taro and grain crops were changed to buckwheat. Such farmland landscapes where buckwheat has been newly introduced was also selected as "important."

18.4.2 Relationships Between Cultural Landscapes and Other Cultural Properties

Based on the Law for the Protection of Cultural Properties (the LPCP), the current protection system is now analyzed from the perspective of monuments.

The LPCP defines five categories of cultural property. The monuments covered in this chapter belong to the third category of the law. Under this law, the Minister of Education, Culture, Sports, Science, and Technology designates important monuments as "historic sites," or "places of scenic beauty and/or natural monuments," and especially important ones as "special historic sites," "special places of scenic beauty," and/or "special natural monuments." The ministry first consults the Council for Cultural Affairs prior to their designation or the annulment of a designation, since these matters require highly technical knowledge and experience.

The guidelines for designation are set out in the criteria for the Designation of a Historic Site, Place of Scenic Beauty, Natural Monument, Special Historic Site, Special Place of Scenic Beauty, and/or Special Natural Monument, as well as by the Committee for the Protection of Cultural Properties, which was set up on May 10, 1951, and revised on March 6, 1997.

"Historic sites" stipulated in law are designated as "important" sites with "archeological remains" that possess high historical and/or scientific value, and consists mainly of "structural relics" and "accompanying remains." Moreover, the criteria refer not only to "structural remains" and accompanying remains, but also to the "scale of the archeological remains" for evaluating the scientific value, indicating that the space occupied by archeological remains is also included. The continuous land area surrounding the archeological remains, combined with the various exterior appearances of the remains, formulates the "historic landscape." A historic site is composed of two factors which are fused, i.e., the historic landscape of the historic site and the landscape of the designated area. The landscape within the designated area and the elements of which it is composed are closely related to the landscapes in the surrounding areas and their elements.

"Places of scenic beauty" include "gardens, bridges, gorges, sea-shores, mountains, and places with beautiful scenery that possess a high artistic value or which deserve great visual appreciation in a region," and in law they are designated as important sites. To comply with the criteria, they should be "essential components of the beauty of Japan, and excellent in terms of scenic beauty, sightseeing, or (natural) scientific value, or in terms of artistic or (man-made) scientific value."

Elements which compose the artistic or aesthetic value of natural "places of scenic beauty" include all the natural and man-made elements that give natural scenery landscapes excellent sightseeing value or scientific value. These include, natural elements which constitute the basis of the artistic and aesthetic qualities of sightseeing spots (climate, weather, landform, geologic features, waters systems, water quality, fauna, and flora), and man-made elements such as historic buildings, structures, and unique land use which show the long history of a specific place as a well-known tourist spot.

Elements composing the artistic or aesthetic value of man-made "places of scenic beauty" include gardens, parks, and bridges which are designed by people for their artistic and aesthetic value. For example, the landform, land division, stonework, planted trees, buildings, and structures are inseparable, and similarly bridges and rivers are also inseparable, so all of these are included. Elements composing man-made "places of scenic beauty" are of two types. The historic sites manifest their value both above-ground and below the ground. In addition, borrowed scenery, such as agricultural land which is located outside a garden area, e.g., ancient Kofun tumuli, should be incorporated into the composition of the garden system as an important component. For example, in Japan, small garden areas are often constructed inside rice paddies, and farmers share such spaces to rest or to meet each other. They should be also included. The artistic and aesthetic value of "places of scenic beauty," whether natural or artificial, lies in organic relations among the above-mentioned elements, and they are visually perceived as scenic beauty by humans.

A "place of scenic beauty" has a cultural value and is perceived as art, or with appreciation. In other words, elements composing a specific land area are perceived as a landscape whose aesthetic value is perceived through the visual sense as "scenic beauty." Therefore, as far as places of scenic beauty are concerned, the perspective or view constitutes an extremely important component in their aesthetic value. In addition, the artistic or aesthetic value of places of scenic beauty presupposes a certain degree of alteration or change in landscape with the passing of time and the vicissitude of the seasons.

Cultural landscapes evaluated from the perspective of places of scenic beauty can be defined as those places that have traditionally been famous as renowned sightseeing spots for a long time, and are also well known in relation to works of art composed of the scenery or landscape of agriculture, forestry, or fisheries which are also highly valuable from the point of view of visual appreciation. Their evaluation should be carried out by clearly distinguishing the elements that are the source of the artistic or aesthetic value. Although not easy to distinguish, the organic relations among elements must be harmonized with the entire landscape of visual appreciation.

18.4.3 Conservation of Cultural Landscapes

First, the characteristics of cultural landscapes must be identified, and then fundamental ideas for their protection will be developed. Later, a plan of the protection system should be proposed, together with guidance for conservation, management, utilization, and promotion.

Cultural landscapes have the seven characteristics shown below.

A. Founded on Traditional Industries and Modes of Life
Cultural landscapes which were formed through interactions with people on the basis of traditional industries and modes of life in agricultural, forestry, and fisheries communities also relate deeply to the contemporary industries and current lifestyles in the region. Therefore, cultural landscapes are significant not only by having continuous land use in the past, but also by being properly maintained in harmony with current activities.

B. Abundant Local Characteristics
Cultural landscapes reflect the unique history and culture of agricultural, forestry, and fisheries communities, and represent the natural features peculiar to the locality. In this regard, they are extremely rich in local characteristics. Being close to the hearts of residents who were born, grew up, and live in the region, they generally symbolize an image of the home town or the spiritual proto-landscape. In addition, by representing a unique form of land use that has been inherited from generation to generation exclusively in that region, many cultural landscapes also serve as important spiritual centers for local citizens.

C. Cyclic Change
Cyclic change is an essential element of cultural landscapes which are constantly evolving in concurrence with the cycle of the traditional industries and lives in agricultural, forestry, and fisheries communities. The diurnal cycle of day and night, and the seasonal or step-wise changes in the processes of production activities and fisheries recur at certain times. Cultural landscapes also alter their forms cyclically and project different appearances. As traditional industries and modes of life have been changing over a period of time, many cultural landscapes have also gradually evolved.

D. Diverse Component Elements and Their Organic Relations
Cultural landscapes are composed of various types of tangible and intangible elements, and are characterized by organic relations among these components. For example, the terraced paddies with levees and stonewalls are assisted by tangible components such as the water supply drainage system, reservoir ponds, rivers, and protection forests which affect the water sources. These diverse elements and the organic relations among them constitute an extremely important value. One example is the diverse communities of fauna and flora inhabiting a particular area, and building the rich ecosystems (Iiyama et al. 2005). Furthermore, various types of work to maintain and manage these tangible elements, or to celebrate an abundant harvest or a large catch of fish, are repeatedly carried out by the people in the region. All of these are important intangible elements that compose cultural landscapes.

E. Diversity in Landscape Structure
Because it is affected by natural features, the structure of cultural landscapes is extremely diverse. The landscape of terraced paddies carries a group of terraces with sloped walls which unfold around a valley. The landscape of continuous farmland extends over the slopes of the pediment area. A compound landscape has rice paddies and farmland spread over a vast plain or basin, with rivers, hills,

and settlements scattered among them. The landscape of fisheries and fishing grounds expands limitlessly over the surface of the sea or lake as far as the eye can see. These landscapes change depending on the size of the entire area in which it unfolds, and its continuity with the surrounding space.

F. Biodiversity and Habitats
Interventions by people, which are repeated through agriculture, forestry, and fisheries, cause a certain degree of disturbance to the ecosystem, and this can conserve a rich biodiversity and range of habitats. On the other hand, the water features of paddies and water channels, and also of Satoyama (Nakagoshi and Hong 2001), as green spaces are not only the habitats of flora and fauna, but are also ecological corridors for plants and animals. Therefore, some of the cultural landscapes provide precious habitats for endangered species.

G. Two Types of Cultural Landscape
Cultural landscapes can be divided into two types: those representing extremely high value by themselves, and those whose value comes from being united with other monuments that those landscapes encompass. Both types occupy a continuous and inseparable space, and are closely related to each other. However it is appropriate to evaluate them separately.

The conservation guidance for cultural landscapes should be designed in full consideration of the following seven points which correspond to the seven characteristics above.

A. Appropriate Harmony with Agriculture, Forestry, and Fisheries, and Traditional Culture
In order to maintain and protect cultural landscapes in an appropriate manner, and to build upon the inherited mechanisms of traditional industries and modes of life in agricultural, forestry, and fisheries communities, it is necessary to continue the agriculture, forestry, and fisheries on which they were founded in a manner which meets the living standards of the contemporary generation. Therefore, proper harmonization between measures for the conservation of traditional culture and those for the development and promotion of agriculture, forestry, and fisheries is an essential perspective. In addition, it is necessary to respect the traditional industries and modes of life, and give due consideration to the transmission of their essential value while building a consensus with local people, and to consider moving on to new methods, always with a long-term perspective.

B. Need for Consensus Building with Local People and Stakeholders
When people appreciate the history and local culture, the beauty and value in landscapes closely associated with traditional industries and modes of life will be discovered. It is important to re-evaluate such attitudes when considering the conservation of cultural properties. Cultural landscapes, which represent the natural features unique to the region, demand the devoted commitment of local communities for their protection. Therefore, it is important for local residents to fully recognize the value of the relevant cultural landscape, and make continuous efforts

for its conservation based on a mutual consensus. It is also important for governmental organs, NPOs, and NGOs to provide constant assistance to their efforts, so that people will learn the significance of discovering the value of the landscape through their daily lives. In particular, local governments need to have concrete policies to conserve cultural landscapes while maintaining traditional industries and lifestyles, as well as to provide opportunities to raise the awareness of the residents. The authorities of areas with cultural landscapes need to actively implement policies to support conservation measures in their jurisdiction.

C. Conservation Accommodating Appropriate Degrees of Change Caused by Time
Cultural landscapes constantly change in accordance with the cycle of traditional industries and modes of life, and in many cases they are evolving with the passage of time. Therefore, their protection measures need to evolve in step with the evolution of the environment. Where changing social circumstances which encompass the traditional industries and lifestyles drive the traditional mechanisms supporting cultural landscapes, it would be necessary to consider flexible measures, including the introduction of a new mechanism which could replace the traditional system. In such cases, it is essential to have regular and frequent communication with the local people who are directly involved in the traditional industries and cultures.

D. Protection with a Focus upon Tangible and Intangible Components and Their Organic Relations
Cultural landscapes not only consist of individual tangible components, but also contain intangible components rising from management activities for the industries, lifestyles, and customs in the region. Their value also consists of organic relations among all the components, including the ecosystems built by communities of fauna and flora. Therefore, comprehensive measures for evaluation and protection with the focus on these factors are necessary.

E. Protection with a Focus on the Diversity of Landscape Structure
The diversity of the landscape structure reflects how local residents perceive and recognize these landscapes (Kamada and Nakagoshi 1996). The extent and spatial size of landscapes as perceived by humans can be identified as the scope of the cultural landscapes that can be viewed daily through the various activities of agriculture, forestry, and fisheries. Therefore, it is important to have an accurate understanding of the characteristics of the landscape structure vis-à-vis the scope and spatial size of cultural landscapes as perceived by residents, and to define the scope of the cultural landscapes to be conserved appropriately.

F. Protection with a Focus on the Role of Biodiversity and the Range of Habitats
Since cultural landscapes are manifestations of how the land is shared by people and nature, their protection must be carried out with sufficient consideration of the diverse communities of animals and plants in the area. It is necessary to remove obstacles hindering the growth and habitations of plants and animals, ensure that water features and green spaces are appropriately maintained, and explore other technical options for creating desirable living environments for fauna and flora.

G. Comprehensive Cultural Landscape Conservation
In examining the conservation system for cultural landscapes, it is essential to have a comprehensive objective of protecting both cultural landscapes which represent high value and those whose value comes from being united with the monuments. With respect to compound landscapes (category IV), it is important to protect the value as a whole, including not only the cultural landscapes, but also the areas existing around or between them.

18.4.4 Conservation of Important Areas

It is expected that a conservation system for cultural landscapes would be multi-layered. To protect both a cultural landscape and its surrounding and related areas, all governments at national and municipal levels should cooperate to apply the regulations and restrictions for protection, regardless of the difference in responsibilities.

The surrounding areas of places under the existing protection system, as well as important places, will be evaluated as cultural landscapes while making it possible to conserve them by making the best use of the existing system in the Law for the Protection of the Cultural Properties (LPCP). At the next level, it would be possible to establish a new system in which, as is the case with the existing preservation areas for groups of historic buildings, the local government first establishes legally binding protection measures by passing appropriate ordinances, and the national government makes the final selection. On the other hand, the use of a registered cultural properties system is another option, whose major goal is to raise awareness and reach out to the public, rather than even more regulations.

However, as many cultural landscapes represent natural features as well as social customs unique to particular localities, there will be cases which cannot be dealt with by the conventional monument designation system, because that is designed to protect features of high historic, scientific, artistic, and aesthetic value through strong regulatory measures. At the same time, it should be noted that many cultural landscapes cover a vast area where a large number of landowners are involved, and therefore the introduction of a zone protection scheme is the most suitable way to provide assistance under appropriate regulations.

For important areas in particular, consideration should be given to continued efforts to designate those that are eligible for protection under the existing system of monuments. However, it is necessary to start a discussion on necessary amendments to the law with a view to making it possible to establish a new system in which local governments set up their own ordinances for zone protection measures. To realize this, the agreement of local people is necessary, with the national government providing the necessary support.

Since cultural landscapes possess the above seven characteristics, there could be some cases which cannot be dealt with under the conventional designation system. Therefore, it is necessary to consider establishing a new protection system under the LPCP. Consideration of a new protection system should be given with full attention to the following two points. (1) The purposes and objectives of the existing cultural properties protection system should be fully respected, and duplication or contradiction between a new system and the conventional system must be avoided by clarifying the scopes and classifications of the protection. (2) Conservation of cultural landscapes is not possible without the consensus of local people and the support of local governments. In recognition of this, unlike the conventional designation system, a mechanism should be incorporated in which local governments have the central role of controlling alterations to the landscapes properly, and the national government provides support when necessary.

Possible options for the new protection system, such as a two-layered scheme, could be suggested in the existing designation of preservation areas for groups of historic buildings under the LPCP. First, local governments, according to their ordinances, should establish protection measures. Second, based on applications by the relevant local governments, the national government selects what could be temporarily called important cultural landscape conservation areas.

In this case, local governments will be required to conduct detailed surveys in and around the candidate areas for selection by the national government as important cultural landscape conservation areas, and prepare a plan for their conservation, management, utilization, and promotion. Therefore, a system in which the national government aids local governments in taking any necessary measures will also be needed. At the same time, the system will also be required to generate support from the national government for local governments when they carry out regular repair work to the principal elements of the relevant cultural landscape, and other measures for their preservation and utilization in the important cultural landscape conservation areas selected by the national government.

18.4.5 Conservation Management Plan and Utilization Promotion Plan

When conservation measures for cultural landscapes are designed, it is necessary to carry out detailed surveys in accordance with the framework indicated in this chapter. Based on the results of the survey, it will be possible to clearly define the scope of the conservation. Following this, it is necessary to consider the nature and components of the relevant cultural landscapes and the organic relations among these components, in order to present the most suitable plans for conservation and management as well as utilization and promotion.

To begin with, the basic principles of the conservation and management of the zone to be covered by the protection measures need to be presented based on concrete methods of preservation and protection for each component of the cultural landscape under consideration. Thus, regulatory measures that are necessary for proper conservation and management must be identified. Secondly, the basic principles of utilization and promotion, including efforts to raise awareness of the value of cultural landscapes, utilization to reactivate the local economy, the training of human resources, and the allocation of responsibilities among the agents implementing the protection and the governments need to be put forward in accordance with these aims. It is vital to clearly illustrate the desirable future image of the relevant cultural landscapes, and the specific methods to be used to attain it. Through the process of preparing such plans for preservation and management and for utilization and promotion, it will be possible to build a consensus for the preservation of the cultural landscapes among stakeholders, including local people.

18.4.6 Efforts for Promoting an Appreciation of the Value of Cultural Landscapes in Daily Lives

The discovery of beauty and the importance of the landscapes surrounding daily life is a cultural attitude. People will gain mental strength or find joy in life in familiar landscapes or in the landscapes of their home town visited after a long absence. Local people must recognize the value of the cultural landscape and the importance of passing it on to future generations. Ideally, they should find pride in their own home town and improve their own lives. It is particularly important for people working in agriculture, forestry, and fisheries to know the value of cultural landscapes, and to recognize their own role in the protection of cultural landscapes through their daily work. Therefore, local governments should work closely with local people and various relevant organizations to foster a variety of creative projects, and help local people to re-identify the unrealized value of their cultural landscapes, which are firmly rooted in local lives, and to learn more about their essential value. In such activities, special efforts should be made so that younger generations in particular can understand that cultural landscapes, which have been traditionally inherited from generation to generation, are compelling examples of land use at a certain period in history, and that they have significance in our contemporary efforts to explore how to enrich our lives in harmony with the natural environment.

On the other hand, in order to pass on cultural landscapes to future generations, it is essential to preserve and transmit the techniques for their proper maintenance. In this regard, it is necessary to provide opportunities for local people to realize the significance of the traditional technologies and skills of agriculture, forestry, and fisheries, and to train the people who succeed to them.

18.4.7 Utilization and Promotion for Revitalization of the Local Economy

Since many cultural landscapes are unique and eloquent of the history, culture, industries, and lifestyles of the localities, it is important to utilize and promote them as the face of the community. This is particularly important for terraced paddies and farmlands which were developed on steeply sloping land. It can be presumed that this development would not have been possible without tremendous labor and pain, and because of that they are the heritage which represents the history of the local economy (Iiyama et al. 2005). However, it is possible that some people may hesitate to agree to protect them as the face of their community. Therefore, it is necessary to explore ways of utilization and promotion which would enable local people to understand the value of the cultural landscapes because they have been built at the cost of such strenuous labor, and to see the significance of taking active measures to protect them, as well as to be able to feel proud of them.

The ownership which is being implemented in many well-maintained terraced paddies is a mechanism by which local governments enable people living in urban areas to rent terraced paddies and experience their cultivation with the cooperation of local farmers, which is a good example of terraced paddies being utilized as a means of exchanges between rural villages and urban cities. Some farmers are running farm tourism and rent-a-farm schemes to attract tourists, using the Internet to advertise cultural landscapes and to make direct sales of their produce on-line. When carrying out such activities, it is necessary to consider cultural landscapes as tourist resources, and to explore active measures for their utilization while paying due attention to maintaining a balance with preservation. With regard to the various activities for re-invigorating local economies which are being carried out throughout Japan, it is necessary that local people, local governments, and relevant organizations work together to provide support to people who visit for observation, agricultural experience, or participation in various activities with a view to enhancing mutual efforts to promote inter-/intra-regional cultural exchanges.

18.4.8 Lead Operators and Building of Human Resources

Since cultural landscapes are characteristically associated with the history and culture unique to each locality, and represent unique natural features, local people living there with a direct interest in the conservation of the relevant cultural landscape should undertake the leading role in their protection. In this regard, it is essential to promote approaches for preservation and utilization in which every resident in the locality can participate. It is also necessary to expand the circle of participation in the conservation of cultural landscapes by actively contacting NPOs, NGOs, and other private organizations. These bodies should take action for the protection of these cultural landscapes by carrying out various programs related to revitalizing the local economy. At the same time, local and national governments need to strengthen their support measures.

The training of human resources to undertake leading roles in the conservation of cultural landscapes at the regional level is necessary. For organizations such as residents associations and cooperatives, NPOs, and NGOs, which can undertake the protection of cultural landscapes, training people who can lead protection activities is a common and important issue. It is also necessary to make efforts to create circumstances in which people from other regions can easily participate and undertake agriculture, forestry, and fisheries, because these are the foundations of the formation of cultural landscapes. In particular, those who have lived in cities for a long time will frequent rural areas when they develop a wish to settle in areas with cultural landscapes surrounding their daily lives, and there are even some who want to find a job in agriculture, forestry, fisheries, or many other traditional industries. Therefore, it is important to establish a local mechanism to accommodate these people and train them to undertake the protection of cultural landscapes in the future.

In order to utilize and promote cultural landscapes, the appropriate and active participation of every citizen living in the locality is foremost. Local people should join groups such as residents associations and cooperatives, so that NPOs and NGOs can cooperate with them according to their local purposes and specific objectives. Local governments and the ministries and agencies concerned should provide active support to such activities. The role of local governments in the appropriate conservation and utilization of cultural landscapes is especially significant, and the national government needs to provide support and assistance. It is essential that the divisions/bureaus of local government which are in charge of cultural properties and the ACA should cooperate in the preservation and utilization of the value of cultural landscapes. The divisions/bureaus of local government which are in charge of agriculture, forestry, and fisheries and the MAFF cooperate closely, and share their roles in revitalizing these local industries. They thus promote their products through mutual communication and promotion of the industries. At the same time, it is also necessary to explore collaborative measures which could contribute to the conservation of cultural landscapes, such as support for the conservation and repair of component elements of cultural landscapes, reinforcement of the existing direct subsidy payment system for communities in semi-mountainous regions, and even tax exemption.

18.5 Registration of Cultural Landscapes

18.5.1 Re-Categorization of Cultural Landscapes

After the publication of the committee's report, it was necessary to realize the registration of cultural landscapes at the ACA. Categories II and III in Table 18.2 were judged to require reconsideration before the registration commenced. The ideas about worship, inspiration, and climate/weather in category II are not always associated with daily human activities and the existing visual culture. These were therefore removed from Table 18.2.

Category III was divided into new sub-categories I.6 and I.7. The original sub-categories I.1, rice paddy, and I.2, farmland landscapes, were combined into the new sub-category I.1, agriculture landscapes. Finally, the official registration categories were established (Table 18.3). The local government must be a landscape

Table 18.3 Official registration standards for important cultural landscapes

Category	Sub-category	Items
I	1	Agriculture landscapes such as paddy fields and dry farmland
	2	Grassland landscapes such as meadows and pastures
	3	Forestry landscapes such as timber production, and for the prevention of disasters
	4	Aquatic cultural landscapes on sea coasts, such as nursery rafts and larva nurseries
	5	Artificial aquatic landscapes such as irrigation ponds, water ways, and ports
	6	Mining and industrial landscapes such as mines, quarries, and factories
	7	Landscapes on roads and open spaces for transportation and views
	8	Residential landscapes with hedgerows and trees, etc.
II	Combination of I-1 to I-8	

administrative organization. Then it can propose the designation of important cultural landscapes. At the same time, this organization has to formulate a plan for developing a good cultural landscape. The actual registration of important cultural landscape sites started in 2005. Considering long-term/continuous registration, additional registration was permitted for the expansion of cultural landscapes.

18.5.2 Registration of Important Cultural Landscapes Sites in July 2009

From January to July 2006, 11 important cultural landscapes sites in 15 local municipalities were designated (Fig. 18.1). One site is located in Hokkaido, five are in Honshu, two in Shikoku, and three in Kyushu. They are briefly introduced below in order from north to south (A to K).

A. Cultural landscapes in the Saru River watershed with Ainu tradition and modern development, Hiratori town, southern Hokkaido (selected in July 2007; Categories I-2, 3, 5, 7, 8, II; Fig. 18.1a).
   Both the traditional culture of the Ainu people of Hokkaido and their Westernization after the Meiji era have contributed to create these landscapes. Their elements spread widely in the watershed, with modern infrastructures built by the national government. The Ainu were the original inhabitants of Hokkaido and adjacent islands, including the Kurile Islands. Over the centuries, the Ainu people have mixed with the Japanese, and there are few full-blooded Ainu people living today. Their living style has played a leading role in the development of the cultural landscape. A Chise (Ainu: house) is exhibited along with other museums such as the Ainu Culture Museum, the Historical Museum of the Saru River, and the Munro Museum. Neil Gordon Munro (1863–1942)

Fig. 18.1 Important cultural landscapes: (**a**) The Saru River watershed in Hiratori; (**b**) rural landscape in Ichinoseki; (**c**) lakeside landscape in Oumihachiman; (**d**) an excursion in a reed lake in Oumihachiman; (**e**) terraced farmland in Uwajima; (**f**) the Shimanto River watershed; (**g**) rice paddies irrigated by the Tsuujun Irrigation Waterway in Yamato; (**h**) the urban area in Kanazawa (under investigation)

was a Scottish physician and anthropologist. Residing in Japan for almost 50 years, he was notable as one of the first Westerners to study the Ainu people of Hokkaido.

B. Grasslands on Arakawa Highland in Toono, the northeastern part of Toono city, Iwate Prefecture (selected in March 2008, addition in February 2009; Category I-2).
The grasslands are characterized by horses and horse breeding. While taking advantage of the grasslands extending throughout the plateau, unique pasturing has long been practiced, and horses are grazed as a core industry of the region. One of the characteristics is that horses are fed in the open fields in summer, and sheltered inside barns located in the communities during the winter season. Along with the designation of the grasslands, the Arakawa-komagata Shinto Shrine, where horses are worshiped as deities, was also designated in February 2009, accompanied by its related facilities.

C. Rural landscape in Hontera, Ichinoseki, Ichinoseki city, Iwate Prefecture (selected in July 2006; Categories I-1, 8, II; Fig. 18.1b).
Hontera was originally a manor in the territory owned by the Chusonji Temple, and the area was in a farming village where paddy fields were developed in the middle age. Rice is still actively cultivated today, and scenes of the rural landscapes are preserved in drawings from the thirteenth and fourteenth centuries. The traditional agricultural infrastructure has been continuously restored, and thus has succeeded in handing the history of rice paddies in Japan from generation to generation. Many visitors come from outside the village to experience agriculture. After it was selected as an important cultural landscape, the village's rice fields were improved, and the landscape was modified to give a taste of the village life. Various official programs with financial assistance have been carried out to create a traditional village farm landscape.

D. Lakeside in Oumihachiman, Oumihachiman city, Shiga Prefecture (selected in January 2006, additions in July 2006 and July 2007; Categories I-1, 3, 5, 8, II; Fig. 18.1c, d).
The designation as an important cultural landscape site was the first in Japan. While reeds grow on the islands of Japan's largest lake, Lake Biwako, the landscape is represented by lakesides and wide rice fields. The traditional industry of utilizing reed materials is actively continued today. In the old days, a marshy spectacle was the typical lowland landscape, and the current landscape was created through development of the marsh. Sightseeing on the waterways provides spectacles of an ancient Japanese landscape. The residents are working to maintain the surrounding Satoyama. Because of the success of the resident's positive scenic conservation activities, and the progress of their conservation work, the area has expanded twice, resulting in 354 ha being additionally designated.

E. Lakeshore landscape in Takashima, Takashima city, Shiga Prefecture (selected in March 2008; Categories I-5, 7).
The north shore of Lake Biwako has received no help from recent developments, so the prototype lakefront village has been continuously restored and

maintained since the Edo era. The waterway developed in the village is used daily. It was historically a transportation node for both the land and the lake areas. The traditional fishing practice has continued, and traditional villages are intentionally preserved. Thus, the quality of the cultural landscape has been continuously enhanced.

F. Cultural landscapes in Uji, Uji City, Kyoto Prefecture (selected in February 2009; Categories I-1, 5, 6, 7, 8, II).
Green tea is a traditional drink of Japanese people, but in the Heian era, only high-ranking people were able to consume it. After entering the Muromachi era, tea cultivation actively started in Uji. The tea produced ranged from the standard to the exclusive. This was possible because the technology of highly valued tea making already existed in Uji. The city has become the center of the tea production area of Japan since the era of the Warring States. Tea shops line the Omotesando, which is the main path for visitors going to the Byodo-in Temple of the World Heritage, and a traditional form of street has been created. A series of traditional industries is continued and supported by the production and sale of tea. This means that a well-managed tea plantation exists in harmony with the traditional shops in the street, creating the graceful cultural landscapes.

G. Terraced Farmland in Yusu-Minaura, Uwajima city, Ehime Prefecture (selected in July 2007; Category I-1; Fig. 18.1e).
Terraced fields with stone walls extend on the steep slopes of a distant peninsula where the water supply faces difficulties. The terraced fields built up over many years are indeed beautiful. Some terraced fields were once abandoned, but have now been revived and again maintained as farmland thanks to the residents' efforts. Mizugaura village is a community living both on agriculture and fishing, and it has been famous since the Edo era for plowing and fishing. Today, potatoes are produced in the terraced fields. The cultivation of young yellow tail is active in the deep bay. Although the currently designated cultural landscape site is the terraced fields and the village, there is a plan to add the bay where aquaculture is conducted.

H. Cultural landscapes in the Shimanto River watershed, located in five municipalities (Yusuhara, Tsuno, Nakatosa and Shimanto towns, and Shimanto city) in western Kochi Prefecture (Agency for Cultural Affairs 2005) (selected in February 2009; Categories I-1, 3, 4, 5, 7, 8, II; Fig. 18.1f).
The Shimanto River starts from Mt. Irazu-yama (altitude 1 336 m), which belongs to the Shikoku mountain range. After combining with many tributaries, it finally flows into the Pacific Ocean. The total length is 196 km, and it has a watershed of 2 270 km^2.
A laurel forest, which is potentially the natural vegetation in western Japan, and a cedar plantation occupy the upstream area. Steep slopes are covered by terraced paddies. Rice fields also extend across flat land from the middle to lower reaches. Since the river is remote from major cities and has no big dams, it is sometimes referred to as "the last clear river of Japan." Fishing and the production of larva are thriving industries along the river.

Traditional fishing continues along the main and local streams. Of this, fishing of Ayu, Plecoglossus altivelis, is the most actively practiced, and in the Kubokawa area, for example, which is located in the middle reaches, various fishing methods are well conserved. The ruins excavated from the Misato Ruin in the lower reaches in Shimanto city prove that fresh water fishing had been practiced since the Jomon era.

The river also has 47 Chinkabashi (sinking bridges, Fig. 18.1f) including those of some tributaries. Chinkabashi are bridges without parapets, designed in order not to be washed away by floods. The prefecture decided to preserve them as part of the cultural heritage.

The success of the wide-ranging designation of this cultural landscape is due to the Basic Ordinance Concerning Conservation and Promotion of the Shimanto River Basin in Koch Prefecture. This ordinance was enforced in 2001 to promote regional development through the conservation of natural environments and landscape in the basin. In 2002, all local municipalities in the watershed followed suit and enforced a similar ordinance adapted to their own natural and cultural conditions.

I. Terraced paddies in Warabino (selected in March 2008; Category I-1).
The Warabino community is located on the top slope of the Hachiman-dake (altitude 764 m) in Karatsu city, Saga Prefecture. Its vast terraced paddy fields stretch from the community toward the lower slopes, along with the hill slopes of five valleys. The rice terraces cover some 40 ha of mountainside, the height of the rubble masonry walls enclosing each terraced paddy averages 3 m, while the maximum of 8.5 m is greater than all other existing walls. From an examination of excavated ruins, the history of the inhabitants can be dated back to the sixteenth or seventeenth century. According to a statement in a historical document, the first rice paddy had been developed by 1839 in one of the valleys. The terraced paddy landscape has been well preserved thanks to strenuous efforts by the local residents to continue traditional farming practices, and it was therefore chosen as one of the "100 Selections of Japanese Terraced Paddy Fields" by the MAFF.

J. Tsuujun irrigation waterway and terraced paddy landscape on Shiraito Plateau, Yamato town, Kumamoto Prefecture (selected in July 2008, addition in July 2009; Categories I-1, 5, 7, II; Fig. 18.1g).
No water supply was available on the Shiraito Plateau to the south of Mt. Aso, a big active volcano, and therefore rice could not be cultivated. Paddy rice production became possible after the construction of the Tsuujun Irrigation Waterway in the period from 1854 to 1856. This irrigation facility uses rainwater, and was used until very recently because of its excellence. After partial damage of the facility, repair work was undertaken by applying appropriate industrial methods to retain the original structure. A rich fish fauna, including some endangered species, live in the fresh water of the waterway, and there is a wide variety of species. The beautiful terraced paddy landscape extending along the irrigation waterway was first specified as an important cultural landscape, and a year later consecutive neighboring rice fields were added to the designation forming a larger area.

K. Village producing Onta Pottery, Hita city, Oita Prefecture (selected in March 2008; Categories I-1, 5, 6, II).
This village is part of the Ikenotsuru area, with terraced paddies and a Satoyama area with good quality soil. Taking advantage of being in the center of the valley where water is easily obtained from the river, the community also produces the famous Onta Pottery. It has been designated an important cultural landscape. A plan to also designate the peripheral forest, where fuel is collected and used for firing pottery and daily life is ongoing.

18.5.3 Problems of Industrial and Urban Cultural Landscapes

There are landscapes of quarries and mines which are components of industry, and which are used as the sources of building materials for agricultural, forestry, and fisheries communities. Landscapes of mining terraces often exist in unity with the current land use, such as agricultural land. As well as agriculture, forestry, and fisheries, however, there must be many other cultural landscapes of high value which are connected with mining in urban life and industries. Therefore, it is essential to carry out a selection and addition of important areas in these fields. Regarding the protection of cultural landscapes with mining and other urban industries, it is appropriate that the relevant ministries and agencies should cooperate, as well as local governments. Cultural landscapes with mining and other urban industries belong to categories I-6, 7, and 8. By the end of 2009, four important new cultural landscapes will be proposed for designation. One of these is an urban "cultural landscape in Kanazawa city with tradition and culture" (Fig. 18.1h). This is the first site to be proposed for designation in a traditional castle city with a heavily populated urban area.

18.6 Perspectives

18.6.1 Requirements for Designation

Cultural landscapes consist of a combination of their component elements, and demonstrate various functions and the organic relations among them. In addition to the cultural landscapes that represent a unique form of land use and exhibit a high value by themselves, many other cultural landscapes which should be protected are yet to be discovered. In aiming to realize zonal protection, I have proposed that local governments enact the relevant ordinances to protect cultural properties and heritage. Each individual cultural landscape of high value must be under a permission-based regulation system, with flexible control measures such as notification and registration, and which covers the closely related areas which surround them or exist among them.

Compound landscapes, which contain a group of several landscapes of high value, often expand beyond the boundaries of several municipalities or prefectures, and are therefore distributed over an extremely large area. In such cases, protection which is provided only from the viewpoint of cultural properties/heritage, and is limited by the law within each municipality, is not sufficiently comprehensive. In this context, it is essential that the ACA and the local governments in charge of cultural properties should take the lead in ensuring the conservation of all cultural landscapes worth protecting. In these circumstances, the relevant local governments should also cooperate to provide zonal protection to cultural landscapes which are worth protecting. When such an area expands beyond a few municipalities or prefectures, it is necessary for the ACA to take the initiative in developing new frameworks of legal protection regardless of whether it is under the LPCP or not.

Important areas, and other areas identified by the provisional committee, were selected based on the procedures in place at the time, and therefore the selection procedure should not be regarded as already finalized. In particular, it should be noted that some areas of high value were not selected as important sites for various reasons such as the lack of consensus in the relevant communities. Efforts should be made to build a consensus among stakeholders so that these areas can be added to the list of important sites.

Various policies and measures for the conservation of cultural landscapes are already being implemented throughout Japan. In order to assist such efforts, it is necessary to organize a forum in which local residents, NPOs, NGOs, other relevant organs, and local governments are involved. The national government organizations should share their achievements, as well as information about cultural landscape conservation which has been carried out in various places, discuss issues concerning conservation sites, and exchange opinions and ideas among stakeholders at the national level. Local governments can take full advantage of websites on the Internet, and publications including brochures and newsletters, to achieve the same purpose. It is essential for local governments to invite residents, relevant organizations, experts, and researchers to set up a committee and participate in discussions about various conservation issues. It is also important for the related organizations and governments in charge of protection to organize seminars and workshops so that they can communicate with and stimulate each other.

18.6.2 Current Conclusion in the Registration of Cultural Landscapes

Cultural landscapes are heritages which are a rich representation of local characteristics, and the closest manifestation of the images of their home town and the spiritual proto-landscape in the hearts of the Japanese people. As industrial structures and modes of life have changed drastically, the scenery surrounding us has also changed so much that some cultural landscapes are facing extinction. Among these, many are highly valuable man-made works, facing nature unswervingly and directly over a long period of time. At present, we have an absolute responsibility to pass the value of these cultural landscapes on to future generations, and this needs to be recognized without delay.

In addition, national and local governments need to pay attention to the domestic and international trends encompassing cultural landscapes, and to make efforts to ensure that the policies and measures necessary for conservation are implemented to protect their high value. It is our earnest wish as members of the ad hoc committee that citizens, local governments, and other relevant organizations should recognize their roles in the conservation of cultural landscapes, explore deeper collaboration, and develop the basic directions in which the conservation of cultural landscapes must proceed.

Acknowledgments I would like to express my gratitude to the staff and related committee members in the Agency for Cultural Affairs for allowing me to refer to their official reports. I am also grateful to the members of the Nationwide Cultural Landscape Liaison Association for their permission to use the photographs. Further, I must thank the officials of the Kochi Prefecture Government for their assistance in the research of the Shimanto River watershed. The Nara National Research Institute for Cultural Properties organized a new training session for the parties concerned, and I delivered a lecture and participated in the discussion. This was a great opportunity for me to summarize and speak about Japanese cultural landscapes. This chapter was financed by Kensetsukankyo-kenkyujo Co. Ltd., Tokyo, and the GELs Programme of IDEC, Hiroshima University.

References

Agency for Cultural Affairs (2005) Cultural landscapes in Japan. Douseisha, Tokyo (in Japanese)

Committee on the Preservation, Development and Utilization of Cultural Landscapes Associated with Agriculture, Forestry and Fisheries (2003) The report of the study on the protection of cultural landscapes associated with agriculture, forestry and fisheries. Agency for Cultural Affairs, Tokyo

Education Board of Akita Prefecture (ed) (2009) Cultural landscapes. Akita (in Japanese)

Fukamachi K, Oku H, Nakashizuka T (2001) The change of a Satoyama landscape and its causality in Kamiseya, Kyoto Prefecture, Japan between 1907 and 1995. Landsc Ecol 16:703–717

Iiyama N, Kamada M, Nakagoshi N (2005) Ecological and social evaluation of landscape in arural area with terraced paddies in southwestern Japan. Landsc Urban Plan 70:301–313

Kamada M, Nakagoshi N (1996) Landscape structure and the disturbance regime at three rural regions in Hiroshima Prefecture, Japan. Landsc Ecol 11:15–25

Nakagoshi N (1995) Changing cultural landscapes in Japan. In: von Droste B et al (eds) Cultural landscapes of universal value. Gustav Fischer, Jena

Nakagoshi N, Hong SK (2001) Vegetation and landscape ecology of East Asian 'Satoyama'. Glob Environ Res 5:171–181

Nakagoshi N, Kim JE (2007) Landscape changes in Japan based on national grid maps. In: Hong SK et al (eds) Landscape ecological applications in man-influenced areas: linking man and nature systems. Springer, Dordrecht

Nakagoshi N, Ohta Y (1992) Factors affecting the dynamics of vegetation in the landscapes of Shimokamagari Island, southwestern Japan. Landsc Ecol 7:111–119

Nakagoshi N, Hikasa M, Koarai M, Goda T, Sakai I (1998) Grip map analysis and its application for detecting vegetation changes in Japan. Appl Veg Sci 1:219–224

Ohta Y, Nakagoshi N (2006) Landscape changes in the Seto Inland Sea, Japan. Ekologia (Bratislava) 25 Suppl 1:190–200
Takashima city (ed) (2009) Japan's original scenery "Cultural Landscape" 2009. Nationwide Cultural Landscape Liaison Association, Takashima (in Japanese)

Chapter 19
Restoring Central Asian Floodplain Ecosystems as Natural Capital and Cultural Heritage in a Continental Desert Environment

Stefan Zerbe and Niels Thevs

19.1 Introduction

The restoration of degraded and damaged ecosystems has become a challenge for landscape management, nature conservation and sustainable land-use development throughout the world in recent decades (Bradshaw and Chadwick 1980; Jordan et al. 1987; Allen 1992; Harris et al. 1996; Urbanska 1997; Perrow and Davy 2002; Temperton et al. 2004; van Andel and Aronson 2006; Falk et al. 2006; Zerbe and Wiegleb 2009). As early as 1995, Daily reported that about 45% of the terrestrial land surface has a reduced land-use capacity due to non-sustainable land use in the past.

The restoration of floodplain ecosystems in desert environments becomes especially difficult with the increasing limitations of water resources, e.g. as a consequence of over-utilization of water and climate change. This is the case in Central Asia, where increasing populations and subsequent land-use pressures have led to severe environmental problems along the major river systems, such as the Amu Darya, Syr Darya, and Tarim River (Giese et al. 1998; Kuzmina and Treshkin 1997; Treshkin 2001; Feng et al. 2005; Glantz 2005). Some severe environmental problems are soil salinity due to inappropriate irrigation, the destruction of the natural vegetation, the expansion of the deserts, and the alteration and even destruction of the natural dynamics of rivers. As rivers and their floodplains provide many ecosystem services, e.g. the purification of water, combating desertification, the accumulation of carbon, and providing habitats for plants and animals, a particular focus is on their restoration (Mant and Janes 2006; Lüderitz and Jüpner 2009). As a case study, we focus on the Tarim River and its floodplain in Southern Xinjiang, NW China.

S. Zerbe (✉)
Faculty of Science and Technology, Free University of Bozen – Bolzano, Piazza Università 5, P.O. 276, 39100 Bolzano, Italy
e-mail: Stefan.Zerbe@unibz.it

N. Thevs
Institute of Botany and Landscape Ecology, Greifswald University, Grimmer Strasse 88, 17487 Greifswald, Germany

S.-K. Hong, et al. (eds.), *Landscape Ecology in Asian Cultures*, Ecological Research Monographs, DOI 10.1007/978-4-431-87799-8_19, © Springer 2011

We aim at an interdisciplinary approach to the landscape, taking ecological as well as socio-economic aspects into account. Therefore, we describe the natural environment, as well as outline the historical development and cultural and economic background. On the basis of our own investigations, we develop a future strategy for the sustainable development of the Tarim river floodplain.

19.2 Study Site

19.2.1 The Natural Environment

The study site is the Tarim river floodplain in Southern Xinjiang (NW China; Fig. 19.1). Xinjiang is located in Central Asia between the Tibetan Plateau, Mongolia and Kazakhstan. The Tarim Basin is filled with quaternary sediments deposited by the rivers and shifted by winds (Zhou and Chen 1992; Thevs 2007). The huge inland basin, mostly covered by the Taklamakan Desert, declines from 1 300 m above sea level in the west to 750 m above sea level in the east (Thevs et al. 2005), and is surrounded by the Kunlun, Tianshan and Pamir mountains. The climate is extremely arid, because these mountain ranges cut off all humid air currents to the Tarim Basin. The average air temperature is –9°C in January and 25°C in July, thus reflecting a strong continental climate. The average annual sum of precipitation is less than

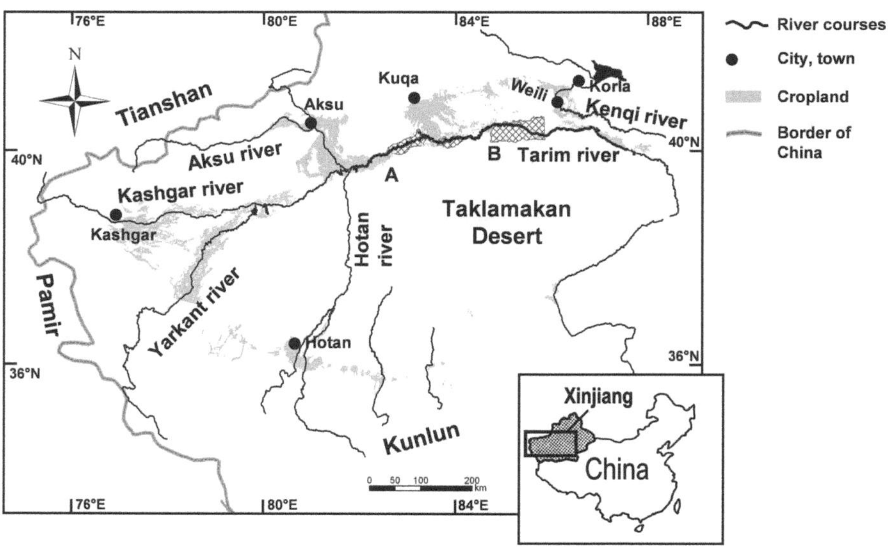

Fig. 19.1 The Tarim Basin with the location of major oases and the locations of the Tarim Shangyou Xayar Nature Reserve (A) and the Tarim Huyanglin Nature Reserve (B); only the Aksu River, the upper and middle reaches of the Tarim River, and the upper reaches of the other river courses are perennial rivers.

25 mm (Hudaberdi and Abdusalih 2002). Consequently, it is clear that the floodplain ecosystem depends heavily on the water supply from river floods or underground water stored during the floods (Wang et al. 1996). Some 75% of the annual run-off of the Tarim River occurs from July to September (Fig. 19.2), due to the glacier melt and summer precipitation in the surrounding mountains (Thevs et al. 2008a, b).

Today, the headwaters of the Tarim River are from the Aksu River. The Yarkant and Hotan Rivers only discharge occasionally during high floods in the Tarim (Giese et al. 2005). The Tarim River flows in a floodplain up to 100 km wide. Along the middle and lower reaches, the Tarim River and the neighbouring Kenqi River share one common floodplain.

Each flood carries huge loads of sediment. Along the middle reaches, the annual sediment load of the Tarim amounts to 13,670,000 t with 37% being transported in August (Xia 1998). Because of the sediments deposited, lateral levees are formed and some stretches of the river course become raised above the floodplain. The back swamps are often lower than the Tarim River itself (Thevs et al. 2008a). Each flood changes the meanders of the river's course. During strong floods, the river courses may break out of their beds and form new courses, and the entire river system can even be changed (Chou 1960), as shown in Fig. 19.3. These river dynamics have lost much of their former force in most places, but still prevail in parts of the Tarim Huyanglin Nature Reserve and the Tarim Shangyou Xayar Nature Reserve at the middle and upper reaches of the Tarim, respectively (Thevs et al. 2008a, b, Fig. 19.1).

About 250 years ago, the main course of the Tarim River was located in a similar position as it is today. The Tarim merged with the Kenqi River, and their common

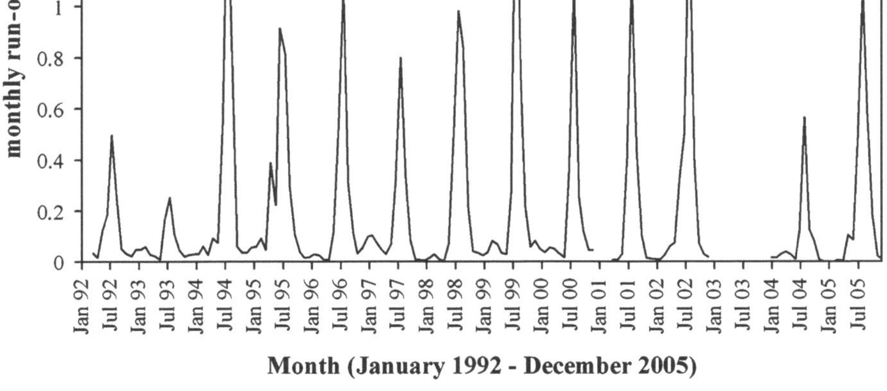

Fig. 19.2 Monthly run-off from the Tarim River from 1992 to 2005 at the Yengi Bazar gauging station, south east of Kuqa City (Thevs 2007)

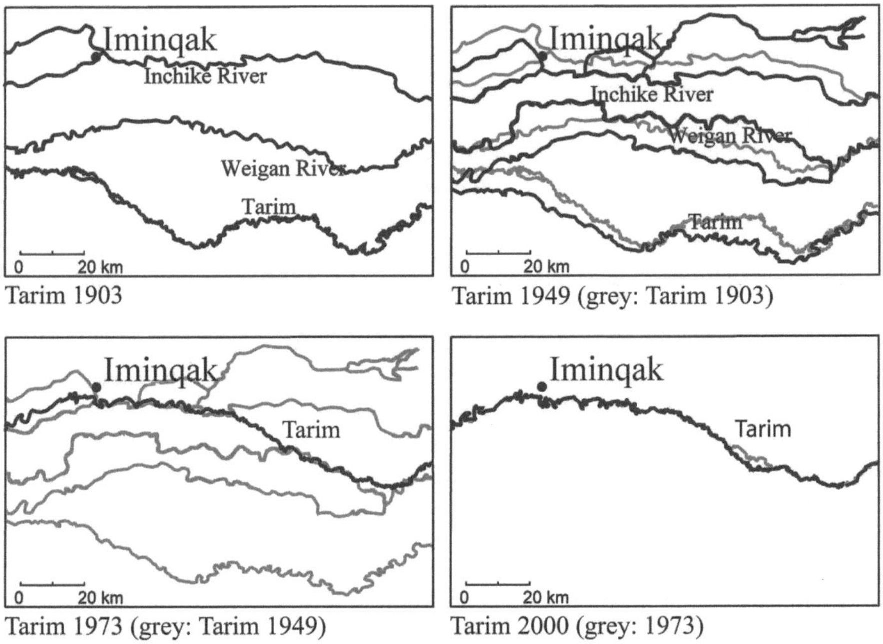

Fig. 19.3 Changes in the course of the Tarim River along the middle reaches according to Hedin (1905), Zhonghua Renmin Gongheguo Guojia Tuciju (1959), Landsat MSS image (path 155, row 31) from 1973/10/20, and Landsat ETM image (path 144 row 31) from 2000/07/06 (Thevs 2007)

main stream drained into Lake Lopnor. Later, the Tarim shifted its course southward, i.e. the old river bed indicated in Fig. 19.3 (Chou 1960; Zhong and Xiong 1999). This was the course of the Tarim as indicated by Hedin (1905).

By 1921, the Tarim River once again diverted more water into Lake Lopnor than into Lake Taitema. The water was diverted through the Kenqi River. This happened when the Tarim River adopted an irrigation channel as its new main course during the summer floods. In 1952, a dam was built in order to separate the Tarim and Kenqi Rivers. The Tarim River then flowed into Lake Taitema through its ancient bed (Chou 1960). Apparently, the river course called "Tarim" today was only one branch of the river 100 years ago. The Tarim River of Hedin's times is today a dry river bed covered with sand dunes (Fig. 19.3).

The soils are described as Arenosols, Fluvisols and Solonchaks (Thevs 2007). Compared with the total number of plant species in Xinjiang, where there are approximately 3 430 vascular plants (Hudaberdi and Xu 2000), Southern Xinjiang is rather poor with approximately 700 species. Xerophytic species contribute to the flora to a large extent. The differentiation and distribution of vegetation depends strongly on the groundwater level and the salt content of the upper soil. Following Hudaberdi and Abdusalih (2002), four major vegetation types can be distinguished in the Tarim Basin (see also Thevs 2005; Thevs et al. 2008b):

Fig. 19.4 Floodplain forest at the Tarim River, Iminqak, Tarim Huyanglin Nature Reserve, approximately 200 km southwest of Korla City (photograph: Thevs, July 2004)

- Floodplain ("Tugai") forests with *Populus euphratica* and *Populus pruinosa* (Fig. 19.4).
- Halophytic shrubs with mainly *Tamarix* species, *Halimodendron halodendron* and *Lycium ruthenicum*.
- Vegetation of the salt desert, with typical species such as *Halostachys caspica* and *Halocnemum strobilaceum*.
- Salt grassland with *Phragmites australis*, *Apocynum venetum* and *Glycyrrhiza inflata* as characteristic species.

19.2.2 Ecology of the Key Species

19.2.2.1 *Populus euphratica*

Populus euphratica is the most dominant forest-building tree along the Tarim River. In the upper reaches of the Tarim and further west only, *Populus pruinosa* mixes into the riparian forests (Wang et al. 1996; Weissgerber 2003). *Populus euphratica* is an obligate phreatophyte, i.e. it continuously contacts the groundwater or the water-saturated zone of the soil (Gries et al. 2003). *Populus euphratica* can contact the groundwater as deep as 12 m, but the electrical conductivity in the water-saturated soil zone must not exceed 8.7 mS/cm (Thevs 2005; Thevs et al. 2008b). After periods of water stress have led to a decrease in growth intensity, annual growth increments increase after an improvement in the water supply (Westermann et al. 2008).

With regard to recruitment, *Populus euphratica* follows two strategies: generative and vegetative recruitment. Generative recruitment depends on flood events and river dynamics (Saito et al. 2002; Weissgerber 2003; Thevs et al. 2008a).

The seeds are light, have pappus-like hairs, and are dispersed by wind and water. The main fruiting period is between July and September, i.e. during the flood period of the Tarim River. Germination and establishment depend on freshly deposited river banks. Optimal germination occurs under conditions of high light, a temperature between 25 and 30°C, and water-saturated soils with a salt content lower than 0.2%. Such sites are created by the annual summer floods and shifting river courses. Germination occurs in lines or narrow strips marking flood water lines on the river banks (China Ministry of Forestry 1990; Liu et al. 1990). After germination, the seedlings invest more in root growth than in shoot growth in order to secure water uptake during the spring and early summer of the following year. If the flood of the year following germination starts too late or does not moisten the soil to the depth of the seedlings' roots, the seedlings cannot survive and establish.

On sites out of the reach of the floods, *Populus euphratica* recruits from root suckers, which emerge from the lateral roots. These clones can cover areas of 4 ha (Bruelheide et al. 2004). As the parent trees supply the root suckers with water until they reach the groundwater by themselves, vegetative recruitment is restricted to sites on which *Populus euphratica* does not suffer water stress; in general, these are sites with a groundwater level not deeper than 8 m.

19.2.2.2 *Phragmites australis*

Phragmites australis is the dominant reed-bed plant along the Tarim River. Furthermore, it frequently occurs as undergrowth in the riparian forests. *Phragmites australis* is an obligate phreatophyte too. Its root system can only exploit groundwater as deep as 3 m. The electrical conductivity in the water-saturated soil zone must not exceed 3.3 mS/cm (Thevs et al. 2007).

Giant reeds have stem heights of 200–400 cm and 15–30 nodes, and are found on submerged or periodically flooded sites. These stands are partly used to harvest reeds as raw materials. Medium-sized reeds have stems 40–200 cm high with 10–20 nodes. These are found on non-flooded sites. Small reeds have stems with an average height of 10–20 cm, and not exceeding 40 cm, and 8–12 nodes, and occur on sites with salt accumulation, i.e. with a salt content of 37.1 g/kg in the upper 30 cm of the soil (Gao and Xu 1995). The latter reeds appear to be sturdy with contracted internodes. The leaves are hard and thorny, and thus resemble claws. Under continuous grazing, the reed shrinks and also assumes this sturdy habit. Thevs et al. (2007) measured average stem heights of 220 cm and 15 cm on a non-grazed and grazed site, respectively. The average number of leaves per metre of stem length was 7 on the non-grazed site and 118 on the grazed site.

According to Haslam (1969, 1970), contracted internodes are due to nutrient deficiency and salt. Consequently, the phenotypes of *Phragmites* reeds seem to correspond with the flood regime and topsoil salinization (Gao and Xu 1995), as well as with the grazing regime (Thevs et al. 2007).

19.2.2.3 *Apocynum venetum*

Apocynum venetum is a herbaceous plant of 1 m height. It forms monospecies stands, is distributed together with herbaceous plants such as *Alhagi sparsifolia*, *Karelinia caspica* or *Glycyrrhiza inflata*, or is part of the undergrowth in the riparian forests. *Apocynum venetum* is distributed on sites with a groundwater level of 6 m on the middle reaches of the Tarim (Thevs et al., unpublished data) and 8 m along the lower reaches (Chen et al. 2006).

19.2.2.4 *Tamarix* sp.

The most common *Tamarix* species along the Tarim River and in the Tarim river floodplain is *Tamarix ramosissima*. It forms monospecies shrub land or occurs as undergrowth in the riparian forests. *Tamarix ramosissima* is distributed from frequently flooded sites to the margins of the Tarim floodplain bordering the Taklamakan Desert, as well as in dune valleys in the desert (Liu 1995).

Tamarix ramosissima is a facultative phreatophyte (Smith et al. 1998), i.e. it can endure phases without groundwater contact. According to Thevs et al. (2008b), *Tamarix ramosissima* is restricted to sites where the groundwater level is not deeper than 12 m, as is *Populus euphratica*. However, *Tamarix ramosissima* has a far higher salt tolerance and was found on sites with an electrical conductivity in the water-saturated soil zone of 25.5 mS/cm.

Tamarix ramosissima, as well as the other *Tamarix* species, stabilizes and accumulates sand. The *Tamarix* species grow up with the dune, which they create by themselves. These dunes are generally 2–5 m high, but can also be as high as 15 m.

The *Tamarix* species also recruit generatively and vegetatively. The seedlings of *Tamarix* are more salt tolerant than those of *Populus euphratica*. Furthermore, *Tamarix* seedlings can survive longer periods without groundwater contact (Liu 1995).

19.2.3 The Cultural History

The Tarim Basin has a long cultural history, which is best known through the silk road, which crossed the Tarim Basin from east to west and was one of the most important trade routes for many centuries (Le Coq 1928; Hedin 1936; Weggel 1985; Fülling 1999). Since the Chinese Han Dynasty around 2000 years ago, goods have been exchanged between the Roman Empire and China. During that time, the oases of today's Kashgar, Kuqa, Bügür (Luntai), Hotan, and Keriya already existed and were inhabited by Indo-European people such as the Tocharians. Their language was related to Sanskrit. During the Tang Dynasty, the Chinese Empire occupied the Tarim Basin from 600 to 650 AD, and this was later followed by the Tibetan Empire (Wood 2005).

The Uyghurs formed their first kingdom in today's Mongolia in 744. This Uyghur Kingdom existed until 840, and had very close socio-economic relationships with the Chinese Tang Dynasty. However, in 840 the Kyrgyz drove the Uyghurs out of their kingdom in today's Mongolia, and all the Uyghur people moved into the Tarim Basin. The Uyghurs mixed with the Tocharian people and adopted their religion and their culture of oasis agriculture (Scharlipp 1992; Soucek 2000).

The Tocharian, Uyghur and Chinese people concentrated in the oases at the foothills of the mountains. The Tarim river floodplain was never densely settled, because the river was too dynamic. Only people like the Lopliks or Abdal lived in the Tarim floodplain. As fishermen and nomads, they were adapted to the dynamic environment of the Tarim river floodplain.

After the Peoples Republic of China was founded in 1949, state farms were built up by the army, the so-called military farms. These farms were founded along the upper and lower reaches of the Tarim River, and served as footholds for the Peoples Liberation Army during the process of occupying Xinjiang. They served as starting points for land-opening campaigns in the Tarim river floodplain on the one hand, and on the other, the farms provided the food supply for the Peoples Liberation Army in Xinjiang (Weggel 1985).

19.2.4 The Socio-Economic Environment of Today

Today in Xinjiang, the East Asian trend of rapidly increasing populations and urban agglomerations (Mackensen 1998; Breitung 2001; Kim et al. 2002) also strongly influences the landscape and land-use development. Consequently, in the past few decades, the traditional oases of silvo-horticultural systems have been transformed into new systems of urban green spaces (Zerbe et al. 2005). Thus, they reflect traces of the cultural identity of the native Uyghur people on the one hand, and the increasing Han-Chinese influence and international trends of urban design and planning on the other. In China, urban development and urban culture reflect the centralized political system controlled to a large extent by the Communist Party.

According to an urban ecological investigation by Zerbe et al. (2005) in Southern Xinjiang, the population sizes of the four largest oases, Aksu, Kashgar, Korla and Hotan, range between approximately 80000 and 213000 inhabitants, with population densities up to 11300 inhabitants per km^2 (Table 19.1). Whilst in the 1940s approximately 75% of Xinjiang's population were Uyghur, the proportion has been reduced to some 45% today. Owing to the official population policy since the late 1950s, the Han-Chinese share of the total population has increased strongly from approximately 7% in the 1950s to about 40% today. The population change in Xinjiang is accompanied by massive shifts in the economy of the oases, i.e. a gradual shift from traditional agriculture and agro-forestry towards an intensive irrigation agriculture mainly based on cotton (*Gossypium* sp.) (Hoppe 1992, 1995; Bohnet et al. 1998, 1999). The traditional land-use system was based on a large variety of cultivated plants (Hoppe 1992), and in particular those which used the water resources in a sustainable way. In Korla (Fig. 19.1), oil and natural gas extraction is the main industry. In the other oases, textile production presents a major source of income (Table 19.1). In the past 20 years

Table 19.1 Characterization of four oases in Southern Xinjiang with regard to city size, population and economic aspects, after Halik (2003) and Zerbe et al. (2005)

Oases	Aksu	Hotan	Kashgar	Korla
City area (km^2)	26.7	10.0	18.5	38.0
Number of inhabitants	213 160	82 000	209 000	208 600
Population density (inhabitants per km^2)	8 000	8 200	11 300	5 500
Uyghur population (%)	41.4	83.0	76.5	30.0
Han-Chinese population (%)	57.2	16.6	22.3	67.0
Main industry and its share of total production (%)	Textile industry (53)	Textile industry (56)	Textile industry (44)	Oil and natural gas (67)
Housing space per person (m^2)	7.5	8.3	8.8	7.9
Traffic area per person (m^2)	6.8	7.2	5.1	12.6
Water consumption per person per day (l)	166	195	162	209
Total annual waste water (million m^3)	15.8	2.8	11.0	22.2

Korla, in comparison with the other oases, has developed most rapidly from a small urban settlement to the most prominent and dynamic industrial centre of Southern Xinjiang. This is indicated by the highest values of traffic area per person (12.6 m^2), water consumption (209 l per person per day) and the amount of waste water per year (22.2 million m^3), as shown in Table 19.1.

Currently, the most prominent land uses in the Tarim floodplain are irrigation agriculture, i.e. mainly cotton and melons, and animal herding. The largest contiguous areas under irrigation fed by the Tarim River are located in the upper reaches and the lower reaches of the Tarim River (Fig. 19.1). The middle reaches of the Tarim remained almost untouched by irrigation agriculture until the beginning of the 1990s. Therefore, the Tarim Huyanglin Nature Reserve and the Tarim Shangyou Xayar Nature Reserve (Fig. 19.1) were founded along the Tarim middle and upper reaches in 1983 and 2006, respectively. However, after 1995, land was also opened up for agriculture in the Tarim Huyanglin Nature Reserve (Hofmann 2006).

Until 2007, the irrigation depended completely on the river run-off of the Tarim. As the Tarim River carries little water during spring and early summer, i.e. when the crops have the highest water demand, huge water reservoirs were constructed during the 1950s (Hoppe 1992).

After 1995, investors, mostly from outside the Tarim Basin, reclaimed land along the middle reaches of the Tarim River, including the Tarim Huyanglin Nature Reserve. Now, the cropland area within the Tarim Huyanglin Nature Reserve as well as along the Aksu River and the Tarim upper reaches continues to increase (Table 19.2).

The pasture system is typically transhumance with goats and sheep as livestock. At the beginning of April, the herds are moved to pasture lands close to the Tarim River to graze there until mid-July or the beginning of August, depending on the onset of the summer flood. From the spring/summer pastures, the herds are then moved to the autumn pastures, which are located in non-flooded areas on the desert

Table 19.2 Agricultural area along the Tarim River in 2000 and 2008, based on mapping from Landsat ETM satellite images (Thevs, unpublished data)

	Area (ha)		Increase from 2000 to 2008	
	Year		(ha)	(%)
Region	2000	2008		
Aksu River and Tarim upper reaches	658 620	796 156	137 536	21
Weili county	42 739	67 710	24 971	58
Tarim Huyanglin Nature Reserve:				
Core area	2 178	5 225	3 047	140
Total	3 346	7 570	4 224	126
Shaya Tarim Shangyou Nature Reserve:				
Core area	444	611	167	38
Total	4 615	5 992	1 377	30

Table 19.3 Productivity and site characteristics of *Phragmites australis* in the Tarim middle reaches (Iminqak)

	Site			
Site parameter	1	2	3	4
Net primary production (above-ground biomass) (t/ha*a)	1.8	0.16	5.1	6.0
Flood duration per year (month)	2	0	1	1
Depth of closed capillary fringe in the soil (m)	1.3	2.5	2.9	2.2
Depth of iron mottles in the soil (m)	0	1.0	–	0
Depth of greyish reduced soil horizon (m)	–	2.7	2.9	2.2
Electrical conductivity at closed capillary fringe (mS/cm)	1.8	2.9	3.3	2.0
Electrical conductivity in the top 1 m of the soil profile (mS/cm)	2.0	52.6	8.5	2.7

Sites: 1 = grazed every year, flooded for 2 months every year; 2 = occasionally grazed, never flooded; 3 and 4 = non-grazed, flooded for 1 month about every second year. For data and site description see Thevs et al. (2007); – indicates none.

fringe. At the beginning of November, the herds move to the winter pastures close to the families' homes. The most important fodder plant is *Phragmites australis* (Table 19.3). The leaves of *Populus euphratica* and *Tamarix* sp. are also used as fodder. *Alhagi sparsifolia* and *Glycyrrhiza inflata*, which belong to the Fabaceae family, serve as the protein supply. If the herders can afford it, they buy corn as additional winter fodder. The fields with cotton stubble are also valued by the herdsmen as good winter pasture.

19.2.5 The Current Situation of the Floodplain Ecosystems

In contrast to the lower reaches of the Tarim River, which has been strongly degraded due to excessive use of the water resources in the past decades (Song et al. 2000; Hai et al. 2006), the Tarim Huyanglin Nature Reserve and the Tarim Shangyou Xayar Nature Reserve, on the middle reaches and the upper reaches, respectively (Fig. 19.1), still have a natural to near-natural floodplain character. In the Tarim

Huyanglin Nature Reserve, natural riparian ecosystems are not only distributed along the Tarim main course, but also along the river branches of the Tarim River's inland delta. Therefore, there is a peculiar mosaic of riparian forests, reeds, shrubland and desert in the Tarim Huyanglin Nature Reserve.

In 2004, dykes were constructed along the Tarim middle and upper reaches within the Tarim River Regulation Program (Zhu et al. 2006). As a result of these dykes, it can be expected that most of the river branches will be cut off from their water supply, so that the groundwater level in the hinterland will drop and the natural ecosystems will degrade in a similar way to what happened along the Tarim lower reaches before the Tarim River Regulation started (Westermann et al. 2008).

Seed germination of the native plants occurs frequently, but they rarely establish successfully. The reason is that one flood allows seed germination, but the flood of the following year is often too low or too late, because during the first days the flood water is diverted into irrigation rather than into the Tarim river bed.

The strongest human influence results from the prevailing system of water management and the large cotton plantations. Excessive use of surface water for irrigation, particularly for cotton fields in the upper and middle reaches of the river, has lowered the groundwater along the lower reaches from 3–5 m in the 1950s to 11–13 m in the 1990s (Chen et al. 2006). In Southern Xinjiang, the amount of water used for the irrigation of cotton fields is 22 500 m^3/ha, which is very high (Fan 1996). Large areas of poplar stands are therefore strongly degraded or completely dead. Furthermore, during spring and early summer in 2007 and 2008, the Tarim River fell dry along the middle and parts of the upper reaches due to lower snow melt run-off (Fig. 19.5).

Fig. 19.5 Dry river bed of the Tarim River at Yengi Bazar, Tarim Huyanglin Nature Reserve, approximately 220 km southwest of Korla City (photograph: Thevs, July 2008)

19.3 Ecosystem Services and Perspectives for Land Use of the Tugai Vegetation

The floodplain of the Tarim River provides many ecosystem services (for a typology, see de Groot et al. 2002) for the local people. The Tugai vegetation provides an important habitat for plant and animal life, and harbours the highest biodiversity in these regions (Thevs 2005; Thevs et al. 2005). In addition, the floodplain vegetation provides environmental benefits such as landscape preservation, wind-breaks, sand fixation and the prevention of sand storms, as well as soil and riverbank protection (Weissgerber 2003). These ecosystem services depend strongly on the key species of the floodplain vegetation, i.e. *Populus euphratica*, *Phragmites australis*, *Apocynum venetum* and *Tamarix ramossissima*.

19.3.1 Populus euphratica

The annual increment of wood of *Populus euphratica,* the main forest-building tree and the natural wood source of the Tarim River, ranges from 1 795 to 3 459 kg/ha*a (Xinjiang Linkeyuan Zaolin Zhisha Yanjiusuo 1989) and from 736 to 1 472 kg/ha*a (Wang et al. 1996) on sites in the Tarim Huyanglin Nature Reserve (Table 19.4). The annual increment of the whole above-ground biomass is 3.17–6.12 t/ha*a (Gries et al. 2005). Thus, *Populus euphratica* serves for biomass and carbon accumulation. Furthermore, the wood can be used as an energy source and construction material (Table 19.4).

In addition, the riparian forests provide shelter for people and livestock against the sun as well as against sand storms. Owing to their transpiration, the local climate

Table 19.4 Productivities and land-use perspectives for the key species of Tugai vegetation

Key species	Productivity (net primary production) above-ground biomass (t/ha*a)	Land-use perspective
Populus euphratica	0.7–3.5, wood (Xinjiang Linkeyuan Zaolin Zhisha Yanjiusuo 1989; Wang et al. 1996) 3.17–6.12, wood and leaves (Gries et al. 2005)	Carbon storage, also with regard to carbon credits; energy source; construction material
Phragmites australis	1.8–6.0 (Thevs et al. 2007)	Carbon storage, also with regard to carbon credits; raw material for paper production; construction material
Apocynum venetum	0.7, standing biomass (unpublished data)	Carbon storage, also with regard to carbon credits; tea; medical applications; fibre
Tamarix ramosissima	1.6–7.2 (Gries et al. 2005)	Carbon storage, also with regard to carbon credits; energy source

in dense riparian forests is considerably cooler than in shrub vegetation or open desert, thus providing areas for recreation.

Along the lower reaches of the Tarim, the annual increment of *Populus euphratica* increased after water was diverted into the river course, although the river course had been dry for nearly 30 years (Westermann et al. 2008). Therefore, *Populus euphratica* floodplain forests have a high restoration potential even after severe degradation.

19.3.2 Phragmites australis

Reed (*Phragmites australis*) is a key species in the floodplains of the streams in Central Asia, e.g. the Tarim River (Thevs 2005, 2007; Thevs et al. 2007), Syr Darya and Amu Darya (Ogar 2003). In these continental-arid regions, where large deserts with sparse vegetation or without any vegetation occur, it serves, for example, as a fodder plant and is intensively grazed by sheep and goats (Thevs 2006), as shown in Fig. 19.6. In addition, it is used as raw material for paper production and mats (Hansmann 2008). The area with reeds has expanded in the past few decades owing to the transformation of former riparian forests into agricultural land as a consequence of land-use intensification. Reeds are distributed along channels and in water reservoirs, as well as in irrigated cropland owing to seepage water.

The annual net primary production of the above-ground biomass on non-grazed sites ranges from 1.8 to 6 t/ha*a compared with 0.16 t/ha*a on grazed sites (Tables 19.3 and 19.4). Therefore, non-grazed reed stands are among the most

Fig. 19.6 Intensively grazed *Phragmites australis* reeds, i.e. site 2 in Table 19.3 (Thevs et al. 2007)

productive ecosystems of the riparian vegetation of the Tarim. Stands with sturdy reeds yielding low year-end biomass can be turned into reeds which yield a much higher biomass. We therefore conclude that biomass harvesting could be an alternative to grazing in terms of sustainable land use (Table 19.4).

19.3.3 *Apocynum venetum*

Apocynum venetum forms meadow-like stands in the floodplain of the Tarim River. The leaves of *Apocynum venetum* are harvested and sold as tea or used as raw material for medicines, e.g. suppressing high blood pressure. The stems have strong fibres, which are considered for textile manufacturing. *Apocynum venetum* is also a fodder plant for grazing. The above-ground biomass reaches 0.7 t/ha (Table 19.4).

19.3.4 *Tamarix ramosissima*

Tamarix ramosissima, as the most prominent *Tamarix* species in the Tarim floodplain, stabilizes floating sand and also stores carbon in its biomass. The annual productivity ranges from 1.55 to 7.15 t/ha*a (Gries et al. 2005). *Tamarix* shrubs form their own dunes and grow higher with the accumulating sand (Fig. 19.7).

Fig. 19.7 Dune profile of a *Tamarix* dune, Iminqak, Tarim Huyanglin Nature Reserve, approximately 200 km southwest of Korla City. The spade is 1 m long (photograph: Thevs, September 2006)

One dune with a diameter of 15 m and a height of 8 m contained 1.8 t of dead wood (data not published). The dead wood had been concentrated in the upper decimeters of the dune, where the wood was too dry to decompose. Thus, *Tamarix* shrublands also accumulate biomass and carbon.

Furthermore, the *Tamarix* shrub communities serve as autumn pasture. Locally, the wood is used as fuel (Table 19.4).

19.4 Stakeholders

The relevant stakeholders are governmental bodies, land users and entrepreneurs, scientists, and non-governmental organizations (NGOs) (cf. Wiegleb and Lüderitz 2009).

19.4.1 Governmental Bodies

The most important governmental bodies with regard to water allocation and ecosystem restoration are the Tarim River Basin Water Resource Commission, the Xinjiang Forestry Administration, and the military farms.

The Tarim River Basin Water Resource Commission was established in 1998 in order to coordinate the water allocation within the whole Tarim watershed, and also to avoid conflicts between prefectures and counties within the watershed. The Tarim River Basin Water Resource Bureau under the Tarim River Basin Water Resource Commission is responsible for enforcing the water allocation and the Tarim River Regulations (Zhang 2006). The Tarim River Basin Water Resource Bureau, based in Korla City, Bayingouleng Prefecture, has subordinate bureaus in all the other prefectures in the Tarim Basin, i.e. Aksu, Kashgar and Hotan (Zhu et al. 2006).

The Xinjiang Forestry Administration is responsible for all nature reserves in the Tarim basin, and it is also in charge of forest protection and management. The everyday management work of the Tarim Huyanglin Nature Reserve and the Tarim Shangyou Xayar Nature Reserve is done by the Nature Conservation Bureaus under the Forestry Administrations of Bayingouleng and Aksu Prefecture, respectively. Nature Conservation Bureaus do not have the power to enforce conservation measures. The Forestry Bureaus at the county level are responsible for the enforcement of conservation measures (Thevs 2007).

The military farms are one of the largest consumers of water along the Tarim River. The military farms are still under the direct rule of the Peoples Liberation Army head quarters in Beijing, and therefore cannot be ruled by the Xinjiang Government.

19.4.2 Land Users and Entrepreneurs

The most prominent land users are farmers and herders. The farmers can be differentiated into small- and large-scale farmers (Hofmann 2006). The small-scale farmers are mostly Uyghur people living in the traditional villages along the Tarim River, such as Tarim Xiang (Xayar County), Karqugha, and Dongkotan (Weili County). The large-scale farmers are mostly Han-Chinese, often from outside the Tarim Basin, who reclaimed land after 1995. The herders are Uyghur people from the villages along the Tarim River.

The most prominent entrepreneurial activities along the Tarim River are cotton processing, oil exploitation and tourism (Table 19.2). The cotton processing only consists of the separation of seeds from fibres and oil production from the seeds. The oil exploitation partly encroaches on areas of natural riparian forests and shrub vegetation. Tourism based on riparian forests started within the Tarim Huyanglin Nature Reserve near Yengi Bazar.

19.4.3 Scientists

All the major research institutions in Xinjiang have carried out research work with regard to ecosystem studies along the Tarim River, i.e. Xinjiang Institute of Ecology and Geography under the Chinese Academy of Sciences, Xinjiang University, Xinjiang Normal University and Xinjiang Forestry Academy. On a local level, the Poplar Research Institute in Korla and the Tarim University in Aral City have to be added to the list of scientific institutes.

The Xinjiang Institute of Ecology and Geography contributed most to the scientific input for the Tarim River Regulations, and carries out monitoring of groundwater and vegetation recovery along the lower reaches of the Tarim (Zhang et al. 2005; Chen et al. 2006). Xinjiang University and Xinjiang Normal University investigated the ecology and land use along the Tarim River (Hoppe et al. 2006). The Xinjiang Forestry Academy, the Poplar Research Institute, and Tarim University carried out applied research projects with regard to afforestation of *Populus euphratica* and *Tamarix* sp. (Xinjiang Linkeyuan Zaolin Zhisha Yanjiusuo 1989). To date, many other institutions throughout the world have done specific investigations or built up international research networks with a focus on the Tarim Basin (Runge and Zhang 2004; Hoppe et al. 2006).

19.4.4 Non-Governmental Organizations

A civil society with NGOs as stakeholders in various fields is just emerging in China (Heberer and Sausmikat 2004). With regard to Xinjiang, there are only two small NGOs which focus on the environment, i.e. Xinjiang Conservation Fund and

Xinjiang Biodiversity Conservation. Both organizations are located in Urumqi and mainly organize student activities, but they also provide environmental education for school teachers and pupils as well as farmers.

19.5 Restoration of the Tarim River Floodplain

Conservation and restoration of the riparian vegetation requires two preconditions. Firstly, a sufficiently high groundwater table must be maintained so that the plants can connect to the groundwater and reach their water supply. Secondly, flood events and river dynamics must prevail as preconditions for generative recruitment and thus the conservation of the genetic diversity of the natural ecosystems.

In order to meet the first condition, the summer floods at least have to be diverted along the whole Tarim River so that the groundwater body is refilled regularly. The river branches along the middle reaches of the Tarim must also be flooded regularly in order to maintain a stable groundwater level in the hinterland of the Tarim River.

The second condition can only be met along particular parts of the Tarim River and requires careful monitoring as well as flood management. If seedling germination occurs on larger areas, the flood water in the following year should be released early so that the seedlings have an opportunity to establish. If the following flood does not start in time or is too low, seedlings should be irrigated during the natural flood period and monitored. Sites with prevailing river dynamics should not be disturbed.

Further land reclamation should be stopped. Along the upper reaches and the Aksu River, newly reclaimed land will decrease river run-off along the middle and lower reaches, while newly reclaimed land along the middle reaches will inevitably depend on groundwater exploitation. Groundwater exploitation will result in a lowering of groundwater levels, which will lead to large-scale vegetation degradation and thus counteract all efforts for vegetation restoration. The groundwater exploitation of the currently existing agriculture has to be limited and monitored in order to avoid lowering groundwater levels and further ecosystem degradation.

Alternative land-use systems based on native plants, e.g. *Phragmites australis* or *Apocynum venetum*, have to be investigated further. While cotton and melons have to be irrigated in the soil surface, the native plants tap the groundwater and thus survive the low water season. They use the groundwater which was stored in the groundwater layer by the preceding summer floods. This groundwater is not prone to losses from evaporation like the water from reservoirs. Furthermore, the water uptake from the groundwater does not lead to soil salinization, while irrigation bears the risk of soil salinization.

In conclusion, sustainable land use and ecosystem restoration require a holistic approach of water allocation and land-use planning, including all stakeholders, and in particular those who are responsible for nature conservation and restoration.

References

Allen EB (1992) Principles of restoration ecology, an integrated approach. Springer, Berlin

Bohnet A, Giese E, Zeng G (1998) Die Autonome Region Xinjiang (VR China) – Eine ordnungspolitische und regionalökonomische Studie (Band I). LIT Verlag, Münster

Bohnet A, Giese E, Zeng G (1999) Die Autonome Region Xinjiang (VR China) – Eine ordnungspolitische und regionalökonomische Studie (Band II). LIT Verlag, Münster

Bradshaw AD, Chadwick MJ (1980) The restoration of land. Blackwell, Oxford

Breitung W (2001) Hongkong und der Integrationsprozess. Basler Beitr z Geographie 48:1–206

Bruelheide H, Manegold M, Jandt U (2004) The genetical structure of *Populus euphratica* and *Alhagi sparsifolia* stands in the Taklamakan Desert. In: Runge M, Zhang XM (eds) Ecophysiology and habitat requirements of perennial plant species in the Taklamakan Desert. Shaker, Aachen

Chen YN, Zilliacus H, Li WH, Zhang HF, Chen YP (2006) Ground-water level affects plant species diversity along the lower reaches of the Tarim river, Western China. J Arid Environ 66:231–246

China Ministry of Forestry (1990) The forests of Xinjiang. Xinjiang Peoples Press, Urumqi (in Chinese)

Chou T (1960) The problem of channel shifting of the middle reaches of the Tarim River in Southern-Sinkiang. In: Murzayev EM, Chou L (eds) Natural conditions of Sinkiang. Chinese Academy of Sciences 1959 and Prirodnyye usloviya Sin'tsziana NAUK 1960, Beijing, Moscow

Daily G (1995) Restoring value to the world's degraded lands. Science 269:350–355

De Groot RS, Wilson MA, Boumans RMJ (2002) A typology for the classification, description and valuation of ecosystem functions and goods and services. Ecol Econ 41:393–408

Falk DA, Palmer MA, Zedler JB (eds) (2006) Foundations of restoration ecology. Island Press, Washington, DC

Fan ZL (1996) Research on the impacts of land utilization to ecology and environment in Xinjiang and the correspondent countermeasures. Qixiang Chubanshe, Beijing (in Chinese)

Feng Q, Liu W, Si JH, Su YH, Zhang YW, Cang ZQ, Xi HY (2005) Environmental effects of water resource development and use in the Tarim River basin of northwestern China. Environ Geol 48:202–210. doi:10.1007/s00254-005-1288-0

Fülling A (1999) Chinas Norden – Die Seidenstrasse. Brackwede, Bielefeld

Gao HY, Xu J (1995) Correlation of growth forms of *Phragmites australis* with soil water and salt. Grassl China 6:20–23 (in Chinese)

Giese E, Bahro G, Betke D (1998) Umweltzerstörungen in Trockengebieten Zentralasiens (West- und Ost-Turkestan): Ursachen, Auswirkungen, Maßnahmen. Steiner, Stuttgart

Giese E, Mamatkanov DM, Wang R (2005) Wasserressourcen und deren Nutzung im Flussbecken des Tarim (Autonome Region Xinjiang (VR China). Zentrum für Internationale Entwicklungs- und Umweltforschung (ZEU), Universität Giessen, Giessen

Glantz MH (2005) Water, climate, and development issues in the Amu Darya Basin. Mitigation and Adaptation Strategies for Global Change 10:1573–1596. doi:10.1007/s11027-005-7829-8

Gries D, Zeng F, Foetzki A, Arndt SK, Bruelheide H, Thomas FM, Zhang XM, Runge M (2003) Growth and water relations of *Tamarix ramosissima* and *Populus euphratica* on Taklamakan desert dunes in relation to depth to a permanent water table. Plant Cell Environ 26:725–736

Gries D, Foetzki A, Arndt SK, Bruelheide H, Thomas FM, Zhang XM, Runge M (2005) Production of perennial vegetation in an oasis-desert transition zone in NW China – allometric estimation, and assessment of flooding and use effects. Plant Ecol 181:23–43

Hai YO, Wai L, Hoppe T, Thevs N (2006) Half a century of environmental change in the Tarim river valley. An outline of the causes and remedies. In: Hoppe T, Kleinschmit B, Roberts B, Thevs N, Halik Ü (eds) Watershed and floodplain management along the Tarim River in China's arid northwest. Shaker, Aachen

Halik Ü (2003) Stadtbegrünung im ariden Milieu – das Beispiel der Oasenstädte des südlichen Xinjiang/VR China, unter besonderer Berücksichtigung ökologischer, sozioökonomischer und kulturhistorischer Aspekte. Berliner Beitr z Umwelt u Entwicklung, vol. 20. TU-Berlin, Berlin

Hansmann P (2008) "They call them Golden Sticks" – socio-economic explorations around the commodity of reed. M.Sc. thesis, University Wageningen, Wageningen
Harris JA, Birch P, Palmer JP (1996) Land restoration and reclamation, principles and practise. Longman, Harlow
Haslam SM (1969) Stem types of *Phragmites communis* Trin. Ann Bot 33:127–131
Haslam SM (1970) Variation of population type in *Phragmites communis* Trin. Ann Bot 34:147–158
Heberer T, Sausmikat N (2004) Bilden sich in China Strukturen einer Zivilgesellschaft heraus? Duisburg Working Papers on East Asian Studies, vol 61. University Duisburg-Essen, Duisburg
Hedin SA (1905) Scientific results of a journey in Central Asia 1899–1902, vol 1: The Tarim River. Lithografic Institute of the General Staff of the Svedish Army, Stockholm
Hedin SA (1936) Die Seidenstrasse. Brockhaus, Leipzig
Hofmann S (2006) Comparative analysis of Uyghur and Han-Chinese farm management along the middle reaches of the Tarim River. In: Hoppe T, Kleinschmit B, Roberts B, Thevs N, Halik Ü (eds) Watershed and floodplain management along the Tarim River in China's arid northwest. Shaker, Aachen
Hoppe T (1992) Chinesische Agrarpolitik und Uygurische Agrarkultur im Widerstreit. Das sozio-kulturelle Umfeld von Bodenversalzungen und -alkalisierungen im nördlichen Tarim-Becken (Xinjiang). Institut für Asienkunde, Hamburg
Hoppe T (1995) Die ethnischen Gruppen Xinjiang: Kulturunterschiede und interethnische Beziehungen. Institut für Asienkunde, Hamburg
Hoppe T, Kleinschmit B, Roberts B, Thevs N, Halik Ü (eds) (2006) Watershed and floodplain management along the Tarim River in China's arid northwest. Shaker, Aachen
Hudaberdi M, Abdusalih N (2002) The features of vegetation and eco-geography in the Taklamaklan Desert. In: Zerbe S, Küchler J, Hamann B (ed) Basic and applied aspects of nature and environmental protection in North China and South Korea with a focus on ecology and socio-economics. Landschaftsentwicklung und Umweltforschung, vol 121. TU Berlin, Berlin
Hudaberdi M, Xu J (2000) Claves plantarum xinjiangensis. Xinjiang University Press, Urumqi (in Chinese)
Jordan WR, Gilpin ME, Aber JD (1987) Restoration ecology: a synthetic approach to ecological research. Cambridge University Press, Cambridge
Kim YM, Zerbe S, Kowarik I (2002) Human impact on flora and habitats of Korean rural settlements. Preslia 74:409–419
Kuzmina ZV, Treshkin SY (1997) Soil salinization and dynamics of Tugai vegetation in the southwestern Caspian Sea region and in the Aral Sea coastal region. Eurasian Soil Sci 30:642–649
Le Coq A (1928) Die buddhistische Spätantike in Mittelasien. Reimer, Berlin
Liu MT (1995) Comprehensive research about the genus *Tamarix* including its propagation on large areas and its utilization. Lanzhou University Press, Lanzhou (in Chinese)
Liu PJ, Zhang L, Fan CQ (1990) The *Populus euphratica* resources along the Tarim river. In: Liang K, Liu PJ, Liang K, Liu PJ (eds) Investigation of resources and the environment along the Tarim river through remote sensing. Science, Technique, and Documentation Press, Beijing, in Chinese
Lüderitz V, Jüpner R (2009) Renaturierung von Fließgewässern. In: Zerbe S, Wiegleb G (eds) Renaturierung von Ökosystemen in Mitteleuropa. Springer Spektrum Akad. Verlag, Heidelberg, pp 95–124
Mackensen R (1998) Bevölkerungsdynamik und Stadtentwicklung in ökologischer Perspektive. In: Sukopp H, Wittig R (ed) Stadtökologie, pp 49–79
Mant J, Janes M (2006) Restoration of rivers and floodplains. In: van Andel J, Aronson J (eds) Restoration ecology. The new frontier. Blackwell, Oxford, pp 141–157
Ogar NP (2003) Vegetation of river valleys. In: Rachkovskaya EI, Volkova EA, Khramtsov VN (ed) Botanical geography of Kazakhstan and middle Asia (desert region). Komarov Botanical Institute of Russian Academy of Sciences, Saint Petersburg, Institute of Botany and Phytointroduction of Ministry of Education and Science of Republic Kazakhstan, Almaty, Institute of Botany of Academy of Sciences of Republik Uzbekistan, Tashkent
Perrow MR, Davy AJ (ed) (2002) Handbook of ecological restoration, vol 1, 2. Cambridge University Press, Cambridge

Runge M, Zhang XM (eds) (2004) Ecophysiology and habitat requirements of perennial plant species in the Taklamakan Desert. Shaker, Aachen

Saito Y, Shiraishi S, Tanimoto T, Yin L, Watanabe S, Ide Y (2002) Genetik diversity of *Populus euphratica* population in Northwestern China determined by PAPD DNA analysis. New For 23:97–103

Scharlipp W-E (1992) Die frühen Türken in Zentralasien: Eine Einführung in ihre Geschichte und Kultur. Wissenschaftliche Buchgesellschaft, Darmstadt

Smith SD, Deviktt DA, Sala A, Cleverly JR, Busch DE (1998) Water relations of riparian plants from warm desert regions. Wetlands 18:687–696

Song YD, Fan ZL, Lei ZD, Zhang FW (2000) Research on water resources and ecology of the Tarim river. Xinjiang Peoples Press, Urumqi, China (in Chinese)

Soucek S (2000) A history of Inner Asia. Cambridge University Press, Cambridge

Temperton VM, Hobbs RJ, Nuttle T, Halle S (2004) Assembly rules and restoration ecology. Bridging the gap between theory and practice. Island Press, Washington

Thevs N (2005) Tugay vegetation in the middle reaches of the Tarim River – vegetation units and their ecology. Arch Nat Conserv Landsc Res 44:64–84

Thevs N (2006) Grazing in the Tarim Huyanglin nature reserve. In: Hoppe T, Kleinschmit B, Roberts B, Thevs N, Halik Ü (eds) Watershed and floodplain management along the Tarim River in China's arid northwest. Shaker, Aachen

Thevs N (2007) Ecology, spatial distribution, and utilization of the Tugai Vegetation at the middle reaches of the Tarim river, Xinjiang, China. Cuvillier, Göttingen

Thevs N, Halik Ü, Schnittler M, Succow M, Zerbe S (2005) Tugay-Wälder im Tarim-Becken (Xinjiang, NW-China): Gefährdung sowie Schutz- und Revitalisierungsbestrebungen. Forst und Holz 60:63–66

Thevs N, Zerbe S, Gahlert F, Mijit M, Succow M (2007) Productivity of reed (*Phragmites australis* Trin. ex. Staud.) in continental-arid NW China in relation to soil, groundwater, and land use. J Appl Bot Food Qual 81:62–68

Thevs N, Zerbe S, Schnittler M, Abdusalih N, Succow M (2008a) Structure, reproduction and flood-induced dynamics of riparian Tugai forests at the Tarim River in Xinjiang, NW China. Forestry 81:45–57

Thevs N, Zerbe S, Peper J, Succow M (2008b) Vegetation and vegetation dynamics in the Tarim river floodplain of continental-arid Xinjiang, NW China. Phytocoenologia 38:65–84

Treshkin SY (2001) The Tugai forests of floodplain of the Amudarya River: ecology, dynamics and their conservation. In: Breckle SW, Veste M, Wucherer W (eds) Sustainable land use in deserts. Springer, Heidelberg

Urbanska KM et al (1997) Safe sites interface of plant population and restoration ecology. In: Urbanska KM, Webb NR, Edwards PJ (ed) Restoration ecology and sustainable development. International conference on restoration ecology and sustainable development, Thessaloniki

Van Andel J, Aronson J (2006) Restoration ecology. The new frontier. Blackwell, Oxford

Wang SJ, Chen BH, Li HQ (1996) Euphrates Poplar Forest. China Environmental Science Press, Beijing (in English)

Weggel O (1985) Xinjiang – Das zentralasiatische China: Eine Landeskunde. Institut für Asienkunde, Hamburg

Weissgerber H (2003) *Populus euphratica*. In: Schütt P (ed) Enzyklopädie der Holzgewächse. Ecomed, Landsberg/Lech

Westermann J, Zerbe S, Eckstein D (2008) Age structure and growth of degraded *Populus euphratica* floodplain forests in north-west China and perspectives for their recovery. J Integr Plant Biol 50:536–546

Wiegleb G, Lüderitz V (2009) Akteure in der Renaturierung. In: Zerbe S, Wiegleb G (ed) Renaturierung von Ökosystemen in Mitteleuropa. Springer, Spektrum Akad. Verlag, Heidelberg

Wood F (2005) The Silk Road: two thousands years in the heart of Asia. The British Library, London

Xia DK (1998) Silt movement and river course shift of the mainstream of Tarim River. In: Mao DH (ed) The water resources, environment and management of the Tarim river watershed. China Environmental Science Press, Beijing (in Chinese)

Xinjiang Linkeyuan Zaolin Zhisha Yanjiusuo (1989) Research on methods to revitalize *Populus euphratica* forests. Xinjiang Linkeyuan, Urumqi (in Chinese)

Zerbe S, Halik Ü, Küchler J (2005) Urban greening in the oases of continental arid Southern Xinjiang (NW China) – an interdisciplinary approach. Die Erde 136:245–266

Zerbe S, Wiegleb G (eds) (2009) Renaturierung von Ökosystemen in Mitteleuropa. Springer, Spektrum Akad. Verlag, Heidelberg

Zhang JB (2006) Water management issues and legal framework development of the Tarim River Basin. In: Wallace J, Wouters P, Pavakavambwa S (eds) Hydrology and water law – bridging the gap. IWA Publishing, London

Zhang YM, Chen YN, Pan BR (2005) Distribution and floristics of desert plant communities in the lower reaches of Tarim River, southern Xinjiang, People's Republic of China. J Arid Environ 63:772–784

Zhonghua Renmin Gongheguo Guojia Tuciju (1959) Map of Xinjiang. Zhonghua Renmin Gongheguo Guojia Tuciju, Beijing

Zhong W, Xiong HG (1999) Paleo-climatic and environmental development since about 4 ka B.P. and the relation with abandonments of ancient cities in Southern Xinjiang. J Desert Res 19:343–347 (in Chinese)

Zhou ZY, Chen PJ (1992) Biostratigraphy and geological evolution of Tarim. Science Press, Beijing (in Chinese)

Zhu XM, Wu LY, Obul O, Habibulla Ä (2006) The regulation of the Tarim River system. In: Hoppe T, Kleinschmit B, Roberts B, Thevs N, Halik Ü (eds) Watershed and floodplain management along the Tarim River in China's arid northwest. Shaker, Aachen

Part III
Concluding Remarks

Chapter 20
Integrating Nature and Culture in Landscape Ecology

Jianguo Wu

20.1 Introduction

Landscape ecology is an interdisciplinary field that aims to understand and improve the relationship between spatial pattern and ecological processes on a range of scales (Wu and Hobbs 2007b). Although the term was coined in Europe in 1939, landscape ecology was not a recognized scientific field of global research until the 1980s, when remote sensing data and computers became widely accessible to ecologists and geographers. The 1980s was also a time when ecological ideas of spatial heterogeneity and nonequilibrium dynamics flourished, and when landscape ecology was reborn in North America. During the two decades of the 1980s and 1990s, landscape ecology swept through North America like a storm, was rejuvenated in Europe, and reached out to other parts of the world, including Asia and Australia. Today, landscape ecology is a well-established field of study, with the active participation of ecological, geographical, and social scientists from around the world.

It has become a cliché to describe landscape ecology as being dominated by two schools of thought: the European perspective and the North American perspective. At the risk of over-simplification, we may consider the European landscape ecology perspective as having been characterized by a more holistic, humanistic, and society-centered view of landscapes, with a focus on user-inspired and solution-driven research. The North American landscape ecology perspective, on the other hand, has been dominated by a more analytical and biological ecology-centered view of landscapes, with a focus on basic science-oriented and question-driven studies (Wu and Hobbs 2002; Wu 2006). However, caution must be exercised to avoid over-interpretation of such dichotomous characterization. The two perspectives are neither

J. Wu (✉)
School of Life Sciences, School of Sustainability, and Global Institute of Sustainability, Arizona State University, P.O. Box 874501, Tempe, AZ 85287-4501, USA
and
Sino-US Center for Conservation, Energy, and Sustainability (SUCCESS), Inner Mongolia University, Hohhot 010021, China
e-mail: Jingle.Wu@asu.edu

S.-K. Hong, et al. (eds.), *Landscape Ecology in Asian Cultures*, Ecological Research Monographs, DOI 10.1007/978-4-431-87799-8_20, © Springer 2011

inclusive nor exclusive; they are not contradictory but complementary to each other. There are, and should be, other approaches to landscape ecology. For example, one could argue from an Australian landscape ecology perspective that focuses on pragmatic and functional approaches, typically tied in with land management, restoration, and conservation issues (e.g., Ludwig et al. 1997; MacKey et al. 2007).

Is there an identifiable Asian landscape ecology perspective? What contributions have Asian scientists and practitioners made to the development of landscape ecology? What is the state of Asian landscape ecology? What are its future directions? These are likely to be interesting questions to the readers of this book, but they are not the key questions to be addressed in this chapter. I will, however, make a few brief comments here which may be helpful to those who are looking for answers to these questions. A quick literature search suggests to me that much of the landscape ecological research in Asia during the past few decades has taken place in China, Japan, and Korea. China has produced substantially more publications than any other Asian country. For example, Cao et al. (2002) reported that Chinese authors published 619 journal articles and 13 books during the 1990s, of which over 90% were in Chinese. Of course, quantity is not quality – numbers do not always translate into impact. Nonetheless, these statistics are indicative of an exceptionally high level of enthusiasm for landscape ecology in China since the late 1980s. A more recent and comprehensive review of landscape ecology in China is found in Fu and Lü (2006). Although I have not detected a similar trend elsewhere in Asia (at least not on this magnitude), the last few decades have also seen the rapid development of landscape ecology in Japan and Korea, among other countries in this region.

Instead of summarizing all other chapters of this book on Asian cultural landscapes or reviewing the history of Asian landscape ecology in general, I thought that this chapter would be more useful if it presented a more comprehensive picture of landscape ecology in relation to cultural landscapes and sustainability. A broader and ecumenical perspective should foster a better understanding of the idiosyncratic topics covered in this book.

20.2 Evolving Concepts of Landscape and Landscape Ecology

20.2.1 What is a Landscape?

The term "landscape" is a key concept in a number of fields, from social to geographical and ecological sciences. With the rise of landscape ecology in the past few decades, the concept of landscape has achieved a prominent status in the interdisciplinary literature. However, because of the plurality of its origins and interpretations, landscape has acquired various connotations. For example, the same word may refer to a natural landscape, a cultural landscape, a political landscape, an economic landscape, a mental landscape, an adaptive landscape, a landscape view, landscaping, or landscape painting (Mitchell 2000; Tress and Tress 2001).

Even within the field of landscape ecology, the word "landscape" has different meanings, and the differences usually hinge on the spatial scale and the contents of a landscape. For example, landscape has been defined as a kilometers-wide geographic area (Forman 1981; Forman and Godron 1986) which corresponds to a "human-scale" landscape. This is the scale at which the field of landscape ecology was originally developed in Europe, and at which most landscape studies have been conducted around the world ever since. The human-scale landscape, in general, seems to coincide well with geographic units such as watersheds and urban regions (Forman 1995), as well as spatial domains of human perception (Gobster et al. 2007). Thus, it resonates with the public, the decision makers, and researchers who are conscious of the environmental setting in which they live, work, and engage in recreation.

However, many other landscape ecologists have treated landscape as a multi-scale or hierarchical concept, meaning that a landscape is a spatially heterogeneous area that may be of various sizes depending on the subject of study and the research questions at hand (Urban et al. 1987; Pickett and Cadenasso 1995; Turner et al. 2001). In this case, landscape is an "ecological criterion" (Pickett and Cadenasso 1995), and its essence does not lie in its absolute scale, but in its internal heterogeneity. Different plant and animal species perceive, experience, and respond to spatial heterogeneity at different scales, and patterns and processes in landscapes tend to have different characteristic scales (Kotliar and Wiens 1990; Wu and Loucks 1995; Wu et al. 2006). Thus, a hierarchical concept of landscape, also encompassing the human-scale of course, is both sensible and necessary. Clearly, one does not need to consider a landscape of tens of square kilometers to study how grassland vegetation patterns affect the movement of beetles (Wiens and Milne 1989) or is affected by gophers (Wu and Levin 1994).

The contents that constitute a landscape vary greatly in landscape ecological research. For simplicity, the components of a landscape may be classified as tangible versus intangible, and biophysical versus cultural. This is not intended to represent a dichotomous view, but rather a continuum within which a variety of components coexist. Tress and Tress (2001) proposed a "trans-disciplinary landscape concept" that encompasses five dimensions: (1) landscape as a spatial entity, (2) landscape as a mental entity, (3) landscape as a temporal dimension, (4) landscape as a nexus of nature and culture, and (5) landscape as a complex system. Landscape ecological studies have often focused on some but not all of these dimensions. Evidently, the concept of landscape provides a meeting ground for a number of disciplines, including archeology, ecology, geography, geology, history, landscape architecture, and regional economics. To achieve its interdisciplinary and trans-disciplinary goals, landscape ecology needs to appreciate and integrate the multi-faceted perspectives on the culture–nature/people–place relationships that are offered by these diverse disciplines.

20.2.2 *What is Landscape Ecology?*

The definitions of landscape ecology are also diverse, although they are not quite as numerous as those of landscape. Images can be powerfully inspiring, and this is especially true to someone who has a special interest in landscape patterns. Partly inspired by the conspicuous spatial patterns revealed in aerial photographs, the German geographer and botanist Carl Troll (1939) coined the term "landscape ecology" and defined it later as "the study of the main complex causal relationships between the life communities and their environment in a given section of a landscape" (Troll 1968, 1971). Carl Troll's training and research in multiple disciplines endowed him with the abilities to synthesize across, and innovate at the interface between, different fields. He was trained as a botanist, did his doctoral dissertation in plant physiology, and then spent decades working on the climatic, geological, geographical, and ecological aspects of various landscapes in Europe, South America, and Africa. It is easy to understand why Troll could simultaneously appreciate the then-new idea of an "ecosystem" put forward by Arthur Tansley (1935), as well as the great potential for geospatial analysis presented by aerial photography. As a result of his attempt to integrate the "vertical" ecological approach with the "horizontal" geographical approach, a new field of study was born.

In the past few decades, landscape ecology has acquired a number of definitions, which are all in some way related to Carl Troll's original definition. For example, Zonneveld (1972) defined landscape ecology as "an aspect of geographical study which considers the landscape as a holistic entity, made up of different elements, all influencing each other." He advocated that the landscape should be studied as the "total character of a region," and not "in terms of the separate aspects of its component elements" (Zonneveld 1972, 1989). This holistic landscape perspective continues and culminates in the work of Naveh (1991), who described landscape ecology as the study of "the total spatial and functional entity of natural and cultural living space" (also see Naveh 1982; Naveh and Lieberman 1984; Naveh 2000).

Some key ideas of contemporary landscape ecology, such as patch dynamics (Levin and Paine 1974; Pickett and Thompson 1978; Burgess and Sharpe 1981) and the patch–corridor–matrix model (Forman and Godron 1981, 1986) began to emerge in North America in the late 1970s, apparently with little connection to their European root. The early ideas of landscape ecology in North America were inspired by the theory of island biogeography (MacArthur and Wilson 1967), with an explicit focus on spatial heterogeneity. The first major communication between North American and European landscape ecologists occurred in 1981, when five American ecologists (including Forman, Golley, and Sharpe) attended the 1st International Congress on Landscape Ecology in The Netherlands. Two years later, 25 ecologists (23 American, 1 Canadian, and 1 French) gathered at Allerton Park, Illinois, USA, to discuss the nature and future directions of landscape ecology. The report of this historic meeting, published in the following year (Risser et al. 1984), became an important guide to budding landscape ecologists in North America.

Why was such a discussion necessary after landscape ecological research had been practiced for more than 40 years in Europe? The answer seems clear from Forman (1983): "What theory explains the spatial heterogeneity of energy, nutrients, water, plants, and animals at the level of a landscape, the setting in which we live? Alas, none." To develop such a landscape theory, broader scales that encompass multiple ecosystems need to be considered, and horizontal interactions have to be a focus of study. Thus, Forman and Godron (1981, 1986) defined landscape ecology as the study of the structure (spatial relationships among the distinctive landscape elements), function (flows of energy, materials, and species among landscape elements), and dynamics (temporal change in landscape structure and function) of landscapes. The main theme of landscape ecology in North America, with an unmistakable focus on spatial heterogeneity, was set out in Risser et al. (1984):

> Landscape ecology focuses explicitly upon spatial pattern. Specifically, landscape ecology considers the development and dynamics of spatial heterogeneity, spatial and temporal interactions and exchanges across heterogeneous landscapes, influences of spatial heterogeneity on biotic and abiotic processes, and management of spatial heterogeneity.

Is landscape ecology a sub-discipline of ecology? Certainly the semantics of the term suggest that it is. In fact, many ecologists do consider landscape ecology to be a branch of ecology (e.g., Turner et al. 2001), and most ecology programs of major research universities world wide now offer courses in landscape ecology. However, Zonneveld (1972) indicated that landscape ecology was not part of biological sciences, but a branch of geography. In fact, Risser et al. (1984) contemplated three ways that landscape ecology might be viewed: as an intersection of many disciplines, as a separate discipline, or as a branch of ecology. They concluded that only the first option was "intellectually and practically the most persuasive." In addition, "viewing landscape ecology as an interdisciplinary field of research avoids the issue of which discipline 'owns' landscape ecology" (a problem that may have hindered the healthy development of some interdisciplinary fields, such as human ecology, for which geography, sociology, and anthropology have all claimed ownership) (Risser et al. 1984). Reflective of the collective view of the group of 25 participants, likely with some internal heterogeneity, the Allerton workshop report clearly recognized the importance of the multi-dimensionality of landscapes and the cross-disciplinarity of landscape ecology:

> A major forcing function of landscapes is the activity of mankind, especially associated cultural, economic, and political phenomena. ... Landscape ecology is not a distinct discipline or simply a branch of ecology, but rather is the synthetic intersection of many related disciplines that focus on the spatial–temporal pattern of the landscape
>
> (Risser et al. 1984).

Today, a general consensus seems to have emerged that landscape ecology is not simply an academic discipline, but rather a highly interdisciplinary field of study (Wu and Hobbs 2002). Landscape ecology is an interdisciplinary and trans-disciplinary science that focuses on the relationship between spatial pattern and ecological processes across scales. The goal of landscape ecology is not only to understand this relationship, but also to influence it so as to help achieve landscape sustainability.

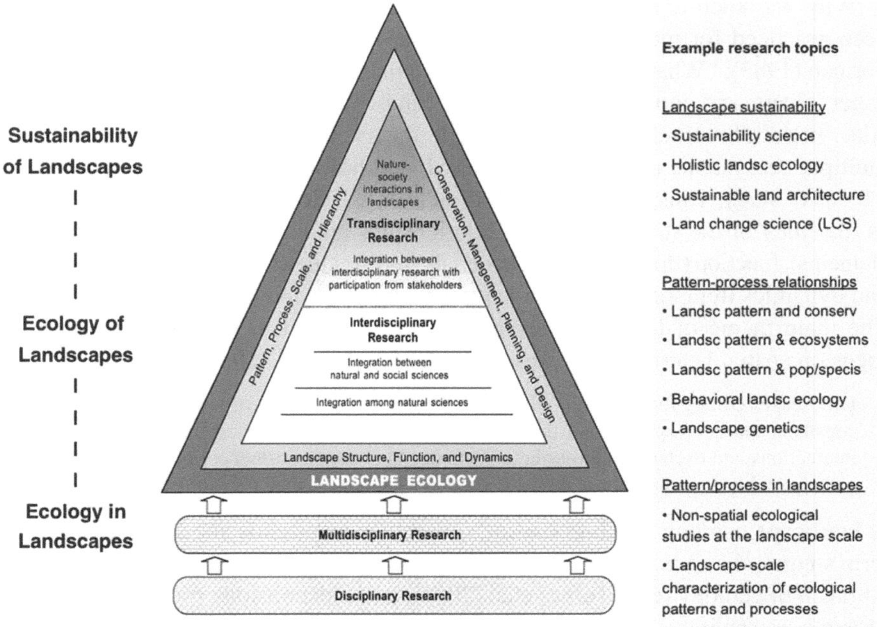

Fig. 20.1 A schematic representation of a pluralistic and hierarchical framework for landscape ecology (modified from Wu 2006; Wu and Hobbs 2007a, b)

In an attempt to integrate the various connotations, Wu and Hobbs (2007b) defined landscape ecology as the integration of the science and art of studying and influencing the relationship between spatial pattern and ecological processes on multiple scales (also see Wu 2006). The "science" of landscape ecology focuses on the theoretical basis for understanding the formation, dynamics, and effects of spatial heterogeneity, whereas the "art" of landscape ecology reflects the humanistic and holistic perspectives necessary for integrating ecology, design and planning, socio-economics, and management practices. Wu (2006, see also Wu and Hobbs 2007b) put forward a pluralistic and hierarchical framework that facilitates synergistic interactions between biophysical/pattern–process and holistic/humanistic perspectives in landscape ecology (Fig. 20.1). The "hierarchical" view here recognizes the varying scope and degree of cross-disciplinarity in landscape ecological studies, whereas the "pluralistic" view stresses the importance of different disciplines and perspectives. This pluralistic and hierarchical framework implies that all the five dimensions of landscape, as discussed in Tress and Tress (2001), are important in landscape ecological studies.

20.3 Landscape of Culture and Culture of Landscape

20.3.1 Cultural Landscapes and People–Landscape Relationships

As discussed earlier in this chapter, the term "landscape" in landscape ecology has various meanings ranging from predominantly biophysical to emphatically holistic and humanistic. In the landscape ecology literature, however, even the "humanistic" definitions are usually much more concerned with contemporary socio-economic processes than with long-term interactions between culture and nature in particular landscapes. The cultural dimension of landscape has not been completely ignored in landscape ecology (especially in Europe), but more emphasis is needed.

"Landscape gives identity to place" and "landscape is where past and present meet" (Phillips 2007). Human geographers may think of landscape as "a work of human labor" or "an activity" of dynamic interactions between people and place (Mitchell 2000). As such, a landscape may also be considered as "a form of ideology" or "a way of carefully selecting and representing the world so as to give it a particular meaning," and thus it can be "an important ingredient in constructing consent and identity" (Mitchell 2000). If one subscribes to the aforementioned holistic and interdisciplinary definition of landscape ecology, such cultural characteristics of landscapes have to be important to the science and practice of the field. Thus, the topic of "cultural landscape," which reflects the interactive relationship between culture and nature in a geographic area, is quite relevant to landscape ecology. The meaning of a cultural landscape is much richer than simply a human-altered setting such as a farm or a city.

The term "cultural landscape," like "landscape," also has various connotations. It has been a fundamental concept in geography since its first use in Germany in the 1890s, when the German geographer Friedrich Ratzel (1895–1896) defined it as "landscape modified by human activity," as opposed to the primeval natural landscape (Jones 2003). The term was introduced to English-speaking countries in the 1920s by the American geographer Carl O. Sauer, who made it the central concept of the Berkeley school of geographic thought (Jones 2003). In his seminal publication, *The morphology of landscape*, Sauer (1925) wrote:

> The cultural landscape is fashioned from a natural landscape by a cultural group. Culture is the agent, the natural are the medium, the cultural landscape is the result.

Since the 1960s, the concept of cultural landscape has been widely used in human geography (of which cultural geography is a part), anthropology, environmental management, and other related fields (Sauer 1925; Webb 1987). A major burst of interest in cultural landscapes took place in the early 1990s, known as the period of "the rise of cultural landscapes" (Jacques 1995).

One of the major factors that contributed to the recent popularity of the term on a global scale was the adoption of cultural landscapes in the International Convention for the Protection of the World's Cultural and Natural Heritage (often referred to as the World Heritage Convention) by the United Nations Educational, Scientific, and Cultural Organization (UNESCO) in 1992. The World Heritage Convention was established in 1972 to recognize and protect the world's natural and cultural heritage of "outstanding universal value," and in 1992 it became the first international legal instrument to recognize and protect cultural landscapes (http://www.whc.unesco.org/en/culturallandscape). The Operational Guidelines for the Implementation of the World Heritage Convention states that:

Cultural landscapes are cultural properties and represent the 'combined works of nature and of man' ... They are illustrative of the evolution of human society and settlement over time, under the influence of the physical constraints and/or opportunities presented by their natural environment and of successive social, economic and cultural forces, both external and internal. ... The term 'cultural landscape' embraces a diversity of manifestations of the interaction between humankind and its natural environment (UNESCO (United Nations Educational 1996).

Three categories of cultural landscape are included in the World Heritage Convention: (1) "clearly defined landscapes designed and created intentionally by humans," which include mainly garden and parkland landscapes, (2) "organically evolved landscapes" resulting from successive interactions between local people and their natural environment (including "relict" and "continuing" landscapes), and (3) "associative cultural landscapes" that have powerful religious, artistic, or cultural associations with the natural elements (Table 20.1). These categories cover landscapes that are profoundly transformed by human actions (designed and created landscapes) as well as those that carry significant cultural values primarily in an intangible way (associative cultural landscapes). This implies that culture and nature are not mutually exclusive, and that cultural landscapes do not have to be entirely created by humans.

As of 2010, 66 cultural landscapes have been included in the World Heritage List (Table 20.2). Although the cultural landscape definition by the World Heritage Convention does not exclude urban landscapes, the sites selected so far are predominantly rural, with only a small number of urban and industrial areas included. Also, a glance at the World Heritage List reveals that there is an evident imbalance in terms of the global geographical representation, as European countries have a disproportionately greater number of selected sites. In particular, Europe has 37 (56.1% of the total), Asia 15 (22.7% of the total), and Africa 9 (13.6% of the total). China has only one, and the United States has none. Sirisrisak and Akagawa (2007) identified "the political and economic stability in each state party" as a major contributing factor to this imbalance. Other factors related to the selection process must have played a role as well.

Cultural landscapes have also been recognized by national programs around the world. For example, in 1988, the United States National Park Service (NPS) formally identified cultural landscapes as a type of cultural resource to be protected in the NPS Management Policies (Page et al. 1998). The NPS defined a cultural landscape as "a geographic area, including both cultural and natural resources and the wildlife or domestic animals therein, associated with a historic event, activity, or person or exhibiting other cultural or esthetic values" (Page et al. 1998). The NPS

Table 20.1 Categories and definitions of cultural landscapes in the World Heritage Convention (UNESCO 1996)

Category	Definition
Clearly defined landscape designed and created intentionally by humans	A landscape designed and created intentionally by humans. This embraces garden and parkland landscapes characteristically constructed for esthetic, social, and recreational reasons, which are often (but not always) associated with religious or other monumental buildings and ensembles
Organically evolved landscape 1. Relict or fossil landscape 2. Continuing landscape	A landscape that results from an initial social, economic, administrative, and/or religious imperative and has developed its present form by association with, and in response to, its natural environment. Such landscapes reflect that process of evolution in their form and component features. They fall into two sub-categories 1. *Relict (or fossil) landscape*: a landscape in which an evolutionary process came to an end at some time in the past, either abruptly or over a period However, its significant distinguishing features are still visible in material form 2. *Continuing landscape*: a landscape which retains an active social role in contemporary society which is closely associated with the traditional way of life, and in which the evolutionary process is still in progress. At the same time, it exhibits significant material evidence of its evolution over time
Associative cultural landscape	A landscape with definable powerful religious, artistic, or cultural associations with the natural element rather than material cultural evidence, which may be insignificant or even absent

Table 20.2 Cultural landscape inscriptions on the World Heritage List as of 2010 (data from UNESCO, http://whc.unesco.org/en/culturallandscape)

Region	Number of inscriptions	Percentage of the total number of inscriptions
Europe	37	56.1
North America	0	0
Asia and the Pacific	15	22.7
Africa	9	13.6
Latin America and the Caribbean	4	6.1
Arab States	1	1.5
Total	66	100

cultural landscapes fall into four general categories: historic sites, historic designed landscapes, historic vernacular landscapes, and ethnographic landscapes (Table 20.3).

Table 20.3 Categories and definitions of cultural landscapes recognized and protected by the National Park Service of the United States (Page et al. 1998)

Category	Definition
Historic site	A landscape which is significant for its association with a historic event, activity, or person. Examples include battlefields and houses of presidents
Historic designed landscape	A landscape which is significant as a design or work of art. A landscape which was consciously designed or laid out by a master gardener, landscape architect, architect, or horticulturist according to a design principle, or by an owner or other amateur according to a recognized style or tradition. A landscape which has a historical association with a significant person, trend, or movement in landscape gardening or architecture, or a significant relationship to the theory or practice of landscape architecture. Examples include parks, campuses, and estates
Historic vernacular landscape	A landscape whose use, construction, or physical layout reflects endemic traditions, customs, beliefs, or values. Expresses cultural values, social behavior, and individual actions over time. A landscape which is manifested in physical features and materials and their interrelationships, including patterns of spatial organization, land use, circulation, vegetation, structures, and objects. It is a landscape whose physical, biological, and cultural features reflect the customs and everyday lives of people. Examples include rural villages, industrial complexes, and agricultural landscapes
Ethnographic landscape	A landscape containing a variety of natural and cultural resources that associated people define as heritage resources. Examples include contemporary settlements, such as the Martin Luther King Jr. National Historic Site, New Orleans neighborhoods, and the Timbisha Shoshone community in Death Valley. Small plant communities, animals, and subsistence and ceremonial grounds are often components

All these connotations of cultural landscapes are rooted in the definitions of Ratzel (1895–1896) and Sauer (1925), with further elaborations and extensions (e.g., the associative cultural landscapes in the World Heritage Convention). However, the degree of human modification or "fashioning" beyond which a natural landscape should be regarded as a cultural landscape is subjective, and has been a point of debate and a source of confusion. On the one hand, cultural landscapes have often referred only to agricultural or rural landscapes that occur between the natural and urban landscapes (Jones 2003). For example, Plachter (1995) advocated a "functional definition" that only includes landscapes in which culture and nature have mutually shaped one another and still do, with modern metropolitan landscapes explicitly excluded. On the other hand, the term has also been used to include all landscapes that are influenced by human activities and human values (Jones 2003). As a result, some have questioned the usefulness of the term based on the argument that landscapes untouched by humans no longer exist in reality.

For instance, Phillips (1998) argued that "Since there are cultural aspects to practically every landscape on earth, it follows that practically all landscapes are cultural landscapes." One conclusion from such an argument is to abandon the term altogether. However, this does not have to be the case, as the vagueness of the meaning of cultural landscape is not "bane or boon" but "both bane and boon." As Rowntree (1996) stated, "This etymological elusiveness [of cultural landscape] is both a liability and asset; to some, the notion of a cultural landscape is an appropriate bridge between space and society, culture and environment, while to others its definitional fluidity weakens the concept and disqualifies it from serious analytical usage." Indeed, this dialectical, rather than binary, property characterizes many terms that are essential to landscape ecology, including patch, disturbance, resilience, sustainability, and the word "landscape" itself. Geography has a long history of studying human–environment relationships, and a number of perspectives have been developed, with different research cores and methodologies that reflect a varying degree of affinity to either natural sciences or the humanities (Turner 1997).

20.3.2 An Asian Perspective on the Culture–Nature Relationship

One of the most far-reaching Asian philosophies about the relationship between culture and nature is the ancient Chinese philosophy known as the "Unity of Man with Nature" ("天人合一"), which has had a widespread influence in Asia and beyond. The Unity of Man with Nature is the unifying theme of several ancient Chinese philosophies and cultural traditions, and is consistent with the most central tenet of Taoism – that people should be in harmony with the rhythms of nature (Ji 2007; Chen and Wu 2009). According to scholars of oriental cultures, the Unity of Man with Nature was the quintessential theme shared by dominant ancient Asian cultures (e.g., Chinese and Indian), and has been described as the greatest contribution of Chinese culture to humanity (Ji 2007). In today's terminology, the Unity of Man with Nature means that human activities, including their architectural creations, should be integrated within natural patterns and processes so that sustainability can be achieved.

Reflective of the Unity of Man with Nature philosophy, Feng–Shui theory (风水理论) – the theory of Feng (wind) and Shui (water) – consists of a set of empirical principles that integrate biophysical landscape features with cultural traditions and religious beliefs to guide the practice of selecting and designing dwellings and burial spaces (Hong et al. 2007; Ji 2007; Chen and Wu 2009). Feng–Shui theory was originally developed based on Taoist Yin–Yang dualism, Five-Element theory, and Eight-Trigram theory. Its main premise is that the human–environment relationship (or the fate of the occupant of a space) can be influenced either positively or negatively by manipulating the Qi (the vital force or energy) that drives all change. As the conceptual basis for both the Five-Element and Eight-Trigrams theory, Yin–Yang dualism emphasizes the balance between natural and anthropogenic forces as well as the harmony and eternity of the whole. The Five-Element theory further articulates how the essential elements of the world are related to each other, and how they can be arranged properly to achieve sustainability. Related to the Five-Element theory, the theory of Eight Trigrams deals with more components that make up the world, and is commonly used as a tool in Feng–Shui practices (Chen and Wu 2009).

A well-known landscape model of the Unity of Man with Nature philosophy is the "Peach Blossom Spring" ("桃花源") ideal, originally described by a celebrated Chinese poet some 1 500 years ago, which vividly portrays an ecologically unspoiled landscape with mountains, water, and fertile land where people integrate themselves harmoniously with their natural environment. This ideal reflects people's desire to be closely connected with nature in order to seek peace and minimize disruptive interactions with the outside world. The philosophy of the Unity of Man with Nature is probably best illustrated in traditional cultural landscapes, such as gardens and farming systems, in China and certain other Asian countries (e.g., Korea and Japan). China is the "mother of gardens" (Wilson 1929), and early Chinese gardens began to appear about 2 000 years ago, mainly as "the gardens of literati" or "scholar's gardens" (Chen and Wu 2009). These gardens were created by combining the concepts from Chinese landscape paintings with poems of idealized bucolic settings. These gardens had neither the rudimentary fabrics of folk dwellings nor the symbolic features of a power hierarchy and social rites often explicit in feudalistic governmental architecture. In general, oriental architecture has a time-honored history of developing structures in concert with natural landscapes using wood as the primary construction material, and emphasizing the proper flows of energy and natural rhythms of the environment. This seems in contrast to the long tradition in Western landscape architecture of creating more permanent monuments with stones and mortar as the main construction materials, which demonstrate human perseverance.

Our perception and understanding of the relationship between people and nature are often influenced by our philosophical roots and cultural traditions. Both classical Western and oriental thinkers meditated on the philosophy of nature and its relationship to humanity. Emerging from this period of classical thought, however, the Western and Eastern perspectives on the natural environment began to diverge. For example, while traditional Chinese culture continued to embrace the power of nature to influence and inform humans, Western culture reacted more audaciously to it. Eastern philosophy emphasized a greater sense of harmony with nature, whereas in Europe there was a stronger emphasis on "taming" nature. In other words, the traditional Western philosophy of nature was based on a one-sided relationship between people and nature: humans are influenced by nature, react to nature, and then find ways to tame nature through technology and policy. Thus, culture and nature were perceived as being separate and conflicting. Such a philosophy represents the historical antecedent to the modern technocratic approach to economic development that has been adopted around the world in the past century. As Phillips (1998) stated: "The separation of culture and nature – of people from the environment which surrounds them – which has been a feature of Western attitudes and education over the centuries, has blinded us to many of the interactive associations which exist between the world of nature and the world of culture." Its influence can be felt even in the way the environment has been studied: "most of our intellectual weapons in the environmental area – from prehistoric fire debates to projections of climate change – have maintained a separation of humans and nature" (Head 2008).

While the ancient Chinese philosophy of the Unity of Man with Nature seems much in tune with the sustainability theme of our time, the environmental movement in the West, which started in the 1960s, had a major role in promoting human values for integration, rather than separation, between culture and nature. Even before that, Aldo Leopold (1949), in his landmark book *A Sand County Almanac*, clearly recognized the problems with the conquering-nature tradition, and promoted "a state of harmony between man and land" with his new land ethic: "The land ethic simply enlarges the boundaries of the community to include soils, waters, plants, and animals, or collectively: the land. …… In short, a land ethic changes the role of *Homo sapiens* from conqueror of the land community to plain member and citizen of it." The eminent American landscape architect Ian McHarg (1969) advocated the "design with nature" approach, which echoed the philosophy of the Unity of Man with Nature. With passion and clarity, he wrote: "Let us then abandon the simplicity of separation and give unity its due. Let us abandon the self-mutilation which has been our way and give expression to the potential harmony of man–nature." After a long period of divergent developments, Eastern and Western cultures now seem to be beginning to converge on a shared recognition and vision – the harmony between culture and nature – of sustainability.

20.4 Connecting Culture and Nature in Landscape Ecology

20.4.1 Emphasizing the Cultural Dimension in Landscape Ecology

If landscape ecology is to achieve its goal of understanding and improving the relationship between spatial pattern and ecological processes, it must explicitly connect culture with nature or people with a place in particular landscapes. In general, landscape ecologists are much more familiar with the physicality than the culture of landscapes. Nonetheless, as discussed earlier, the cultural dimension of landscapes has always been a part of landscape ecology since its inception, particularly in Europe. In recent years, the need to reconnect culture with nature has increasingly been recognized by landscape ecologists around the world.

For example, following the European tradition of landscape ecology illustrated by numerous studies, most noticeably in Germany and The Netherlands, Naveh (1982, 1998) has repeatedly stressed the necessity and importance of cultural landscapes, suggesting that cultural landscapes should encompass all landscapes created and modified by humans. Farina (2000) advocated the use of the cultural landscape as a model for integrating ecology with economics, because they are "geographic areas in which the relationships between human activity and the environment have created ecological, socio-economic, and cultural patterns and feedback mechanisms that govern the presence, distribution, and abundance of species assemblages." In the case of Farina (2000), cultural landscapes referred only to traditional cultivated landscapes. Tress et al. (2001) stated that: "The perceived division between nature and culture has dominated the academic world. In the case of landscapes, this divide is counter-productive and must be overcome since all landscapes are multidimensional and multifunctional."

The dichotomous characterization of European versus North American perspectives may suggest that the latter focuses only on the biophysical aspects of landscapes, but this is not true. The importance of cultural aspects and the inseparability of culture and nature in human-dominated landscapes were also recognized in the nascent stage of North American landscape ecology. This was made clear in the ground-breaking book by Forman and Godron (1986): "To understand why a landscape looks as it does, we cannot limit ourselves to the natural or physical environment. We must also understand human influences and culture. ... In a landscape with people, the human role and the role of nature may be alternatively emphasized but cannot be disentangled." However, this vision has not been adequately implemented in research practice in North America in the past 30 years. As Nassauer (1995) noted:

> Culture changes landscapes and culture is embodied by landscapes. Both aspects of this dynamic are encompassed by landscape ecology, but neither has been examined sufficiently to produce cultural theory within the field. ... American landscape ecology has entered the cultural realm with its vocabulary and in environmental policy, but cultural effects on landscapes have been more assumed than examined. Research in landscape ecology has not focused on culture despite its centrality to the field.

Landscape ecology, like landscapes themselves, is changing. In North America and other parts of the world, landscape ecology has evidenced a rapid increase in a research emphasis on the integration between the culture and nature of landscapes in recent years. A fundamental reason for this surge of interest is the realization, increasingly shared by landscape ecologists around the world, that the world has been on an unsustainable trajectory, particularly since the Industrial Revolution, and that landscape ecology can and must contribute to sustaining our landscapes and the world (Wu 2006; Naveh 2007; Fu et al. 2008; Wu 2008; Barrett et al. 2009; Musacchio 2009b; Wu 2010). One example of recent studies on this topic is the special issue of *Landscape Ecology*, the flagship journal in the field, which was published in 2009 with the title: "The ecology and culture of landscape sustainability: emerging knowledge and innovation in landscape research and practice" (Musacchio 2009a). However, much needs to be done to reconnect culture and nature in landscape ecology. To move forward, "We must formulate ecological questions by considering cultural possibilities, and we must formulate cultural questions by considering ecological processes" (Nassauer 1997). To formulate such interdisciplinary questions, the four general principles of the culture and nature relationship articulated by Nassauer (1995) should be helpful.

1. Human landscape perception, cognition, and values directly affect the landscape and are affected by the landscape.

2. Cultural conventions powerfully influence landscape patterns in both inhabited and apparently natural landscapes.
3. Cultural concepts of nature are different from scientific concepts of ecological function.
4. The appearance of landscapes communicates cultural values.

Landscape ecology needs more integrated studies that consider cultural landscapes as co-evolved holistic systems of culture and nature. False separations of humans from nature may adversely affect the quality of our research and practice (Head 2008). In our attempt to integrate culture and nature in landscapes, we need to fully recognize the necessity and opportunities of taking pluralistic and ecumenical approaches in landscape ecological research (Wu et al. 2006; Wu and Hobbs 2007a). No single perspective or approach is sufficient to understanding human–environment relationships (Turner 1997). At the same time, collaborations between natural and social sciences, which are designed to synthesize and integrate diverse perspectives, are crucial. Diversity is not divergence. Diversity is a basis for innovation, whereas divergence is more a cause for distraction. After all, the usefulness of pluralism is predicated on the effectiveness of building bridges among research cores with different perspectives (Turner 1997).

20.4.2 Understanding the Diversity of Cultural Landscapes

Cultural landscapes are diverse; cultural landscapes are idiosyncratic; cultural landscapes carry the legacies of the past and foster possibilities for the future. Therefore, to formulate either ecological questions of culture or cultural questions of ecology, it is necessary to recognize the diversity of landscapes with different degrees of human intervention in particular cultural settings. To this end, it is useful to recall the five landscape types classified by Forman and Godron (1986), which constitute a landscape modification gradient.

1. Natural landscape – without significant human impact.
2. Managed landscape – where native species are managed and harvested.
3. Cultivated landscape – with villages and scattered patches of natural or managed ecosystems.
4. Suburban landscape – a town and country area with a heterogeneous patchy mixture of residential areas, commercial centers, cropland, managed vegetation, and natural areas.
5. Urban landscape – with remnant managed park areas scattered in a densely built up matrix.

Forman and Godron's (1986) classification can be complemented or refined by considering characteristics more directly related to the resilience and self-regenerative capacities of the system (Walker and Salt 2006). For example, Naveh (1998) proposed that cultural landscapes should include semi-natural and managed multi-functional landscapes (e.g., protected areas, parks, recreation areas), traditional agricultural landscapes, rural and suburban landscapes, and urban landscapes. He also articulated that these different types of cultural landscapes can be distinguished based on their energy inputs and self-organizing and regenerative capacities through the photosynthetic conversion of solar energy: (1) "solar-powered" semi-natural and managed landscapes, ranging from protected areas and traditional agricultural landscapes to contemporary organic farming systems; (2) "intensive agro-industrial" landscapes, including modern agricultural systems that are heavily subsidized by fossil energy; (3) "technosphere" landscapes, including rural, suburban, and urban–industrial landscapes that are supported primarily by fossil energy, with all internal natural regenerative capacities lost (Naveh 1998). Such landscape gradients provide a broader framework based on which different cultural landscapes can be compared, idiosyncratic studies can be synthesized, and thus our understanding of landscape sustainability can be improved.

20.4.3 Learning About Sustainability from Cultural Landscapes

Based on the discussion in previous sections, I argue that the concept of a cultural landscape is useful and effective, especially when it is used in the context of a landscape modification gradient. Biophysical forces create, alter, and maintain landscapes, but humans have played a rapidly increasing role in the processes of landscape development during the past century. In today's human-dominated earth system, almost all landscapes around the world have been somewhat influenced, and even "domesticated," by anthropogenic processes (Kareiva et al. 2007). Humans now appropriate about 24% of the Earth's terrestrial net primary productivity (Haberl et al. 2007), and have directly influenced 83% of the world's land area through agriculture, urbanization, and associated activities (Kareiva et al. 2007). There are still landscapes, on increasingly smaller scales, that may be called natural or semi-natural. It is evident, however, that the major objects of landscape ecological research are cultural landscapes.

Scholars who study landscapes from either ecological or cultural perspectives seem to agree on the importance of the landscape on an operational scale in the study and practice of sustainability. For example, Forman (1990) argued that human-scale landscapes, as a spatial scale for the study and practice of sustainable development, have significant advantages over broader scales such as the continent. Forman (1995) further pointed out that to deal with "the paradox of management," i.e., that actions tend to be more effective at local scales, whereas success often needs to be achieved at broader scales, "management and planning for sustainability at an intermediate scale, the landscape or region, appears optimum." The ordinary elements of human landscapes (e.g., forests, crop fields, urban land cover, residential areas, streams, and streets) also resonate well with human perception and thus facilitate decision making (Nassauer 1997; Gobster et al. 2007). From a cultural geographer's perspective, Phillips (1998) commented that cultural landscapes are "places which can demonstrate that talk of sustainable development can be more than rhetoric."

"Cultural landscapes often reflect specific techniques of sustainable land-use, considering the characteristics and limits of the natural environment they are established in, and a specific spiritual relation to nature" (UNESCO, United Nations Educational 1996). As well as contemporary cultural landscapes such as agricultural and urban landscapes, traditional cultural landscapes should also be emphasized in landscape ecological studies. Such landscapes are the products of long-term co-evolution between culture and nature, and there is much to be learnt from them. Good examples include the rice terrace landscapes in the northern Philippines, the Iberian agri-silvo-pastoral landscapes of the montado and dehesa, the Scandinavian grazed deciduous woodlands, the puszta of Hungary, and the sheep grazed downlands of southern Britain (Stanners and Bourdeau 1995; Phillips 1998). Many Asian countries are rich in such traditional cultural landscapes, some of which are discussed in other chapters of this book. Cultural landscapes that have survived for hundreds of years must have some sustainable land management strategies and techniques that can contribute to our abilities to develop and maintain sustainable landscapes in future. Even those that have disappeared may still provide us with valuable insights.

For example, based on a review of lessons from history, Forman (1995) observed that water problems, soil erosion, high population density, war, and a decline in exports are key attributes associated with decreased sustainability, whereas cultural cohesion, low population density, an export–import trade, the overall level and arrangement of the resource base, religious cohesion, varied links with adjacent areas, and a major irrigation or dike system are key attributes associated with increased sustainability. Selman (2007) suggested three propositions as a basis for assessing the sustainability of cultural landscapes: (1) "cultural landscapes are sustainable if they are regenerative," (2) "landscape sustainability is characterized by ecological integrity and cultural legibility," and (3) "regenerative landscapes are distinguished by feedback loops leading to an accumulation of cultural and ecological assets." Forman (1990) postulated that "for any landscape or major portion of a landscape, there exists an optimal spatial configuration of ecosystems and land uses to maximize ecological integrity, achievement of human aspirations, or sustainability of an environment." More detailed studies need to be carried out to further test these observations, propositions, and hypotheses. This represents a promising future direction not only for landscape ecology, but also for sustainability science.

20.5 Conclusions

Landscape ecology is now a well-established interdisciplinary field of study, which is evidenced by several characteristics. These include an evolving but identifiable system of concepts, theories, principles, methods, and applications, a hierarchy of professional organizations consisting of international associations and regional and local chapters, a reputable flagship journal as both a platform and a barometer of the development of the field, the adoption of educational and training programs by major universities and research institutes around the world, and an increasing number of publications in main-stream scientific journals which indicate its recognized status as well as its expanding impacts on related disciplines. One may argue, however, that these characteristics do not constitute a complete set of necessary and sufficient criteria that qualify landscape ecology to be a well-established "discipline" in the strict sense of the word. This deficiency may be attributed to the lack of consensus on a set of clearly articulated research questions and goals, as well as a systematic methodology for the field. Indeed, in the past few decades, some have been concerned with the diversity and divergence of concepts and ideas in landscape ecology, and others have worried about its loss of identity as a field of study. While such concerns are common with rapidly developing fields, landscape ecology is not a "discipline," but rather an interdisciplinary and trans-disciplinary science. The state of landscape ecology today is stronger than ever; its relevance to science and society is clearer than ever; and its future looks brighter than ever.

Although landscape ecology has come of age, it is not yet a mature science that is capable of achieving its trans-disciplinary goals. The most important and challenging goal of all involves providing a theoretical basis, developing a set of systematic methodologies, and demonstrating successful applications through place-based studies, in order to understand, manage, and design sustainable landscapes. To achieve this goal, as I have discussed in this chapter, landscape ecology must reconnect culture with nature, and unite people with place in both theory and practice. Cultural landscapes will be the main objects in future ecological landscape studies. Although they are common, the divisions between culture and nature, between society and environment, and between people and place are not based on reality, but on human perception. While such divisions are useful and even necessary as we try to simplify complexity or to reveal mechanistic details, any artificial separation of constituents without a holistic unifying framework tends to obstruct, not construct, a genuine understanding of complex adaptive systems such as landscapes. This is especially important when our research questions are about landscape sustainability.

To landscape ecologists, there is much to be learnt from human geography and other social sciences, there is much to be gained by integrating analytical and holistic approaches within the field, and there is much to be studied of contemporary and traditional cultural landscapes! However, as we expand the spectrum of our research interests, embrace a greater complexity of landscapes, and reach a higher level of trans-disciplinarity, we must not forget the quintessential characteristics of landscape ecology – the emphasis on spatial heterogeneity and associated spatially explicit methodology – which underlies the original definition by Carl Troll, and which has become the cornerstone of landscape ecology today. These are not merely some unique features that distinguish this field from others; more importantly, they provide landscape ecology with a special capacity for tackling complex real-world problems.

Acknowledgments I would like to thank the following students and colleagues of mine for discussions on issues related to Asian landscape ecology: Cheng Li, Junxiang Li, Yu Tian, Qi Yang, and Ting Zhou. It is always a great pleasure to discuss sustainability and philosophical issues with Tong Wu, who also provided valuable comments on the manuscript of this chapter. My research in landscape ecology and sustainability has been supported by grants from the National Science Foundation (Central Arizona–Phoenix Long-Term Ecological Research: DEB 9714833 and DEB-0423704; Biocomplexity/CNH: BCS-0508002), the US Environmental Protection Agency (Science to Achieve Results (STAR) program: R827676-01-0), and the National Natural Science Foundation of China and the Chinese Academy of Sciences (multiple collaborative grants).

References

Barrett TL, Farina A, Barrett GW (2009) Aesthetic landscapes: an emergent component in sustaining societies. Landsc Ecol 24(8):1029–1035

Burgess RL, Sharpe DM (eds) (1981) Forest Island dynamics in man-dominated landscapes. Springer, New York

Cao Y, Xiao DN, Li XZ, Hu YM (2002) Literature analysis and research progress of landscape ecology in China in the 1990s. J For Res 13:98–102

Chen X, Wu J (2009) Sustainable landscape architecture: implications of the Chinese philosophy of "unity of man with nature" and beyond. Landsc Ecol 24:1015–1026

Farina A (2000) The cultural landscape as a model for the integration of ecology and economics. Bioscience 50(4):313–320

Forman RTT (1981) Interaction among landscape elements: a core of landscape ecology. In: Tjallingii SP, de Veer AA (eds) Perspectives in landscape ecology: contributions to research, planning and management of our environment. Pudoc, Wageningen, pp 35–48

Forman RTT (1983) An ecology of the landscape. Bioscience 33:535

Forman RTT (1990) Ecologically sustainable landscapes: the role of spatial configuration. In: Zonneveld IS, Forman RTT (eds) Changing landscapes: an ecological perspective. Springer, New York, pp 261–278

Forman RTT (1995) Land mosaics: the ecology of landscapes and regions. Cambridge University Press, Cambridge

Forman RTT, Godron M (1981) Patches and structural components for a landscape ecology. Bioscience 31:733–740

Forman RTT, Godron M (1986) Landscape ecology. Wiley, New York

Fu B, Lü Y (2006) The progress and perspectives of landscape ecology in China. Prog Phys Geogr 30:232–244

Fu B, Lü Y, Chen L (2008) Expanding the bridging capability of landscape ecology. Landsc Ecol 23(4):375–376

Gobster PH, Nassauer JI, Daniel TC, Fry G (2007) The shared landscape: what does aesthetics have to do with ecology? Landsc Ecol 22(7):959–972

Haberl H, Erb KH, Krausmann F, Gaube V, Bondeau A, Plutzar C, Gingrich S, Lucht W, Fischer-Kowalski M (2007) Quantifying and mapping the human appropriation of net primary production in earth's terrestrial ecosystems. Proc Natl Acad Sci USA 104:12942–12947

Head L (2008) Is the concept fo human impacts past its use-by date? The Holocene 18:373–377

Hong S-K, Song I-J, Wu J (2007) Fengshui theory in urban landscape planning. Urban Ecosyst 10:221–237

Jacques D (1995) The rise of cultural landscapes. Int J Herit Stud 1:91–101

Ji X (2007) Ji Xianlin on Chinese culture. China Books, Beijing

Jones M (2003) The concept of cultural landscape: discourse and narratives. In: Palang H, Fry G (eds) Landscape interfaces. Kluwer, Dordrecht, pp 21–51

Kareiva P, Watts S, McDonald R, Boucher T (2007) Domesticated nature: shaping landscapes and ecosystems for human welfare. Science 316(5833):1866–1869
Kotliar NB, Wiens JA (1990) Multiple scales of patchiness and patch structure: a hierarchical framework for the study of heterogeneity. Oikos 59:253–260
Leopold A (1949) A Sand County Almanac. Oxford University Press, New York
Levin SA, Paine RT (1974) Disturbance, patch formation and community structure. Proc Natl Acad Sci USA 71(7):2744–2747
Ludwig J, Tongway D, Freudenberger D, Noble J, Hodgkinson K (1997) Landscape ecology, function and management: principles from Australia's rangelands. CSIRO, Collingwood
MacArthur RH, Wilson EO (1967) The theory of island biogeography. Princeton University Press, Princeton
MacKey BG, Soulé ME, Nix HA, Recher HF, Lesslie RG, Williams JE, Woinarski JCZ, Hobbs RJ, Possingham HP (2007) Applying landscape-ecological principles to regional conservation: the Wildcountry Project in Australia. In: Wu J, Hobbs R (eds) Key topics in landscape ecology. Cambridge University Press, Cambridge, UK, pp 192–213
McHarg IL (1969) Design with nature. Natural History Press, Garden City, NY
Mitchell D (2000) Cultural geography: a critical introduction. Blackwell, Oxford
Musacchio LR (2009a) The ecology and culture of landscape sustainability: emerging knowledge and innovation in landscape research and practice. Landsc Ecol 24(8):989–992
Musacchio LR (2009b) The scientific basis for the design of landscape sustainability: a conceptual framework for translational landscape research and practice of designed landscapes and the six Es of landscape sustainability. Landsc Ecol 24(8):993–1013
Nassauer JI (1995) Culture and changing landscape structure. Landsc Ecol 10(4):229–237
Nassauer JI (ed) (1997) Placing nature: culture and landscape ecology. Island Press, Washington, DC
Naveh Z (1982) Landscape ecology as an emerging branch of human ecosystem science. Adv Ecol Res:188–237
Naveh Z (1991) Some remarks on recent developments in landscape ecology as a transdisciplinary ecological and geographical science. Landsc Ecol 5:65–73
Naveh Z (1998) Ecological and cultural landscape restoration and the cultural evolution towards a post-industrial symbiosis between human society and nature. Restor Ecol 6:135–143
Naveh Z (2000) What is holistic landscape ecology? A conceptual introduction. Landsc Urban Plan 50:7–26
Naveh Z (2007) Landscape ecology and sustainability. Landsc Ecol 22(10):1437–1440
Naveh Z, Lieberman AS (1984) Landscape ecology: theory and application. Springer, New York
Page RR, Gilbert CA, Dolan SA (1998) A guide to cultural landscape reports: contents, process, and techniques. U.S. Department of the Interior, National Park Service, Washington, DC
Phillips A (1998) The nature of cultural landscapes – a nature conservation perspective. Landsc Res 23:21–38
Phillips A (2007) International policies and landscape protection. In: Benson JF, Roe M (eds) Landscape and sustainability, 2nd edn. Routledge, New York, pp 84–103
Pickett STA, Cadenasso ML (1995) Landscape ecology: spatial heterogeneity in ecological systems. Science 269:331–334
Pickett STA, Thompson JN (1978) Patch dynamics and the design of nature reserves. Biol Conserv 13:27–37
Plachter H (1995) Functional criteria for the assessment of cultural landscapes. In: von Droste B, Plachter H, Rossler M (eds) Cultural landscapes of universal value – components of a global strategy. UNESCO, Gena, pp 380–392
Risser PG, Karr JR, Forman RTT (1984) Landscape ecology: directions and approaches. Illinois Natural History Survey Special Publ. 2, Champaign
Rowntree LB (1996) The cultural landscape concept in American human geography. In: Earle C, Mathewson K, Kenzer MS (eds) Concepts in human geography. Rowman & Littlefield, Lanham, pp 127–159
Sauer CO (1925) The morphology of landscape. Publications in Geography (Berkeley: University of California), vol 2, pp 19–53

Selman P (2007) Landscape and sustainability at the national and regional scales. In: Benson JF, Roe M (eds) Landscape and sustainability, 2nd edn. Routledge, New York, pp 104–117
Sirisrisak T, Akagawa N (2007) Cultural landscape in the world heritage list: understanding on the gap and categorization. City and Time 2:2. (online) URL: http://www.ct.ceci-br.org
Stanners D, Bourdeau P (1995) Europe's environment: the DobriS assessment. European Environment Agency, Copenhagen
Tansley AG (1935) The use and abuse of vegetational concepts and terms. Ecology 16(3):284–307
Tress B, Tress G (2001) Capitalising on multiplicity: a transdisciplinary systems approach to landscape research. Landsc Urban Plan 57:143–157
Tress B, Tress G, De'camps H, d'Hauteserre A-M (2001) Bridging human and natural sciences in landscape research. Landsc Urban Plan 57:137–141
Troll C (1939) Luftbildplan and okologische bodenforschung. Zeitschraft der Gesellschaft fur Erdkunde Zu Berlin 7–8:241–298
Troll C (1968) Landschaftsokologie. In: Tuxen R (ed) Pflanzensoziologie und Landschaftsokologie. Berichte das 1963 Internalen Symposiums der Internationalen Vereinigung fur Vegetationskunde. Junk, The Hague, pp 1–21
Troll C (1971) Landscape ecology (geoecology) and biogeocenology – a terminology study. Geoforum 8(71):43–46
Turner BL II (1997) Spirals, bridges and tunnels: engaging human–environment perspectivers in geography. Ecumene 4:196–217
Turner MG, Gardner RH, O'Neill RV (2001) Landscape ecology in theory and practice: pattern and process. Springer, New York
UNESCO (United Nations Educational S, and Cultural Organization), (1996) Operational guidelines for the implementation of the world heritage convention. UNESCO, Paris. http://whc.unesco.org/archive/opguide05-annex3-en.pdf
Urban DL, O'Neill RV, Shugart HH (1987) Landscape ecology: a hierarchical perspective can help scientists understand spatial patterns. Bioscience 37:119–127
Walker B, Salt D (2006) Resilience thinking: sustaining ecosystems and people in a changing world. Island Press, Washington, DC
Webb M (1987) Cultural landscapes in the National Park Service. The Public Historian 9:77–89
Wiens JA, Milne BT (1989) Scaling of 'landscape' in landscape ecology, or, landscape ecology from a beetle's perspective. Landsc Ecol 3:87–96
Wilson EH (1929) China: mother of gardens. Stratford, Boston
Wu J (2006) Landscape ecology, cross-disciplinarity, and sustainability science. Landsc Ecol 21(1):1–4
Wu J (2008) Making the case for landscape ecology: an effective approach to urban sustainability. Landsc J 27(1):41–50
Wu J (2010) Urban sustainability: an inevitable goal of landscape research. Landsc Ecol 25:1–4
Wu J, Hobbs R (2002) Key issues and research priorities in landscape ecology: an idiosyncratic synthesis. Landsc Ecol 17:355–365
Wu J, Hobbs RJ (eds) (2007a) Key topics in landscape ecology. Cambridge University Press, Cambridge
Wu J, Hobbs R (2007b) Landscape ecology: the-state-of-the-science. In: Wu J, Hobbs R (eds) Key topics in landscape ecology. Cambridge University Press, Cambridge, UK, pp 271–287
Wu J, Levin SA (1994) A spatial patch dynamic modeling approach to pattern and process in an annual grassland. Ecol Monogr 64(4):447–464
Wu J, Loucks OL (1995) From balance-of-nature to hierarchical patch dynamics: a paradigm shift in ecology. Q Rev Biol 70:439–466
Wu J, Jones KB, Li H, Loucks OL (eds) (2006) Scaling and uncertainty analysis in ecology: methods and applications. Springer, Dordrecht, The Netherlands
Zonneveld IS (1972) Land evaluation and land(scape) science. International Institute for Aerial Survey and Earth Sciences, Enschede, The Netherlands
Zonneveld IS (1989) The land unit – a fundamental concept in landscape ecology, and its applications. Landsc Ecol 3(2):67–86

Index